FUZZY LOGIC AND INTELLIGENT SYSTEMS

INTERNATIONAL SERIES IN INTELLIGENT TECHNOLOGIES

Prof. Dr. Dr. h.c. Hans-Jürgen Zimmermann, Editor
European Laboratory for Intelligent
Techniques Engineering
Aachen, Germany

Other books in the series:

Applied Research in Fuzzy Technology
by Anca L. Ralescu

Analysis and Evaluation of Fuzzy Systems
by Akira Ishikawa and Terry L. Wilson

FUZZY LOGIC AND INTELLIGENT SYSTEMS

EDITED BY

Professor Hua Li
Texas Tech University
Lubbock, Texas, USA

Professor Madan Gupta
University of Saskatchewan
Saskatoon, Canada

SPRINGER SCIENCE+BUSINESS MEDIA, LLC

Library of Congress Cataloging-in-Publication Data
Fuzzy logic and intelligent systems / edited by Hua Li, Madan Gupta.
p. cm. -- (International series in intelligent technologies ; 3)
ISBN 978-94-017-3773-9 ISBN 978-0-585-28000-4 (eBook)
DOI 10.1007/978-0-585-28000-4
1. Neural networks (Computer science) 2. Fuzzy systems.
3. Expert systems (Computer science) I. Li, Hua, 1956-
II. Gupta, Madan. III. Series.
QA76.87.F89 1995
629.8'36--dc20 95-16748
CIP

Printed on acid-free paper.

CONTENTS

PREFACE xiii

1 IMPROVEMENT OF FUZZY CONTROL METHODS
M. Mizumoto 1
1 INTRODUCTION 1
2 MULTIPLE FUZZY REASONING 2
3 FUZZY CONTROLS 3
4 FUZZY CONTROLS UNDER VARIOUS AGGREGATION METHODS 5
5 NEW FUZZY REASONING METHODS 6
6 FUZZY CONTROLS UNDER VARIOUS DEFUZZIFIER METHODS 8
7 CONCLUSIONS 11
REFERENCES 11

2 NEURAL NETWORKS AND FUZZY BASIS FUNCTIONS FOR FUNCTIONAL APPROXIMATION
Liang Jin, Madan M. Gupta, Peter N. Nikiforuk 17
1 INTRODUCTION 17
2 STONE-WEIERSTRASS THEOREM AND IMPLICATIONS 19
3 TRIGONOMETRIC FUNCTION NETWORKS 21
4 MFNNs AS UNIVERSAL APPROXIMATORS 27
5 KOLMOGOROV'S THEOREM AND FEEDFORWARD NETWORKS 31

6 APPROXIMATION USING POLYNOMIAL NEURAL NETWORKS 36
7 GAUSSIAN NETWORKS 46
8 UNIVERSAL APPROXIMATION USING FUZZY BASIS FUNCTION NETWORKS 50
9 CONCLUSIONS 55
REFERENCES 56

3 ORDERING FUZZY REAL QUANTITIES
Robert Lowen 69
1 INTRODUCTION 69
2 STOCHASTIC FUZZY REAL NUMBERS 69
3 ORDERING OF THE STOCHASTIC FUZZY REAL LINE 74
REFERENCES 83

4 FUZZY LOGIC CONTROLLERS FOR AIRCRAFT FLIGHT CONTROL
Jia Luo and Edward Lan 85
1 INTRODUCTION 85
2 FUZZY PID CONTROLLER 87
3 FUZZY FLIGHT CONTROL SYSTEM FOR THE X-29 AIRCRAFT 90
4 VALIDATION THROUGH NONLINEAR SIMULATION 106
5 CONCLUSIONS 114
REFERENCES 115

5 A CLASSICAL CONTROLLER: A SPECIAL CASE OF THE FUZZY LOGIC CONTROLLER
Thomas Brehm and Kuldip Rattan 125
1 INTRODUCTION 125
2 FLC ARCHITECTURE AND TERMINOLOGY 126
3 FLC OUTPUT EQUATION DERIVATION 130
4 FLC AS A PIECEWISE LINEAR CONTROLLER 135
5 CLASSICAL CONTROLLER: A SPECIAL CASE OF FLC 138
6 SUMMARY 149

REFERENCES 154

6 REAL TIME FUZZY LOGIC CONTROLLER FOR BALANCING A BEAM-AND-BALL SYSTEM
Nowell Godfrey, Hua Li, Yuandong Ji , William Marcy 157
1 INTRODUCTION 157
2 PROBLEM FORMULATION 158
3 THE DESIGN OF FUZZY LOGIC CONTROLLER 161
4 COMPARISONS TO OTHER CONTROL STRATEGY 167
5 SUMMARY 173
6 ACKNOWLEDGEMENT 174
REFERENCES 174

7 DESIGN OF FUZZY CONTROLLERS BASED ON FREQUENCY AND TRANSIENT CHARACTERISTICS
Kazuo Tanaka 187
1 INTRODUCTION 187
2 FREQUENCY AND TRANSIENT CHARACTERISTICS 189
3 PHASE-LEAD COMPENSATION 190
4 DESIGN EXAMPLES 198
5 APPLICATION TO TANK LEVEL CONTROL 200
6 CONCLUSION 201
REFERENCES 202

8 FUZZY INFERENCE INTEGRATING 3D MEASURING SYSTEM WITH ADAPTIVE SENSING STRATEGY
Koji Shimojima, Toshio Fukuda, Fumihito Arai, and Hideo Matsuura 213
1 INTRODUCTION 214
2 3-D MEASUREMENT SYSTEM 216
3 ADAPTIVE SENSING STRATEGY 225
4 EXPERIMENTS AND RESULTS 226
5 CONCLUDING REMARKS 232

REFERENCES 234

9 ROBOT HAND-EYE COORDINATION BASED ON FUZZY LOGIC

Sukir Kumaresan, Hua Li, and Xing-Min Li 245

1 INTRODUCTION 246

2 SURVEY OF EXISTING TECHNIQUES 247

3 MATHEMATICAL FORMULATION 249

4 ALGORITHM FOR THE COORDINATION BASED ON FUZZY LOGIC 253

5 EXPERIMENTS 257

6 OPERATION UNDER CAMERA DISPLACEMENT 260

7 COMPARISON 264

8 CONCLUSION 266

REFERENCES 267

10 USING FPGA TECHNIQUE FOR DESIGN AND IMPLEMENTATION OF A FUZZY INFERENCE SYSTEM

Donald Hung 271

1 INTRODUCTION 271

2 THE FUZZY INFERENCE ALGORITHM 272

3 FPGA DESIGN CONSIDERATIONS 277

4 DESIGN METHODOLOGY AND IMPLEMENTATION 280

5 TESTING RESULTS 285

6 CONCLUSION AND DISCUSSION 285

REFERENCES 287

11 AN EMPIRICAL ANALYSIS OF ONE TYPE OF DIRECT ADAPTIVE FUZZY CONTROL

Hugues Bersini and Vittorio Gorrini 289

1 INTRODUCTION 290

2 A GRADIENT METHOD FOR SUGENO'S FUZZY SYSTEM 291

3 A SIMPLE DIRECT ADAPTIVE FUZZY CONTROL 293

4 AN ADAPTIVE FUZZY PID 300

5 ON THE EQUIVALENCE BETWEEN RBF AND FUZZY SYSTEMS 302
6 STABILITY AND CONCLUSIONS 305
REFERENCES 307

12 AUTOMATIC OPTIMAL DESIGN OF FUZZY SYSTEMS BASED ON UNIVERSAL APPROXIMATION AND EVOLUTIONARY PROGRAMMING
Mattias Nyberg and Yoh-Han Pao 311
1 INTRODUCTION 311
2 FUZZY SETS AND SYSTEMS 312
3 AUTOMATIC DESIGN OF FUZZY SYSTEMS 317
4 UNIVERSAL APPROXIMATION WITH FUZZY SYSTEMS 321
5 GUIDED EVOLUTIONARY SIMULATED ANNEALING 327
6 WANG'S FUZZY SYSTEM AND GESA 335
7 FUNCTION ESTIMATION: THE GENERALIZED PARITY-2 PROBLEM 337
8 ON FUZZY LOGIC CONTROL: THE INVERTED PENDULUM 342
9 CONCLUSION AND FUTURE WORK 358
REFERENCES 362

13 INTELLIGENT CONTROL USING DYNAMIC NEURAL NETWORKS WITH ROBOTIC APPLICATIONS
Liang Jin, Madan M. Gupta, Peter N. Nikiforuk 367
1 INTRODUCTION 367
2 DESCRIPTION OF DISCRETE-TIME NONLINEAR SYSTEMS 370
3 DYNAMIC RECURRENT NEURAL NETWORKS 374
4 REGULATION OF KNOWN NONLINEAR SYSTEMS USING DRNNS 382
5 STATE REGULATION OF UNKNOWN PLANTS USING DRNNS 388
6 OUTPUT TRACKING OF UNKNOWN PLANTS USING DRNNS 392

7 APPLICATIONS TO ROBOTIC CONTROL 395
8 CONCLUSIONS 401
REFERENCES 402

14 CAMCORDER OPERATION JUDGEMENT USING A NEURAL COMPUTING APPROACH
Kitahiro Kaneda and Paul Wang 409
1 INTRODUCTION 409
2 PROBLEM FORMULATION AND METHODOLOGY 411
3 EXPERIMENTAL RESULTS 416
4 CONCLUSION 418
REFERENCES 418

15 MODEL REDUCTION AND CONTROL OF MULTISTAGE FLASH (MSF) DESALINIZATION PLANTS
Srinivas Ramamurthy,Jayanta Pal, Ak Sinha, Darwish MK Al Gobaisi, Ganti Prasada Rao 423
1 INTRODUCTION 424
2 MSF CONTROL SYSTEMS 424
3 MSF PLANT MODELLING 427
4 REDUCED ORDER MODELLING 428
5 NEURAL NETWORK MODEL 429
6 MODEL REDUCTION 432
7 ROBUST CONTROLLER DESIGN 439
8 CONCLUSIONS 441
REFERENCES 442

INDEX 449

CONTRIBUTORS

M. Mizumoto
Division of Information and
Computer Sciences
Osaka Electro-Communication University
Neyagawa, Osaka 572, Japan

Liang Jin, Madan M. Gupta, Peter N. Nikiforuk
Intelligent Systems Research Laboratory
College of Engineering
University of Saskatchewan
Saskatoon, Saskatchewan
Canada S7N 0W0

Robert Lowen
Departement Wiskunde en Informatica
Universiteit Antwerpen
Froenenborgerlaan 171
2020 Antwerpen, Belgium

Jia Luo and Edward Lan
Department of Aerospace Engineering
University of Kansas
Lawrence, KA 66045, USA

Thomas Brehm and Kuldip Rattan*
Department of Computer Science
and Engineering
*Department of Electrical Engineering
Wright State University
Dayton, OH 45435, USA

Nowell Godfrey, Hua Li, Yuandong Ji*, and William Marcy
Computer Science Department
Texas Tech University
Lubbock, TX 79409, USA
*Department of System Engineering
Case Western Reserve University
Cleveland, OH 44106, USA

Kazuo Tanaka
Department of Mechanical Systems
Engineering
Kanazawa University
2-40-20, Kodatsuno, Kanazawa 920, Japan

Koji Shimojima, Toshio Fukuda, Fumihito Arai, and Hideo Matsuura
Department of Mechano-Informatics
and Systems
Nagoya University
Furo-cho, Chikusa-ku, Hagoya 464-01, Japan

Sukir Kumaresan, Hua Li, and Xing-Min Li*
*Computing Center, The Institute of Textile Engineering
He-Dong, Tianjin, China
Computer Science Department
Texas Tech University
Lubbock, TX 79409, USA

Donald Hung
Department of Electrical Engineering
Gannon University
Erie, PA 16541, USA

Hugues Bersini and Vittorio Gorrini
IRDIA - CP 194/6
Universite Libre de Bruxelles
50, Av. Franklin Roosevelt
1050 Bruxelles, Belgium

Mattias Nyberg and Yoh-Han Pao
Department of Electrical Engineering
and Applied Physics
Case Western Reserve University
Cleveland, OH 44106, USA

Kitahiro Kaneda and Paul P. Wang*
Video Products Development Center and
Fuzzy Logic Research Laboratory
CANON INC., Tokyo, Japan
*Department of Electrical Engineering
Duke University
Durham, NC 27708-0291, USA

Srinivas Ramamurthy, Jayanta Pal, Ak Sinha, darwish Al Gobaisi†and Ganti Rao
Department of Electrical Engineering
India Institute of Technology
Kharagpur 721302, India
†Water and Electricity Department
Government of Abu Dhabi, UAE

PREFACE

Building an intelligent system is challenging work, it can involve many different aspects of learning, adaptation, and control under uncertainty. Neural networks and fuzzy logic have been the major tools for scientists and engineers working in this field. Over the past several years, we have witnessed the rapid growth of utilizing fuzzy logic both in the theory and applications in many different engineering fields. Unlike most of the traditional techniques, fuzzy logic algorithms provide a tool to deal with uncertainty, random disturbances, and yet with relatively modest computational effort. The world is full of uncertainty. Everyday when we drive to work, we may experience different weather conditions and traffic patterns. Everytime we park at the same parking space, but we may not be able to park the car at exactly the same spot as we did the day before. The uncertainty and imprecision seem always come together, at least at many situations, and it seems that the well balanced solution between the uncertain conditions and imprecision, or adaptive decision-making serves our daily life very well. It is then natural to ask how much preciseness or impreciseness can we tolerate in the real world situation? at what cost? Fortunately, fuzzy logic as a tool to deal with uncertainty, random disturbances, and *ill-defined* problems (the problems that can not be well described by closed form equations,) allows us to tackle these problems.

Fuzzy logic algorithms offer many attractive features and they have been gaining popularity for solving many real engineering application problems, especially, the problems or systems which are highly nonlinear, which involve many parameters and many of these parameters are changing or drifting in time. As many people have realized that one of the major obstacles for us to build a real intelligent machine is to deal with random disturbances, to process huge amount of imprecise data, to interact with a dynamically changing environment, and to cope with uncertainty. With the use of neural-fuzzy techniques, it is now possible to attack and solve a class of certain problems by using a customer developed algorithm with an intelligent behavior.

Working in the area of neural networks, fuzzy logic, and their applications, we felt strongly that there is a need to review the state-of-the-art development in this emerging field. During the 1993 IEEE International Conference on Fuzzy Logic jointly held with the 1993 IEEE International Conference on Neural Networks in San Francisco (March 28-April 3, 1993), Kluwer Academic Publishers invited us to put a book together. Since then, we have been working on this book project with the researchers who have made significant contributions to this emerging field. Now we are pleased to present this volume as a result of this work. This edited book consists of fifteen chapters by authors from US, Canada, Japan, Europe, and India. The subjects presented in this book reflect the most recent developments in neural networks, fuzzy logic and their applications in intelligent systems. In addition, the balance between theoretical work and applications makes this book not only suitable for researchers, engineers, but also for graduate students as well.

Without the hard work of many contributors, this book would have not been possible. In particular, it is our pleasure to acknowledge the inspiring contributions that Professor Lofti A. Zadeh has made in fuzzy logic, which has inspired many researchers in the field of building intelligent systems.

During the editorial phase of this book many students from Texas Tech University have shared various responsibilities. In particular, we would like to thank Samuel Huang who helped to coordinate the peer review process, Nowell Godfrey who painstakingly converted some manuscripts to Latex format and helped to plan the layout of tables and figures to arrange them in an orderly fashion. We would also like to thank Xiao-hui Meng, Lazslo Moldovan, and Dongming Liang at Texas Tech University who have helped the preparation of typesetting. The help from Mr. Zachary Rolnik, Senior Editor of Kluwer Academic Publishers, is appreciated.

Hua Harry Li
Texas Tech University
Lubbock, Texas, USA

Madan M. Gupta
University of Saskatchewan
Saskatoon, Saskatchewan, CANADA

1

IMPROVEMENT OF FUZZY CONTROL METHODS

M. Mizumoto

Division of Information and Computer Sciences
Osaka Electro-Communication University
Neyagawa, Osaka 572, Japan

ABSTRACT

This chapter introduces improvement methods for fuzzy controls. At first, we introduce new aggregation operators for obtaining a fuzzy set of control actions. A number of new fuzzy reasoning methods such as product-sum-gravity method, min-sum-gravity method, (min/product)-max-gravity method and others are also introduced. Finally, we show several defuzzification procedures for obtaining a representative point of a fuzzy set of control actions.

1 INTRODUCTION

A number of studies on fuzzy logic controllers has been reported since Mamdani [3] implemented a fuzzy logic controller on a boiler steam engine. For most of the existing fuzzy logic controllers, the center of gravity method is widely used as a defuzzification method which decides an actual control action in a fuzzy set of control actions aggregated by using max operator from fuzzy sets inferred from fuzzy control rules.

In this chapter, we propose several new fuzzy control methods and compare control results of these methods. At first, new aggregation operators are introduced which obtain a fuzzy set of control actions. If averaging operators of arithmetic mean, dual of geometric mean and dual of harmonic mean are used as the aggregation operators, better control results can be obtained compared to the case of using max operator as in the "min-max-gravity method" by Mamdani [3]. In this connection, by introducing t-norms and t-conorms we propose new kinds of fuzzy logic reasoning methods which include product-sum-gravity method, min-sum-gravity method, (min-product)-max-gravity method, bounded product-bounded sum-gravity method. Then we introduce several defuzzifier procedures for obtaining a representative point of a fuzzy set of control actions. Finally, we show that two defuzzification methods called height method and area method provide better control results than that of the widely used center of gravity method.

2 MULTIPLE FUZZY REASONING

We shall consider the following multiple fuzzy reasoning form with several fuzzy rules combined with "else",

$$\begin{array}{rccccl} Rule1: & A_1 \; and \; B_1 & \Rightarrow & C_1 & else & \\ Rule2: & A_2 \; and \; B_2 & \Rightarrow & C_2 & else & \\ & \cdots\cdots & & & & \\ Rule\; n: & A_n \; and \; B_n & \Rightarrow & C_n & & \\ Fact: & & x_0 \; and \; y_0 & & & \\ \hline Cons: & & C' & & & \end{array} \tag{1.1}$$

where $A_i, i = 1, \cdots, n$ are fuzzy sets in X; B_i in Y; and C_i, C' in Z and $x_0 \in X, y_0 \in Y$. Fuzzy rule $\lceil A_i$ and $B_i \Rightarrow C_i \rfloor, i = 1, \cdots, n$ is defined as

$$\mu_{A_i \; and \; B_i \Rightarrow C_i}(x, y, z) = \mu_{A_i}(x) \wedge \mu_{B_i}(y) \wedge \mu_{C_i}(z) \tag{1.2}$$

where $\wedge$ stands for min.

The inference result C_i' inferred from the fact $\lceil x_0$ and $y_0 \rfloor$ and fuzzy rule $\lceil A_i$ and $B_i \Rightarrow C_i \rfloor$ is given as

$$\mu_{C_i'}(z) = \mu_{A_i}(x_0) \wedge \mu_{B_i}(y_0) \wedge \mu_{C_i}(z) \tag{1.3}$$

The consequence C' of (1.1) at x_0 and y_0 is given as follows by interpreting "else" union ($\cup$).

$$C' = C'_1 \cup C'_2 \cup \cdots \cup C'_n \tag{1.4}$$

$$\mu_{C'}(z) = \mu_{C'_1}(z) \vee \mu_{C'_2}(z) \vee \cdots \vee \mu_{C'_n}(z) \tag{1.5}$$

where $\vee$ stands for max.

The method of obtaining a singleton z_0 which is a representative point for the resulting fuzzy set C' of (1.4) is called a *deffuzification method.* For example, the point which has the largest membership grade of C' can be taken as the desired singleton. The method of taking the center of gravity of C' as the desired singleton z_0 is widely used in the fuzzy controls and is called the *center-of-gravity method*, which is given as

$$z_0 = \frac{\int z \cdot \mu_{C'}(z)dz}{\int \mu_{C'}(z)dx.} \tag{1.6}$$

The above fuzzy reasoning method obtained by using (1.3)-(1.6) is known as Mamdani's method [3] and referred as *"min-max-gravity method"* (as illustrated Figure 1)

3 FUZZY CONTROLS

We shall consider a plant model G(s)=$e^{-2s}/(1+20s)$ with first order delay, which is used in the discussion in the next section. Fuzzy control rules for the plant model are shown in Table 1 [7, 8] and interpreted as:

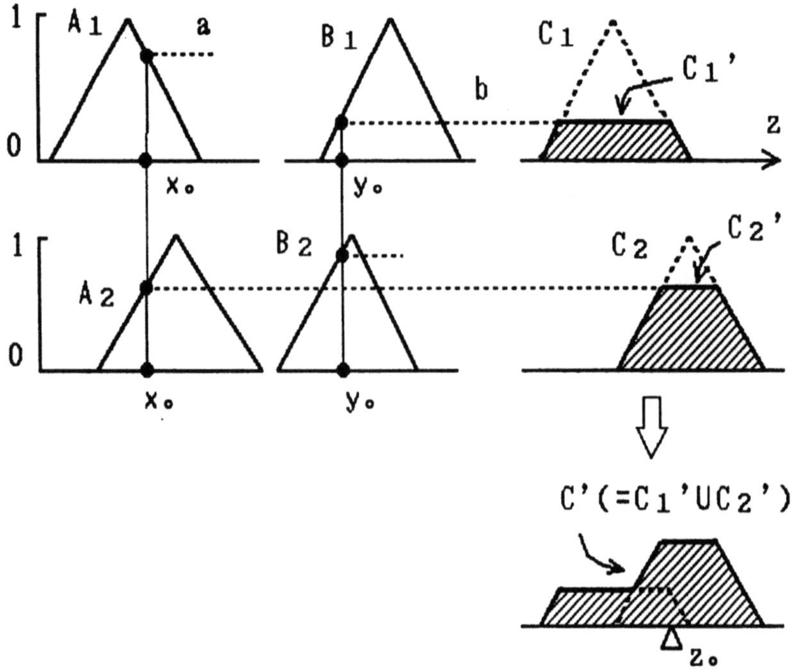

Figure 1 Min-max-gravity method using equation (1.3), (1.5) and (1.6)

Δq		Δe						
		NB	NM	PS	ZO	PS	PM	PB
	NB				PB			
	NM				PM			
	NS				PS			
e	ZO	PB	PM	PS	ZO	NS	NM	NB
	PS				NS			
	PM				NM			
	PB				NB			

Table 1 Fuzzy control rules: e, Δe $\Rightarrow$ Δq

$$\begin{aligned} R_1 :& \quad e \; is \; NB \; and \; \Delta e \; is \; ZO \Rightarrow \quad \Delta q \; is \; PB \\ R_2 :& \quad e \; is \; NM \; and \; \Delta e \; is \; ZO \Rightarrow \quad \Delta q \; is \; PM \\ & \quad \cdots\cdots\cdots\cdots\cdots\cdots \\ R_{13} :& \quad e \; is \; ZO \; and \; \Delta e \; is \; PB \Rightarrow \quad \Delta q \; is \; NB \end{aligned} \tag{1.7}$$

where e is error and Δe is change in error and Δq is change in action. NB, NM, $\cdots$, PB are fuzzy sets as shown in Figure 2 (width W=6). When $e = e_0$ and $\Delta e = \Delta e_0$ are given to a fuzzy controller as premises of (1.7), each control rule $R_i(i = 1, \cdots, 13)$ infers a fuzzy set C' for Δ q by using

$$\mu_{C'}(\Delta q) = \mu_{A_i}(e_0) \wedge \mu_{B_i}(\Delta e_0) \wedge \mu_{C_i}(\Delta q) \tag{1.8}$$

where A_i, B_i and C_i are fuzzy sets given in Figure 2.

The resulting fuzzy set C' is obtained by taking the union ($\bigcup$) of C' as shown in (1.4). The actual change in action $\Delta q = \Delta q_0$ to the plant is obtained as the center of gravity (1.6) of the fuzzy set C' which is aggregated from the fuzzy sets $C_1', \cdots, C_{13}'$ inferred from each fuzzy control rule (1.7) given singletons e_0 and Δe_0 by use of (1.8).

4 FUZZY CONTROLS UNDER VARIOUS AGGREGATION METHODS

As was seen in the definition of C' of (1.4), "max" ($= \vee$) operator is usually used as in (1.5). This section indicates the improvement of fuzzy control results when aggregation operators such as sum, arithmetic mean, dual of geometric mean and dual of harmonic mean are used to aggregate C' from $C_1', C_2', \cdots, C_n'$ instead of using max operator.

As a generalization of (1.5), the aggregation of C' from $C_1', C_2', \cdots, C_n'$ can be given by using a certain addition operation [+].

$$\mu_{C'}(z) = \mu_{C_1'}(z)[+]\mu_{C_2'}(z)[+]\cdots[+]\mu_{C_n'}(z) \tag{1.9}$$

It is possible to use sum operation a+b [5, 6], algebraic sum $a + b - ab(= 1 - (1 - a)(1 - b)$, bounded-product $1\wedge$ $(a + b)$ and, more generally, t-conorms as addition operation [+]. For example, we have

for the sum

$$\mu_{C'}(z) = \mu_{C_1'}(z) + \mu_{C_2'}(z) + \cdots + \mu_{C_n'}(z), \tag{1.10}$$

for the algebraic sum

$$\mu_{C'}(z) = 1 - (1 - \mu_{C'_1}(z)) \cdots (1 - \mu_{C'_n}(z)), \tag{1.11}$$

for the bounded sum

$$\mu_{C'}(z) = 1 \wedge [\mu_{C'_1}(z) + \cdots + \mu_{C'_n}(z)]. \tag{1.12}$$

Moreover, we can also use averaging operations [4] as aggregation operators. For the arithmetic mean

$$\mu_{C'}(z) = \frac{\mu_{C'_1}(z) + \cdots + \mu_{C'_n}(z)}{n}. \tag{1.13}$$

For the dual geometric mean

$$\mu_{C'}(z) = 1 - \sqrt[n]{(1 - \mu_{C'_1}(z)) \cdots (1 - \mu_{C'_n}(z))}. \tag{1.14}$$

For the dual harmonic mean

$$\mu_{C'}(z) = 1 - \frac{n}{\frac{1}{1-\mu_{C'_1}(z)} + \cdots + \frac{1}{\mu_{C'_n}(z)}}. \tag{1.15}$$

Now we shall show control results when using various aggregation operators. It is found from Figure 2(a) that the width W of fuzzy sets NB, NM, $\cdots$, PB is 6 as in Figure 2. In Figure 2(b), the averaging operators of arithmetic mean, dual of geometric mean and dual of harmonic mean are shown to obtain better control results than the max operator in case of W=10. However, the aggregation operators such as algebraic sum and bounded-sum do not offer so good control results.

It is noted that the same control results are observed in the case of using sum and arithmetic mean as aggregation operators since the same center of gravity is obtained under these operations.

5 NEW FUZZY REASONING METHODS

Given the definition of the aggregation of C' of (1.9), we can use several aggregation operators such as averaging operators and t-conorms other than max as shown in (1.10)-(1.16). This fact indicates the possibility of using appropriate operations in place of $\wedge$ (=min) in (1.3) in the process of obtaining inference result C'.

In general, inference result C' from fact ⌈x_0 and y_0⌋ and fuzzy rule ⌈A_i and $B_i \Rightarrow C_i$⌋ is derived by using two kinds of operations $\star_1$, $\star_2$ in the following:

$$\mu_{C_i'}(z) = [\mu_{A_i}(x_0) \star_1 \mu_{B_i}(y_0)] \star_2 \mu_{C_i}(z). \tag{1.16}$$

As the candidates for operations $\star_1$ and $\star_2$ we can use t-norms such as

$$\begin{aligned}
\text{min}: &\quad a \wedge b &&= min(a,b), && (1.17)\\
\text{algebraic product}: &\quad a \cdot b &&= ab, && (1.18)\\
\text{bounded product}: &\quad a \odot b &&= (a+b-1) \vee 0, && (1.19)\\
\text{drastic product}: &\quad a \wedge b &&= \begin{cases} a \cdots b & = 1 \\ b \cdots a & = 1 \\ 0 \cdots a, & b < 1 \end{cases} && (1.20)
\end{aligned}$$

and averaging operators of

$$\begin{aligned}
\text{harmonic mean}: &\quad \frac{2}{1/a+1/b}, && (1.21)\\
\text{geometric mean}: &\quad \sqrt{ab}. && (1.22)
\end{aligned}$$

It is noted that these averaging operators have 0 when a or b = 0.

We can propose new fuzzy reasoning methods of "product-sum-gravity method" [5, 6] in which algebraic product $\cdot$ is used as $\star_1$ and $\star_2$ in (1.16) and sum operation is used as in (1.10) (see Figure 3). Namely,

$$\mu_{C_i'}(z) = [\mu_{A_i}(x_0) \cdot \mu_{B_i}(y_0)] \cdot \mu_{C_i}(z), \tag{1.23}$$
$$\mu_{C'}(z) = \mu_{C_1'}(z) + \mu_{C_2'}(z) + \cdots + \mu_{C_n'}(z). \tag{1.24}$$

Moreover, we can define *"(min/product)-max-gravity method"* in the following, where $\star_1 = \wedge$ and $\star_2 = \cdot$ (algebraic product).

$$\mu_{C_i'}(z) = [\mu_{A_i}(x_0) \wedge \mu_{B_i}(y_0)] \cdot \mu_{C_i}(z) \tag{1.25}$$

In the same way, we can propose a number of fuzzy reasoning methods for fuzzy controls, say, (1) *product-algebraic sum-gravity method, (2) bounded product-bounded sum-gravity method, (3) drastic product-drastic sum-gravity method* (all of which use dual operators), (4) *bounded product-algebraic sum-gravity method, (5) (product/min)-algebraic sum-gravity method, (6) (min/drastic product)-harmonic mean-gravity method, (7) (product/drastic product)-sum-gravity method* and so on by combining appropriate operations in (1.9) and (1.16).

Figure 4 compares control results by min-max-gravity method, product-sum gravity method, min-sum gravity method and (min/product)-max gravity method. It is found from their computer simulations that product-sum-gravity method gives the best control results comparing with other methods and that min-sum-gravity method and (min/product)-max-gravity method are found to be better than min-max-gravity method.

Figure 5 shows the control results by min-max-gravity method, product-algebraic sum-gravity method and bounded product-bounded sum-gravity method in which the operations used are dual each other. Min-max-gravity method gives also worse control result.

6 FUZZY CONTROLS UNDER VARIOUS DEFUZZIFIER METHODS

In this section we shall discuss several defuzzification methods and show that two defuzzifier methods called height method and area method can get better control results than the widely used center of gravity method.

In the following, we list several defuzzification methods.

6.1 Center-Of-Gravity Method [2, 3]

The center-of-gravity method is widely used in the fuzzy controls. The center of gravity of C' is adopted as the desired singleton z_0 (see (1.6)).

$$z_0 = \frac{\int z \cdot \mu_{C'}(z)dz}{\int \mu_{C'}(z)dx}. \tag{1.26}$$

6.2 Average of Maxima Method [1]

Using average-of-maxima method, the representative point x_0 is obtained as an average of the elements which give the maximal grade in C', that is,

$$z_0 = (\sum_{j=1}^{m} z_j)/m. \tag{1.27}$$

where z_j is an element giving the maximal grade in C' and m is the number of such elements z_j.

6.3 Midpoint-Of-Maxima-Method [3]

The midpoint-of-maxima method is a simplified version of the average-of-maxima Method of 6.2. Instead of taking all elements z_j which give the maximal grade, the smallest elements z' and the largest element z'' among them are picked up and the midpoint of z' and z'' is given as the representative point z_0, that is,

$$z_0 = (z' + z'')/2. \tag{1.28}$$

6.4 Median Method

By the median method, the representative point z_0 is obtained as the point which divides C' into two equal areas.

The above defuzzifier methods of obtaining a representative point z_0 are all derived from the calculation of the fuzzy set C' aggregated from $C'_1, \cdots, C'_n$ as in (1.4) and (1.9). In the following, we shall show several defuzzification methods which use the properties of each $C'_i (i = 1, ..., n)$ rather than C'.

Figure 6 shows fuzzy sets C'_i inferred from fuzzy rules $\lceil A_i$ and $B_i \Rightarrow C_i \rfloor$ given x_0 and y_0 (see (1.4)). The height h_i of C'_i is given as

$$h_i = \mu_{A_i}(x_0) \wedge \mu_{B_i}(y_0) \tag{1.29}$$

and the area of C_i' is denoted as S_i.

Note that we can use alternative operators (1.18)-(1.22) in (1.29) as shown in (1.16).

The representative point (say, the center of gravity) of the conclusion part of C_i of fuzzy rule $\lceil A_i$ and $B_i \Rightarrow C_i \rfloor$ is assumed to be denoted as z_i.

6.5 Height Method [1]

The height method obtains z_0 as a weighted average of the representative points z_i of C_i by the heights h_i of C_i'. Namely,

$$z_0 = \frac{h_1 \cdot z_1 + h_2 \cdot z_2 + \cdots + h_n \cdot z_n}{h_1 + h_2 + \cdots + h_n}. \tag{1.30}$$

This method may be considered as a special case of Sugeno's fuzzy reasoning method [8] in which the conclusion part of a fuzzy control rule is a function or a real number rather than a fuzzy set.

6.6 Maximal Height Method

By maximal height method, a representative point z_j of C_j which corresponds to the màximal height h_j among $h_i (i = 1, \ldots, n)$ is adopted as z_0,

$$z_0 = z_j \ (h_j \text{ the maximal height}). \tag{1.31}$$

6.7 Area Method [4]

The z_0 is obtained as the weighted average of the representative points z_i by the areas S_i of C_i',

$$z_0 = \frac{S_1 \cdot z_1 + S_2 \cdot z_2 + \cdots + S_n \cdot z_n}{S_1 + S_2 + \cdots + S_n}. \tag{1.32}$$

6.8 Maximal Area Method

A representative point z_j of C_j which corresponds to the maximal area S_j among $S_i (i = 1, \ldots, n)$ is selected as z_0, that is,

$$z_0 = z_j \ (S_j \text{ the maximal area}). \tag{1.33}$$

Figure 7 indicates control results under various defuzzification methods when fuzzy sets in Figure 2 are of width W=6 and 8. It is found from computer simulation that the defuzzification methods of Height Method and Area Method obtain better control results than the center of gravity Method which is widely used in the fuzzy control. Defuzzifier methods such as Average of Maxima Method, Midpoint of Maxima Method, Maximal Height Method and Maximal Area Method which use maximal values show bad control results. These defuzzifier methods tend to continue selecting the same Δq_0 owing to the use of maximal values so that the control results by these methods do not converge to the set point 40.

7 CONCLUSIONS

We have proposed several different kinds of fuzzy reasoning methods for fuzzy controls by introducing appropriate operations such as t-norms and t-conorms in place of min and max. Moreover, a number of defuzzification methods are also introduced. Combining these operations and defuzzification methods, we can define new kinds of fuzzy reasoning methods, such as min-sum-height method, product-bounded sum-median method, min-arithmetic mean-area method and so on.

It was found from our discussion and computer simulations that min-max-gravity method by Mamdani is not the only fuzzy reasoning method for fuzzy controls and that there exist a number of fuzzy reasoning methods which get better control results than min-max-gravity methods.

REFERENCES

[1] Brae, M. & Rutherford, D. A., Fuzzy relations in a control setting,

Kybernetes, 7, 185-188, 1978.

[2] Ostergaard, J. J., Fuzzy logic control of a heat exchanger process, Report No. 7601, Danish Technical University, 1976.

[3] Mamdani, E. H., Applications of fuzzy algorithms for conreol of a simple dynamic plant, *Proc. of IEEE*, 12 1, 1585-1588, 1974.

[4] Mizumoto, M., "Pictorial representation of fuzzy connectives, Part I: Cases of t-norms, t-conorms and averaging operators", *Fuzzy Sets and Systems*, 31, 217-242, 1989.

[5] Mizumoto, M., Fuzzy controls by product-sum-gravity method, *Advancement of Fuzzy Theory and Systems in China and Japan* (ed. Liu and Mizumoto), International Academic Publishers, c 1.1- c 1.4, 1990.

[6] Mizumoto, M., Fuzzy controls under product-sum-gravity method and new fuzzy control methods, *Fuzzy Control Systems* (ed. A. Kandel & G. Langholz), 276-294, CRC Press, 1993.

[7] Yamazaki, T. & Sugeno, M., Fuzzy controls, *Systems and Controls*, 28, 442-446, 1984 (in Japanese).

[8] Sugeno, M., *Fuzzy Controls*, Nikkan Kogyo Pub., 1988 (in Japanese).

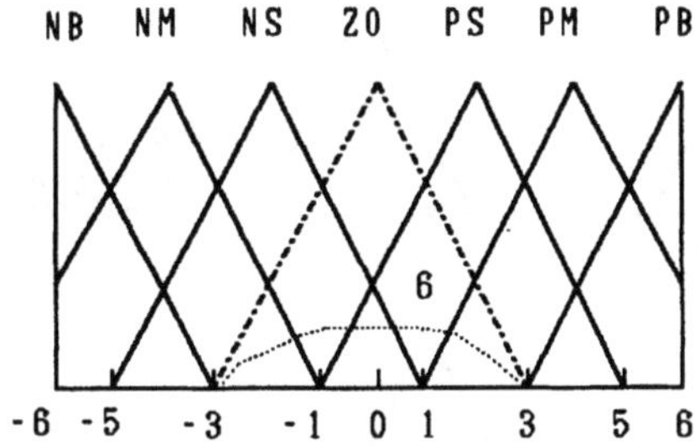

Fuzzy sets of width W = 6

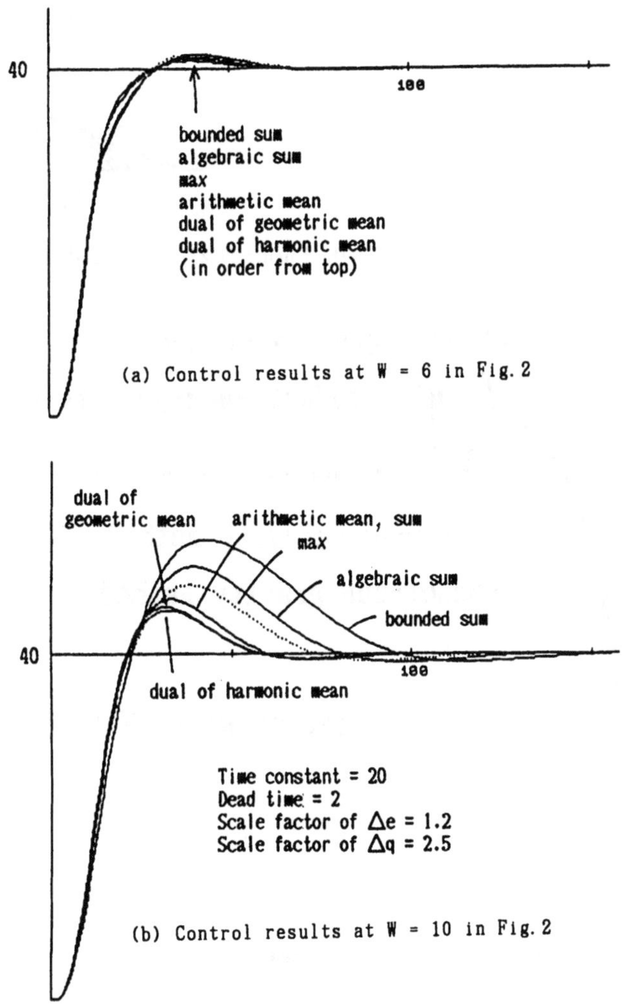

Figure 2 Control results under various aggregation methods: (a) control result at W=6; (b) control result at W=10.

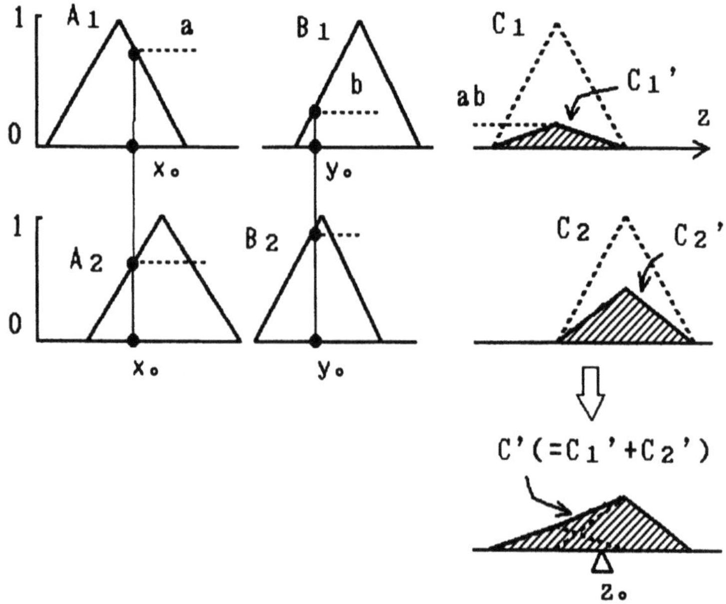

Figure 3 Product-sum-gravity method using (1.23), (1.24) and (1.6)

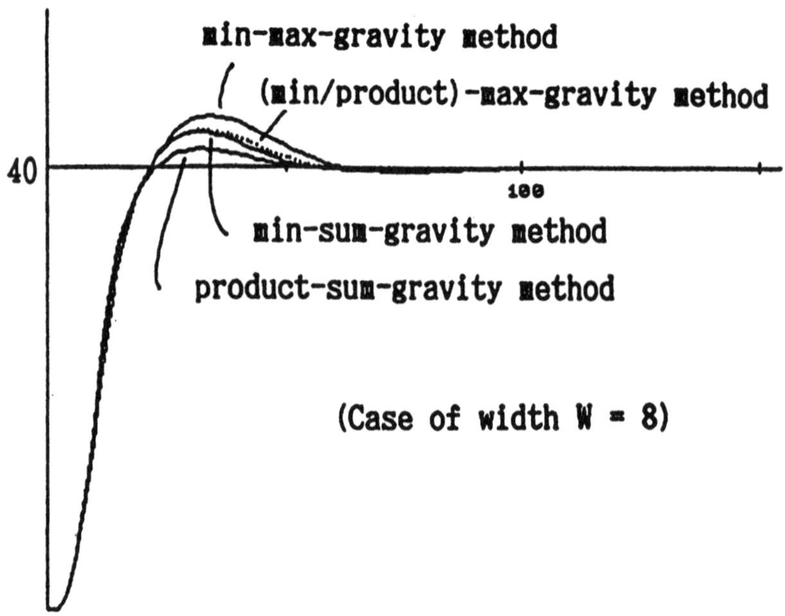

Figure 4 Control results by new fuzzy reasoning methods.

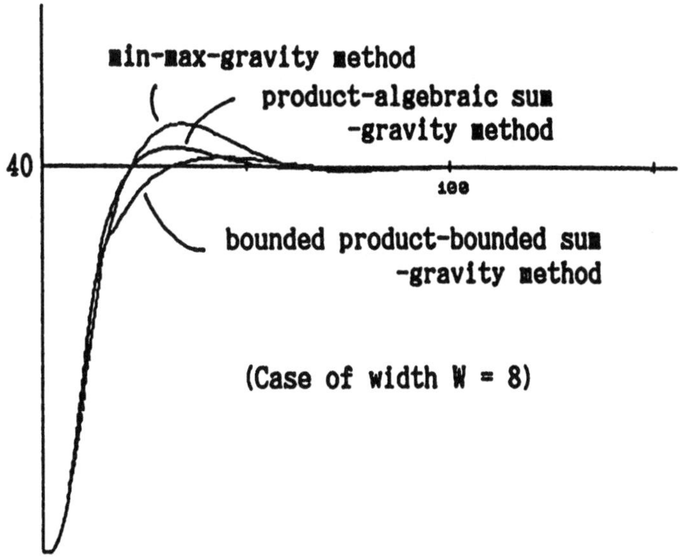

Figure 5 Case of fuzzy reasoning methods with dual operations.

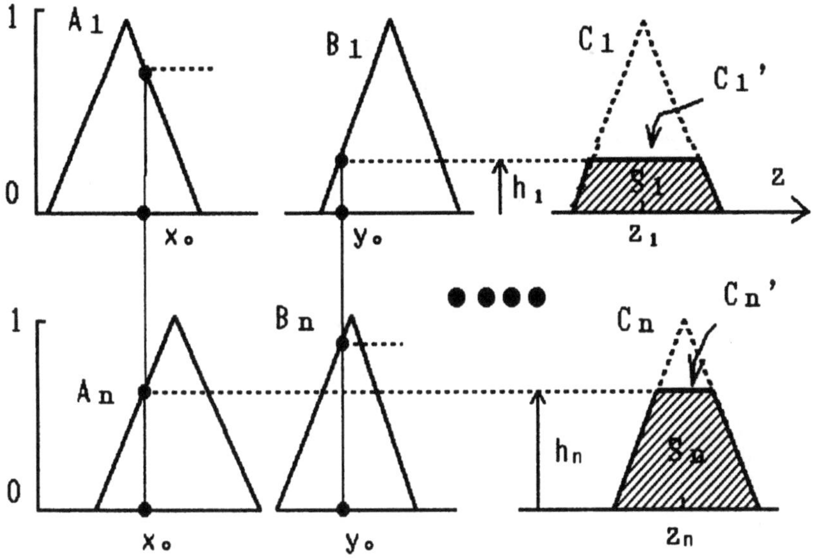

Figure 6 Height h_i and area S_i of inference result C_i'

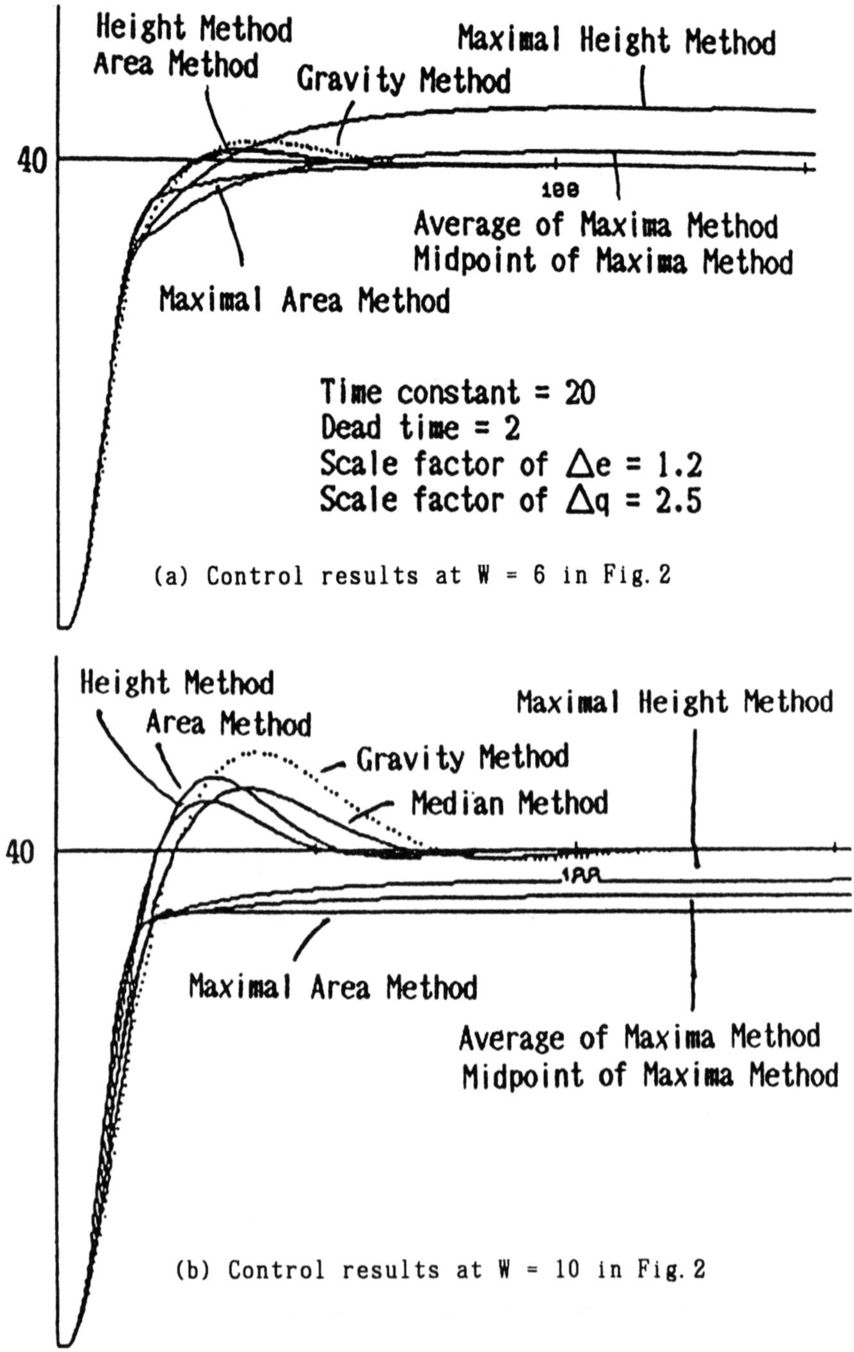

Figure 7 Control results under various defuzzification methods: (a) at W=6 in Figure 2, (b) at W=10 in Figure 2.

2

NEURAL NETWORKS AND FUZZY BASIS FUNCTIONS FOR FUNCTIONAL APPROXIMATION

Liang Jin, Madan M. Gupta, Peter N. Nikiforuk

Intelligent Systems Research Laboratory
College of Engineering, University of Saskatchewan
Saskatoon, Saskatchewan, Canada S7N 0W0

ABSTRACT

Universal approximation capabilities of neural networks and fuzzy basis functions are given in this chapter using the Stone-Weierstrass theorem, Kolmogorov's theorem and functional analysis methods. This study focuses on few commonly-used neural networks such as multilayered feedforward neural networks (MFNNs) with sigmoidal activation functions, trigonometric networks, higher-order neural networks, Gaussian radial basis function networks, and fuzzy basis function networks. The results show that an arbitrary continuous function on a compact set may be approximated to any degree of accuracy by such a neural network or fuzzy system. However, the accuracy of the approximations is strongly related to the design of the learning phases of the network parameters. The theory presented in this chapter provides a theoretical basis for applications to the fields of identification, control and pattern recognition.

1 INTRODUCTION

The function approximation capability of a multilayered feedforward neural network (MFNN) architecture is one of the most exciting properties of the neural structure and has the potential for applications to problems such as system identification, control and pattern recognition. During the past several years, the rigorous investigation of the approximation capabilities of standard multilayered feedforward architectures has received broad research interest. A feedforward network structure may be treated as a rule for computing output values of the neurons in the lth layer using the output values of the $(l-1)$th layer, hence, implementing a class of mapping from the input space $\Re^n$ to the output space $\Re^m$. Of interest here in this study, is what type and how well mappings from $\Re^n$ to $\Re^m$ can be approximated by the network, and how many neural layers and neurons in such layers are sufficient for this approximating process. This issue has been investigated by many authors including Carroll and Dickinson [6], Cybenko [5], Funahashi [7], Gallant and White [8], Hecht-Nielsen [10], and Hornik [15].

For function approximation, both the series expansion approach and the Stone-Weierstrass theorem are very effective analytic tools. However, Hecht-Nielsen [10,11] found the relationship between the Kolmogorov's theorem and the approximation principle of the feedforward networks. Indeed, functional analytic methods have been successfully used to show that feedforward neural structures with at-least one hidden layer are capable of simultaneously approximating continuous functions in several variables and their derivatives if the neural activation functions of the hidden neural units are differentiable [15]. On the other hand, the interplay between neural networks and fuzzy systems, in general, has received a great deal of attention in recent years [22]. Wang and Mendel [36] have shown that fuzzy systems may be expressed as a layered feedforward network. Therefore, building on the theoretical result that fuzzy systems are universal approximations is another interesting problem in the field of neural networks and fuzzy systems.

In this chapter, the universal approximation capabilities of feedforward neural networks are studied mainly using the well-known Stone-Weierstrass theorem. After the basic introduction of this theorem is provided in Section 2, the function approximation capabilities of trigonometric function network structures are discussed in Section 3. Functional approximation capabilities of multilayered feedforward neural networks (MFNNs) are addressed in Section 4. In Section 5, the relationships between the Kolmogorov's theorem and the feedforward neural networks are presented. As alternative structures of feedforward networks, some structures of higher-order neural networks are proposed in Section 6. Gaussian radial basis function networks are discussed in Section 7. The universal approximation using fuzzy basis function networks are discussed in Section 8. The results show that such a fuzzy neural network with Gaussian membership function is a universal approximator. Finally, Section 9 contains some conclusions.

2 STONE-WEIERSTRASS THEOREM AND IMPLICATIONS

In recent years, there have been attempts to find a mathematical justification for the use of the MFNNs. Typical results deal with the possibility, given a network, of approximating any continuous function arbitrarily well. In mathematical terms this means that a function can be computed by a network if it is dense in the space of the continuous functions defined on some subset of $\Re^n$. Next, we will show that the Stone-Weierstrass theorem plays an important role in conducting such an issue of approximation of the feedforward neural networks (Hornik, Stinchcombe and white, [14]; Cotter, [4]). The two equivalent descriptions of this theorem are as follows (Ray, [28]).

Theorem 1 (Stone-Weierstrass Theorem I, Ray, [28]) *Let S be a compact set with n dimensions, and $\Omega \supset C(S)$ be a set of continuous real-valued functions on S, satisfying the following conditions: (a) Identity Function: The constant function $f(\mathbf{x}) = \mathbf{1}$ is in Ω;* (b) *Separability: For any two points $\mathbf{x_1}, \mathbf{x_2} \in \mathbf{S}$ and $\mathbf{x_1} \neq \mathbf{x_2}$, there exists a $f \in \Omega$ such that $f(\mathbf{x_1}) \neq \mathbf{f}(\mathbf{x_2})$;* (c) *Linear subspace: For any f, $g \in \Omega$ and $\alpha \in \Re$, the functions αf and $f + g$ are in Ω;* (d) *Lattice Property: For any $f, g \in \Omega$, the functions $f \vee g = max(f, g)$ and $f \wedge g = min(f, g)$ are in Ω. Then Ω is dense in $C[S]$. In other words, for any $\epsilon > 0$ and any function $g \in C[S]$ there is a function $f \in \Omega$ such that $|g(\mathbf{x}) - \mathbf{f}(\mathbf{x})| < \epsilon$ for all $\mathbf{x} \in \mathbf{S}$.*

This lattice property is somewhat difficult to verify. Consequently, a slightly different statement of this theorem, with respect to the properties of algebraic closure, is sometimes more useful in applications.

Theorem 2 (Stone-Weierstrass Theorem II, Ray, [28]) *Let S be a compact set with n dimensions, and $\Omega \supset C[S]$ be a set of continuous real-valued functions on S satisfying the conditions: (a) Identity Function: The constant function $f(\mathbf{x}) = \mathbf{1}$ is in Ω; (b) Separability: For any two points $\mathbf{x_1}, \mathbf{x_2} \in \mathbf{S}$ and $\mathbf{x_1} \neq \mathbf{x_2}$, there exists a $f \in \Omega$ such that $f(\mathbf{x_1}) \neq \mathbf{f}(\mathbf{x_2})$; (c) Algebraic Closure: For any f, $g \in \Omega$ and α, $\beta \in R$, the functions fg and $\alpha f + \beta g$ are in Ω. Then Ω is dense in $C[S]$.*

Although the Stone-Weierstrass theorem has a potential application to continuous function approximation, many interesting functions, including step functions, are discontinuous. These functions are members of the set of bounded measurable functions that are continuous and bounded functions and have a finite number of discontinuities. Fortunately, the Stone-Weierstrass theorem can be extended to bounded measurable functions by applying the following theorem.

Theorem 3 (Lusin, Ray, [28]) *If g is a measurable real-valued function that is bounded almost everywhere on a compact set $S \supset \Re^n$, then given $\delta > 0$ there is a continuous real-valued function f* on *S such that the measure of the set where f is not equal to g is less than δ*

$$m\left\{ \mathbf{x} : \mathbf{f}(\mathbf{x}) \neq \mathbf{g}(\mathbf{x}), \mathbf{x} \in \mathbf{S} \right\} < \delta.$$

In other words, the minimum total volume of open spheres required to cover the set where $f \neq g$ is less than δ.

Theorem 3 shows that the continuous functions are dense in the space of the bounded measurable functions on a compact set S. Generally, for a compact set $S \supset \Re^n$, the space $L^p[S]$, $1 \leq p < \infty$, which consists of all the real measurable Lebesgue-integrable functions with a finite L^p norm is

$$L^p[S] = \left\{ f(\mathbf{x}) : ||\mathbf{f}(\mathbf{x})||_\mathbf{p} < \infty, \quad \mathbf{x} \in \mathbf{S} \right\}$$

,

where L^p, $1 \leq p < \infty$ norm is defined as

$$||f||_p \equiv \left\{ \int_S |f|^p dx \right\}^{1/p}$$

Therefore, the continuous function space $C[S]$ in the Stone-Weierstrass theorem may be replaced by $L^p[S]$ so that we could consider not only the continuous function approximation problem but also the discontinuous cases.

For the applications of neural networks, we have to assume that S is an arbitrary compact set in $\Re^n$, so the concept of the uniformly denseness given as follows is important. Let $S \subset \Re^n$ be a compact set, and $\Omega \subset C[S]$ be a set of continuous real-valued functions on S. If Ω for arbitrary S is dense in $C[S]$ then Ω is *uniformly dense* in $C[S]$.

The feedforward neural networks as described by some nonlinear mapping from the input pattern space to the output pattern space are said to be the universal approximators in terms of that they are capable of approximating arbitrary nonlinear functions on compact sets to any degree of error. However, implementing such an approximation process fully depends upon an effective weight learning procedure. Based on the principle offered by the Stone-Weierstrass theorem, as another objective of the next several sections, we will design some feedforward neural network structures which might be different from the conventional MFNNs with sigmoidal activation functions and satisfy the Stone-Weierstrass theorem so that they are also universal approximators.

3 TRIGONOMETRIC FUNCTION NETWORKS

Trigonometric functions have been extensively used in the Fourier series systems for representing the functions in the forms of trigonometric series consisting of sines and cosines. In this section, we will show that, as a choice of the nonlinear activation functions in the feedforward neural networks, trigonometric functions, in particular, sines and cosines, may be employed in the hidden neural units so that the resulting networks satisfy the conditions of the Stone-Weierstrass theorem. Also, the studies on the trigonometric function networks in this section provide the preparations for exploring universal approximation of the MFNNs which will be addressed in the next section. Without loss of generality, the case of single output is discussed, however, extension of the results to the networks with multiple outputs is straightforward.

By the basic trigonometric system we mean the system of functions 1, $cos\ x$, $sin\ x$, $cos\ 2x$, $sin\ 2x$, ..., $cos\ nx$, and $sin\ nx$. All these functions have the common period 2π. A two-layered trigonometric network with a single hidden layer is described by the following input-output transfer function

$$y = \sum_{i=1}^{N} u_i \phi_i (\sum_{j=1}^{n} w_{ij} x_j + \theta_i), \tag{2.1}$$

which is simply obtained by replacing the sigmoid function with a trigonometric function in the conventional two-layered neural network as shown in Figure 1. The trigonometric activation function ϕ may be chosen as: i) all $\phi_i(x) = cos(x)$, (cosine network); ii) all $\phi_i(x) = sin(x)$, (sine network); iii) $\phi_i(x) = cos(x)$, or $sin(x)$, (trigonometric network).

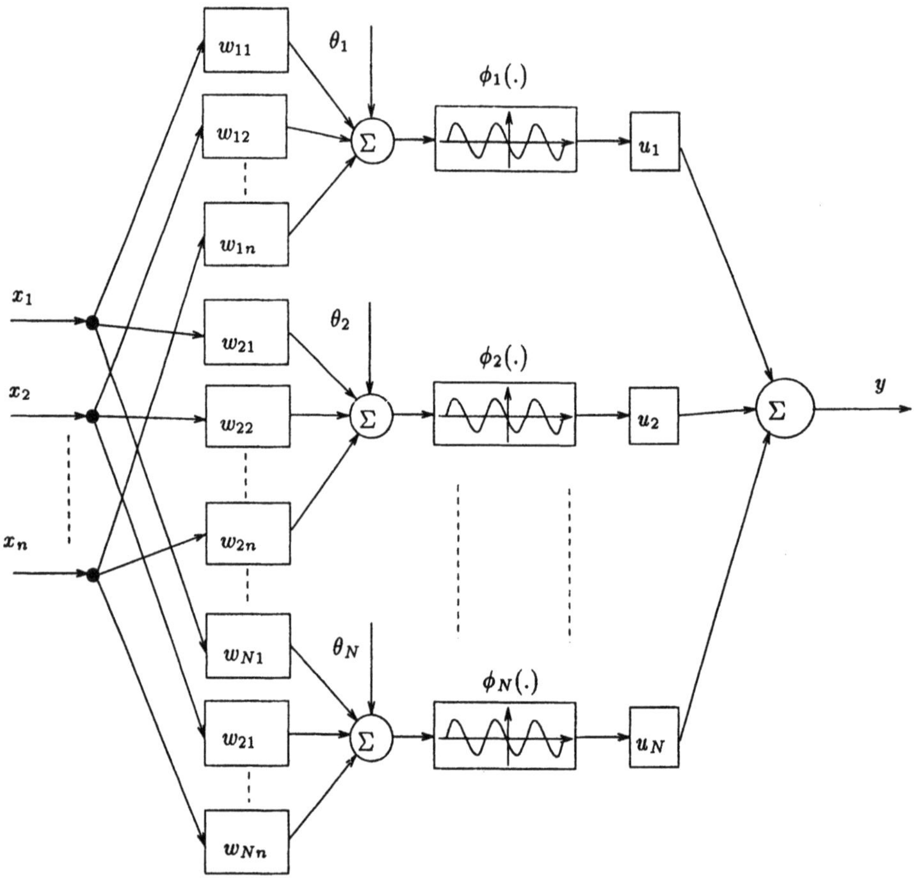

Figure 1 Block diagram of the trigonometric network.

Trigonometric functions can process transforming multiplication into addition since the familiar trigonometric formulas. Hence, the following results of the trigonometric network (2.1) ensure the universal approximation capability.

Theorem 4 *Let Ω be the set of all the functions that can be represented by the trigonometric networks on a compact set $S \supset \Re^n$:*

$$\Omega_N = \left\{ f(\mathbf{x}) = \sum_{\mathrm{i}=1}^{\mathrm{N}} \mathbf{u}_\mathrm{i} \phi(\sum_{\mathrm{j}=1}^{\mathrm{n}} \mathbf{w}_{\mathrm{ij}} \mathbf{x}_\mathrm{j} + \theta_\mathrm{i}) : \mathbf{u}_\mathrm{i}, \mathbf{w}_{\mathrm{ij}}, \theta_\mathrm{i} \in \Re, \mathbf{x} \in \mathbf{S} \right\}$$

$$\Omega = \bigcup_{N=1}^{\infty} \Omega_N$$

Then, Ω is uniformly dense in $C[S]$.

The trigonometric networks are a typical class of the feedforward neural networks with non-sigmoidal functions. A comparison of the classical trigonometric series expansion and the network expression of a continuous function indicates that the trigonometric network is more flexible and useful for many applications since the restriction of the periodic property of the function is removed, and the coefficient of the trigonometric series have to be solved analytically using the function to be approximated while the weights of the network can be determined through a learning process.

The trigonometric activation functions used in the trigonometric networks are well defined on the real axis in the sense of continuity and differentiable property. Having a closer look on these functions in an interval with only half period finds that there exist some similarities such as the characteristics of non-decreasing and bounded between the trigonometric functions and the sigmoidal functions used in the MFNNs. The concept of the squashing functions introduced by Hornik, Stinchcombe and White [14] may generalize the group of the sigmoidal functions which are assumed to be continuous and differentiable. For convenience, we consider the bipolar squashing functions in this text which are formally defined as follows: A function $\psi : \Re \longrightarrow [-1, 1]$ is a squashing function if it is non-decreasing and satisfies

$$\lim_{x \longrightarrow +\infty} \psi(x) = 1, \quad \textit{and} \quad \lim_{x \longrightarrow -\infty} \psi(x) = -1.$$

The squashing functions have at the most countably many discontinuities, they are measurable. The sigmoidal functions which are obviously squashing functions are usually chosen as continuous and differentiable functions. Typical examples of such functions $\sigma(.)$ are

$$tanh(x); \qquad \frac{1-e^{-x}}{1+e^{-x}}; \qquad \frac{2}{\pi}tan^{-1}(\frac{\pi}{2}x).$$

Another useful group of squashing functions include the signum function $sgn(x)$ defined by

$$sgn(x) = \begin{cases} -1, & if \ x < -1 \\ 1, & if \ x \geq 1 \end{cases}$$

and the ramp or saturating function $sat(x)$ defined by

$$sat(x) = \begin{cases} -1, & if \ x \leq -1 \\ x, & if \ -1 \leq x \leq 1 \\ 1, & if \ x \geq 1 \end{cases}$$

which are also given in Figure 2.

The trigonometric functions defined on the whole real axis do not belong to the group of the squashing functions because of their periodic property, however, they may be used to form some new squashing functions as seen in the following discussion.

The Fourier network is a direct extension of the cosine network. It is another two-layered network with a non-sigmoidal function and was proposed by Gallant and White [8] who implemented the Fourier series in the network structure. The activation function in the original Fourier networks was obtained by chopping the sinusoids into half-cycle sections and adding flat fails (Gallant and White, [8]). The resulting function is called a *bipolar sigmoidal cosine activation function.* Fourier networks with a bipolar activation function may be represented, therefore, by

$$y = \sum_{i=1}^{N} w_i \psi(\sum_{j=1}^{n} w_{ij} x_j + \theta_i) \tag{2.2}$$

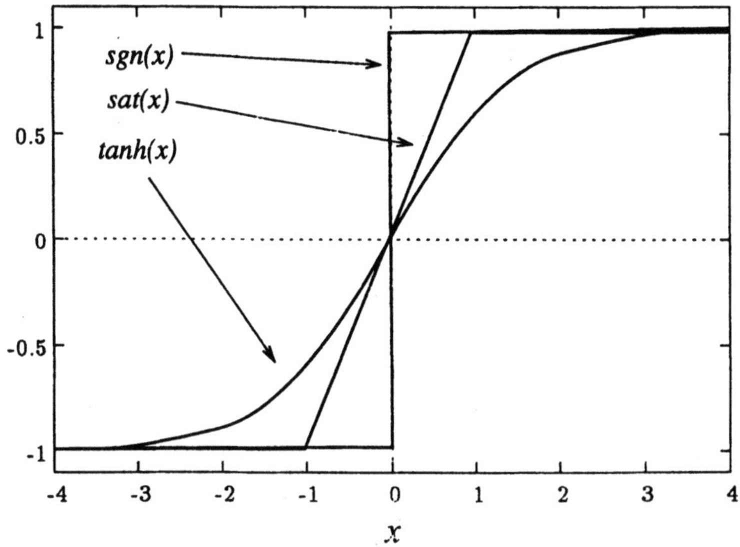

Figure 2 The sigmoidal function $tanh(x)$, signum function $sgn(x)$ and saturating function $sat(x)$.

where $\psi(.)$ is a sigmoidal cosine squashing function, as shown in Figure 3a, with a form

$$\psi(x) = \begin{cases} -1, & x \leq -\pi/2 \\ cos(x + 3\pi/2), & -\pi/2 \leq x \leq \pi/2 \\ 1, & x \geq \pi/2 \end{cases} \tag{2.3}$$

A slightly modified version of the sigmoidal cosine squashing function $\psi(x)$ is a cosine squashing function and is called a *cosig function* [4],

$$cosig(x) = \begin{cases} -1, & x \leq -1/2 \\ cos(2\pi x), & -1/2 \leq x \leq 0 \\ 1, & x \geq 0 \end{cases} \tag{2.4}$$

which is shown in Figure 3b. Corresponding to the cosig activation function, a two-layered cosig network may be given by

$$y = \sum_{i=1}^{N} w_i cosig(\sum_{j=1}^{n} w_{ij} x_j + \theta_i) \tag{2.5}$$

which deals with only a left-half set of functions computed by the Fourier networks (2.2).

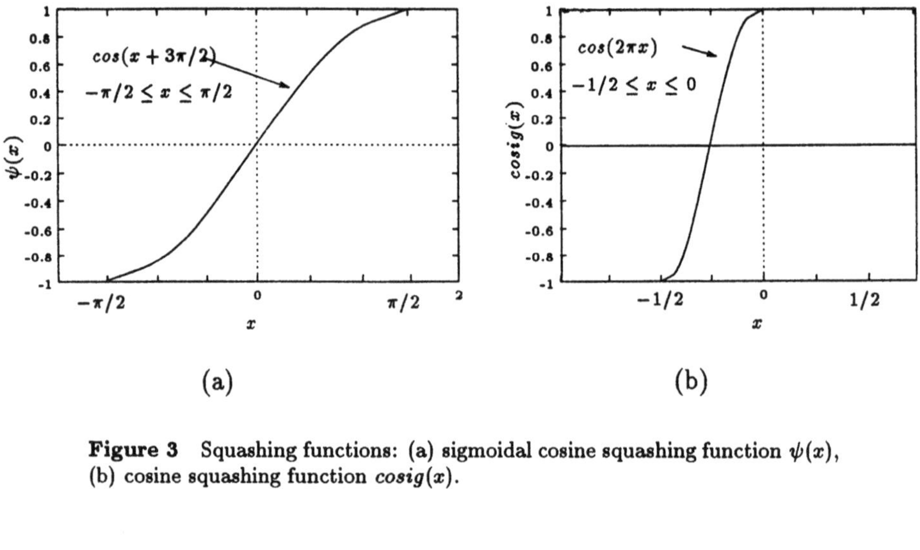

(a) (b)

Figure 3 Squashing functions: (a) sigmoidal cosine squashing function $\psi(x)$, (b) cosine squashing function $cosig(x)$.

Theorem 5 *Let Ω be the set of all functions that can be represented by either the Fourier network or the cosig network on a compact set $S \supset \Re^n$, then Ω is uniformly dense in $C[S]$.*

Proof. See Appendix 1.

4 MFNNS AS UNIVERSAL APPROXIMATORS

The commonly used two-layered feedforward neural network with a continuous sigmoidal function does not satisfy the Stone-Weierstrass theorem because the multiplicative condition fails. Hence, the denseness of such a feedforward neural network can not be immediately implied using the Stone-Weierstrass theorem. Using functional analysis methods, the capabilities of MFNNs may be addressed in either a constructive or non-constructive way. The scope of all these proof methods is too ambition, even if some significant proofs are worth to be reviewed. In particular, Hornik, Stinchcombe, and White's proof [14] used the trigonometric networks as an intermediate tool to conduct the problem, however, the trigonometric networks are not unique choice of the basis functions as pointed out by Blum and Li [1], the Chebyshev polynomials may replace the trigonometric functions as the basis functions. Blum and Li [1] addressed the approximation of real functions by feedforward networks based on the fundamental principle of approximation by piecewise-constant functions. Also, they suggested the difficulties associated with determining two-layered networks. For other approaches to approximation by the feedforward networks see Cybenko [5], Lapedes-farmer, and Funahashi [7]. All of these use semi-linear units and, for the most part, monotonic threshold functions. The proofs are non-constructive or not constructive in a simple way, since they depend on the Fourier transforms, Radon transforms, the Hahn-Banach theorem and so on.

4.1 A Sketch Proof for Two-Layered Networks

The approximation capabilities of the MFNNs will be discussed using the denseness of the cosig network presented in the last section. Hornik, Stinchcombe and White [14] proposed an elegant approach to indirectly conduct the denseness of the space spanned by the two-layered networks with sigmoidal functions in the continuous function space. The first step shows that the a single variable cosine squasher function can be uniformly approximated by a single-input and two-layered network with a sigmoid function. Secondly, one proves that an arbitrary cosig network discussed in the previous section can be uniformly approximated by a two-layered network with sigmoidal function. Finally, the denseness of the space spanned by the cosig network, as known in the last section, implies the denseness of the space of the two-layered networks with sigmoidal functions. Next, we will outline a proof which is mainly based on the Hornik, Stinchcombe and White's idea and is starting from the following lemma with some more readable description.

Lemma 1 *Let* $\sigma : \Re \longrightarrow [-1,1]$ *be a sigmoidal function and* $cosig : \Re \longrightarrow [-1,1]$ *be a cosine squashing function defined in Eq. (2.4). For every* $\epsilon > 0$ *there exists a two-layered network* $f(x) = \sum_{i=1}^{N} u_i\sigma(w_i x + \theta_i)$, x, u_i, w_i, $\theta_i \in R$, *such that* $\sup_{x\in R} |f(x) - cosig(x)| < \epsilon$.

Proof. See Appendix 2.

Lemma 1 not only shows the capability of the two-layered networks for approximating a cosig function being a special class of the squashing functions but also gives an analytic formulation for selecting the number of the hidden units for a desired degree of approximation error since its proof is completed using a constructive way. Next, using the results obtained in Lemma 1, we will show that an arbitrary cosig network may be uniformly approximated by a two-layered network with sigmoidal function.

Lemma 2 *Let* $\mathbf{x} \in \Re^{\mathbf{n}}$, *an arbitrary two-layered cosig network* $g(\mathbf{x}) = \sum_{\mathbf{i=1}}^{\mathbf{N}} \mathbf{u_i}$ $cosig(\sum_{j=1}^{n} w_{ij}x_j + \theta_i)$, $u_i, w_{ij}, \theta_i \in \Re$ *and* $\sigma : \Re \longrightarrow [-1,1]$ *be a squashing function. For every* $\epsilon > 0$ *and an arbitrary compact set* $S \supset \Re^n$, *there is a two-layered feedforward neural network* $f(\mathbf{x}) = \sum_{\mathbf{l=1}}^{\mathbf{\tilde{N}}} \alpha_{\mathbf{l}}\sigma(\sum_{\mathbf{p=1}}^{\mathbf{n}} \beta_{\mathbf{lp}}\mathbf{x_p} + \gamma_{\mathbf{l}})$ *such that* $\sup_{x\in S} |f(\mathbf{x}) - \mathbf{g}(\mathbf{x})| < \epsilon$.

Proof. See Appendix 3.

Lemma 2 indicates that the function space spanned by two-layered networks with sigmoidal functions is uniformly dense in the cosig network function space if both the networks are defined on a compact set. With these preliminary results in hand, the following main theorem may be derived.

Theorem 6 *Let $\sigma : \Re \longrightarrow [-1,1]$ be a sigmoid function and Ω be the set of all the functions that can be represented by the two-layered network on an arbitrary compact set $S \supset \Re^n$:*

$$\Omega_N = \left\{ f(\mathbf{x}) = \sum_{\mathrm{i}=1}^{\mathrm{N}} \mathbf{u}_\mathrm{i}\sigma(\sum_{\mathrm{j}=1}^{\mathrm{n}} \mathbf{w}_{\mathrm{ij}}\mathbf{x}_\mathrm{j} + \theta_\mathrm{i}) : \mathbf{u}_\mathrm{i}, \mathbf{w}_{\mathrm{ij}}, \theta_\mathrm{i} \in \Re, \mathbf{x} \in \mathbf{S} \right\}$$

$$\Omega = \bigcup_{N=1}^{\infty} \Omega_N.$$

Then Ω is uniformly dense in $C[S]$.

Proof. Since the function space spanned by the cosig network is uniformly dense in $C[S]$, and Ω is uniformly dense in the cosig network space by Lemma 2, the theorem is then implied. □

4.2 Approximation using General MFNNs

The approximation capability of the two-layered neural networks with sigmoidal functions is ensured in Theorem 8. However, no information is given on the number of the hidden units to achieve a satisfactory approximation even the continuous functions to be approximated are well-known. On the other hand, one may note that the continuities of the sigmoidal functions are not necessary in the above proof. This leads a natural extension that the sigmoidal functions often used in the conventional neural networks may be replaced with a more general class of the squashing functions in terms of the universal approximation.

An interesting consequence is that the approximation using a two-layered network, as shown in Figure 4 with the hidden units of McCulloch-Pitts, called Mc-P units, is easily implied. This type of feedforward neural networks is said to be the *neural logic networks* and may be obtained by replacing the sigmoidal function $\sigma(x)$ with the signum function as follows

$$f(x) = \sum_{i=1}^{N} u_i sgn(\sum_{j=1}^{n} w_{ij} x_j + \theta_i) \tag{2.6}$$

The network (2.6) consists of the hidden units of threshold elements which deal with a threshold logic on the real input variables $x_1, x_2, \ldots, x_n$. When the input $\mathbf{x}$ is restricted on a compact set in $\Re^n$ the network is capable of approximating any continuous function to a desired degree of accuracy. This conclusion is also summarized in the following corollary.

Corollary 1 *Let $S \supset R^n$ be a compact set and $g \in C[S]$ be a continuous function. For any $\epsilon > 0$, there is a two-layered network consisting of Mc-P units in the hidden layer with the form* $f(\mathbf{x}) = \sum_{i=1}^{N} \mathbf{u}_i \mathbf{sgn}(\sum_{j=1}^{n} \mathbf{w}_{ij}\mathbf{x}_j + \theta_i)$ *such that* $|f(\mathbf{x}) - \mathbf{g}(\mathbf{x})| < \epsilon$, for $\mathbf{x} \in \mathbf{S}$.

Although two-layered networks with Mc-P hidden neurons are capable of approximating arbitrary continuous functions, Blum and Li [1] proved that there is a class of piecewise constant functions which can not be implemented by a two-layered Mc-P network. Therefore, in the direct approach to function approximation, three-layered Mc-P networks with two hidden layers are required in general. Using the results on the two-layered networks, the approximation capabilities of the multilayered feedforward neural networks (MFNNs) may be easily explored.

Corollary 2 *Let $S \supset R^n$ be a compact set and $g \in C[S]$ be a continuous function. For any $\epsilon > 0$, there is a MFNN with arbitrary hidden layers and the sigmoidal functions which approximates g uniformly on S with error $< \epsilon$.*

Proof. See Appendix 4.

Corollary 2 gives the results of the approximation capabilities of general MFNNs with sigmoidal functions. In fact, the neural activation functions in MFNNs may be relaxed to any continuous, bounded and non-constant functions (Hornik, [12]).

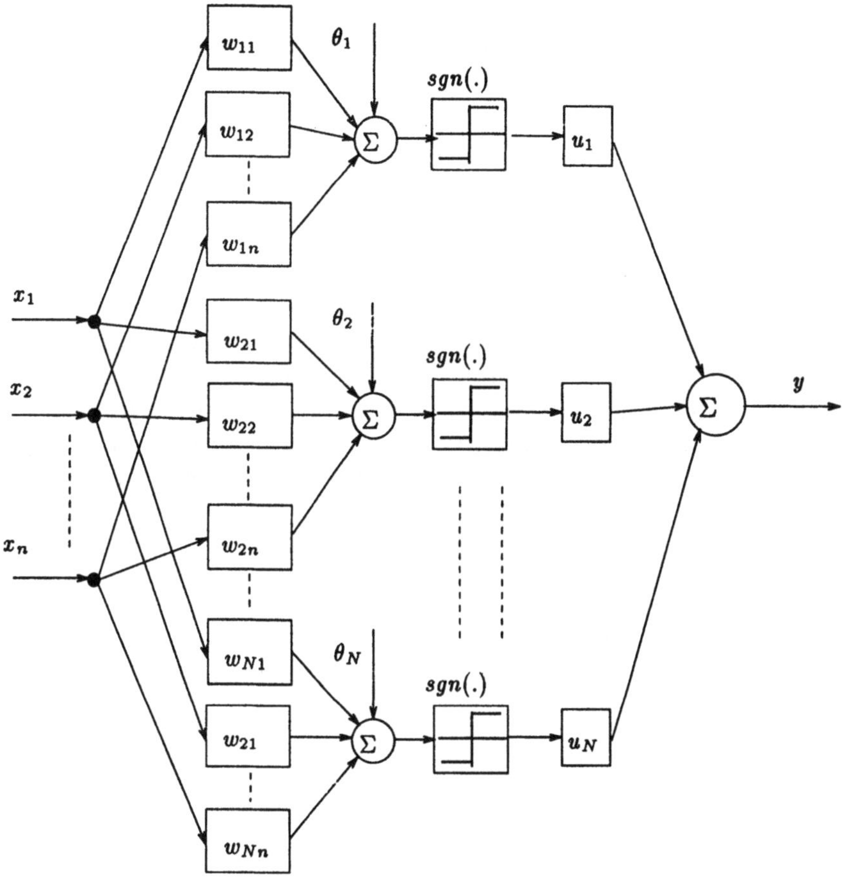

Figure 4 Two-layered network with the Mc-P hidden units.

5 KOLMOGOROV'S THEOREM AND FEEDFORWARD NETWORKS

The applications of the Kolmogorov's superposition theorem which concerns the representation of continuous functions defined on an n-dimensional cube by sums and superpositions to feedforward neural networks were first studied by Hecht-Nielsen [10], [11]. The study gives an existence of an exact implementation of every continuous function in a structure of three-layered networks. As one of the pioneers in the field of neural networks, Hecht-Nielsen gave some interpretation of the approximation principle of the Kolmogorov's theorem in terms of feedforward neural networks before some more practical achievements of universal approximation capabilities of the feedforward networks were develop independently by Cybenko [5], Funahashi [7], and Hornik, Stinchcombe and White [14]. Most recently, Sprecher [32] presented some new results which may be viewed as a stronger version of the Hecht-Nielsen's results. However, Girosi and Poggio [27] pointed out that the one-variable functions constructed by Kolmogorov [21], and the later improvements by Lorentz [25] and Sprecher [32], are far from being any of the type of functions used in feedforward neural networks.

Let $I = [0,1]$ denote the closed unit interval and $I^n = [0,1]^n$ $(n \geq 2)$ be the Cartesian product of I. The superposition theorem of Kolmogorov (1957) establishes that for each integer $n \geq 2$ there are $n \times (2n+1)$ continuous monotonically increasing functions h_{pq} and $(2n+1)$ continuous functions g_q which can be used to represent exactly every real-valued continuous function $f : I^n = [0,1]^n \longrightarrow \Re$. The original statement of Kolmogorov can be given as follows.

Theorem 7 (Kolmogorov's Superposition Theorem) *There exist a set of increasing continuous functions $h_{pq} : I = [0,1] \rightarrow \Re$ so that each continuous function f on I^n can be written in the form*

$$f(\mathbf{x}) = \sum_{\mathbf{q=1}}^{\mathbf{2n+1}} \mathbf{g_q}\left(\sum_{\mathbf{p=1}}^{\mathbf{n}} \mathbf{h_{pq}}(\mathbf{x_p})\right) \tag{2.7}$$

where g_q are properly chosen continuous functions of one variable.

Kolmogorov's theorem shows that any continuous function of several variables can be represented exactly by means of a superposition of continuous functions of a single variable and the operation of addition. Moreover, the functions h_{pq} are universal for the given dimension n; they are independent of a given function f. Only the functions g_q are specific for the given function f. Using the language of feedforward neural networks, we may explain Kolmogorov's theorem as follows: Any continuous function defined on an n-dimensional cube can be implemented exactly by a two-layered feedforward

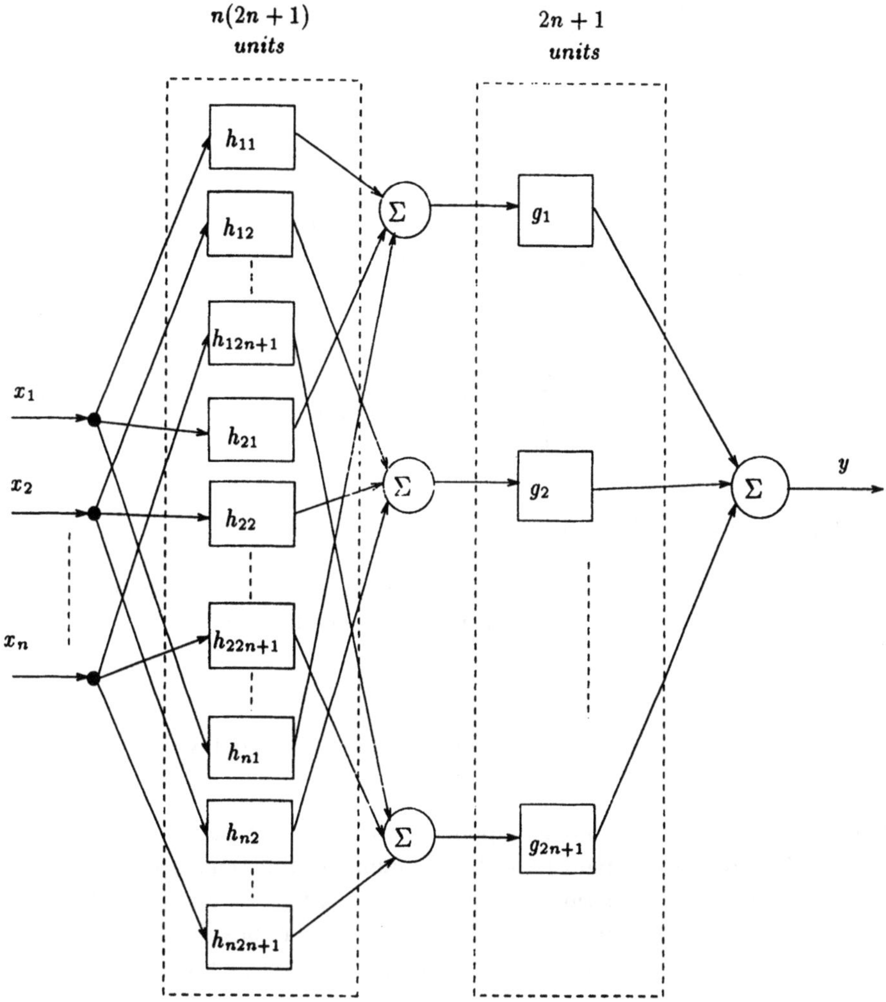

Figure 5 The block diagram representing the Kolmogorov's superposition theorem using a two-layered network structure.

network as shown in Figure 5, which has $n(2n+1)$ units with the increasing continuous functions $h_{pq} : I \to R$ in the first hidden layer and $(2n+1)$ units with the continuous functions g_q in the second hidden layer.

The main improvements to the original Kolmogorov's theorem concentrate on the possibility of replacing the function g_q by a single function g (Lorentz, [25]) and of transforming h_{pq} into $l_p h_1$ (Sprecher, [32]). Let H be the space with the uniform norm consisting of all the nondecreasing continuous functions on the closed interval $I = [0, 1]$ and $H^k = H \times \ldots \times H$ be the kth power of space H. Kahane modified Kolmogorov's theorem using the following results.

Theorem 8 *Let l_p $(p = 1, \ldots, n)$ be a collection of rationally independent constants. Then for quasi every collection $\{h_1, \ldots, h_{2n+1}\} \in H^{2n+1}$ any function $f \in C(I^n)$ can be represented on I^n in the form*

$$f(\mathbf{x}) = \sum_{\mathbf{q=1}}^{\mathbf{2n+1}} \mathbf{g}\Big(\sum_{\mathbf{p=1}}^{\mathbf{n}} \mathbf{l_p h_q(x_p)}\Big) \tag{2.8}$$

where g is a continuous function.

In order to give a geometric interpretation of Theorem 8, consider the mapping of I^n into a $(2n + 1)$-dimensional space defined by

$$y_q = l_1 h_q(x_1) + \ldots + l_n h_q(x_n), \qquad q = 1, \ldots, 2n + 1 \tag{2.9}$$

This is a continuous one-to-one mapping. Otherwise, two points of I^n would exist, which are not distinguishable by the family of functions $y_q(x_1, \ldots, x_n)$, $q = 1, \ldots, 2n + 1$. All functions which are representable by Eq. (2.9) would coincide at these two points, and Eq. (2.9) would be impossible for some functions $f \in C[I^n]$. Indeed, since I^n is compact, its image under mapping is

$$T = \left\{ y = (y_1, \ldots, y_{2n+1}) : y_q = \sum_{p=1}^{n} l_p h_q(x_p), x \in I^n \right\} \tag{2.10}$$

which is also compact, and the mapping (2.9) is a homomorphism between I^n and T. It implies that there exists a one-to-one relationship between all the continuous functions $f(x_1, \ldots, x_n)$ on I^n and all the continuous functions $F(y_1, \ldots, y_{2n+1})$ on T. Therefore, Theorem 8 can be rewritten as follows : There exists homeomorphic embeddin (2.9). from I^n into the $(2n+1)$-dimensional Euclidean space $\Re^{2n+1}$; that is $y_q : I^n \rightarrow \Re$, $q = 1, \ldots, 2n+1$, so that each continuous function F on the image space T of I^n has the form

$$F(y_1, \ldots, y_{2n+1}) = \sum_{p=1}^{2n+1} g(y_p) \tag{2.11}$$

More recently, an improved version of Kolmogorov's theorem due to Sprecher [32] was represented by Hecht-Nielsen [11] as a result concerning the existence of feedforward neural networks. The theorem follows:

Theorem 9 (Kolmogorov's Mapping Neural Network Existence Theorem, Hecht-Nielsen, [11]) *Given any continuous function* $\mathbf{f} : \mathbf{I^n} \longrightarrow \Re^{\mathbf{m}}$ *with* $n \geq 2$, $\mathbf{f}(\mathbf{x}) = \mathbf{y}$. *Then* $\mathbf{f}$ *can be implemented exactly by the following network*

$$f_i(\mathbf{x}) \quad = \quad \sum_{k=1}^{2n+1} g_i(z_k) \tag{2.12}$$

$$z_k = \sum_{j=1}^{n} \lambda^k h(x_j + k\epsilon) + k \tag{2.13}$$

where the real constant λ *and the continuous real monotonically increasing function* h *are independent of* f *although they do depend on* n. *The real and continuous functions* g_i *are dependent on the function* f_i and ϵ. *The constant* ϵ *is a rational number* $0 \leq \epsilon \leq \delta$, *where* δ *is an arbitrary chosen positive constant.*

The proof of the above theorem may be completed using the result of Sprecher [32] to each of the m coordinates of $\mathbf{y}$ separated. As shown in Figure 6, the implementation given in the above theorem is a two-layered neural network having n processing units in the input layer, $(2n+1)$ processing units in the hidden layer which receive the input $\mathbf{x}$ and create the outputs z_1, z_2, ..., z_{2n+1}, and m processing units in the output layer which give the outputs $y_1, y_2, \ldots, y_m$.

Kolmogorov's theorem provides only a structure of three-layered feedforward networks which can represent exactly arbitrary continuous functions. No further results concerning the network functions g and h_q have been obtained yet. The proof of the theorem is not constructive, so it does not show us how to select these quantities. It is strictly an existence theorem. This becomes a main limitation for the application of Kolmogorov's theorem. As expected by Hecht-Nielsen [11], a potentially high-payoff challenge is to discover an adaptive mechanism whereby the g_i's could self-organize themselves in response to incoming $\mathbf{x}$ and $\mathbf{y}$ vector pairs. However, exact network expressions of arbitrary continuous functions are very attractive for the function approximation issue. In addition, some progress on the applications of Kolmogorov's theorem for constructing multilayered neural networks has been made during the last several years (Kurkova, [23]; Sprecher, [3]). In particular, the work on the estimation of the number of the hidden units of a three-layered network using Kolmogorov's theorem presented by Kurkova [23] is one of the interesting examples for these applications.

6 APPROXIMATION USING POLYNOMIAL NEURAL NETWORKS

6.1 Higher-Order Neural Networks

As seen previously, a conventional neuron in the MFNN has only a linear correlation between the input vector and the synaptic weight vector. This correlation was described as a sort of synaptic operations. To capture higher order nonlinear properties of the input pattern space, recently, the extensive attempts have been made by Rumelhart [29]; Giles and Maxwell; Softky and Kammen [34]; Xu, Oja and Suen [37]; Taylor and Coombes [35] towards developing architectures of the neurons which are capable of capturing not only the linear correlation of the input patterns and the weights but also higher-order correlations between the components of the input pattern. Higher-order neural networks have been proven to have good computational, storage, pattern recognition, and learning properties and are realizable in

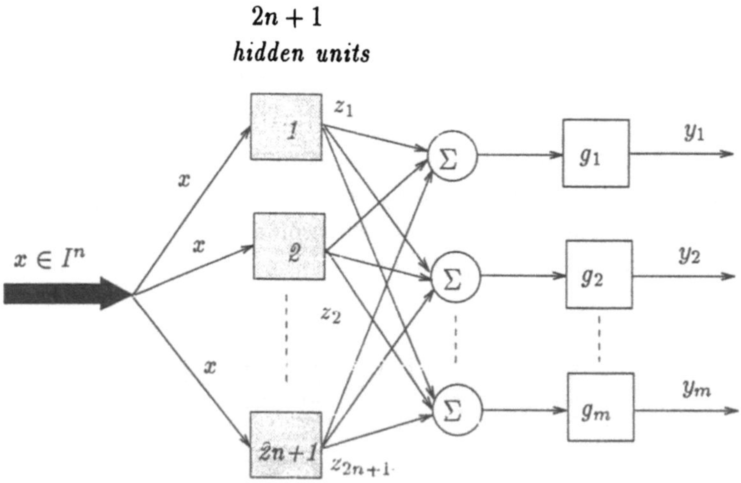

Figure 6 A schematic representation of the Kolmogorov's mapping neural network existence theorem proposed by Hecht-Nielsen, [11].

hardware (Taylor and Coombes, [35]). The regular polynomial networks which contain the higher-order correlations of the input components satisfy the Stone-Weierstrass Theorem, but the number of weights required to accommodate all of the higher-order correlations increases exponentially with the number of the inputs. The basic building block of such a *higher-order network* (HON), also called higher-order neural unit (HONU), is the Nth order processing unit whose output y is given by

$$y = \phi(z) \tag{2.14}$$

$$z = w_0 + \sum_{i_1}^{n} w_{i_1} x_{i_1} + \sum_{i_1,i_2}^{n} w_{i_1 i_2} x_{i_1} x_{i_2} + \ldots + \sum_{i_1,\ldots,i_N}^{n} w_{i_1 \ldots i_N} x_{i_1} \ldots x_{i_N} \tag{2.15}$$

where $\mathbf{x} = [\mathbf{x_1}\ x_2\ \ldots\ x_n]^T$ is the input vector, y is an output, and $\phi(.)$ is a strictly monotonic activation function such as a sigmoidal function whose inverse exists and is denoted as $\phi^{-1}(.)$. The summation for the kth order correlation is taken on a set $C(i_1 \ldots i_j)$ $(1 \leq j \leq N)$ which is a set of the combinations of j indices $1 \leq i_1, \ldots, i_j \leq n$ defined by

$$C(i_1 \ldots i_j) = \left\{ < i_1 \ldots i_j >: 1 \leq i_1 \ldots i_j \leq n, \quad i_1 \leq i_2 \leq \ldots \leq i_j \right\}$$

where $1 \leq j \leq N$. The introduction of the set $C(i_1 \ldots i_j)$ is used to absorb the redundant terms due to the symmetry of the induced combinations. In fact, Eq. (2.15) is a truncated Taylor series with some adjustable coefficients. The Nth order neural unit needs a total of

$$\sum_{j=0}^{N} \left(\begin{array}{c} n+j-1 \\ j \end{array} \right) = \sum_{j=0}^{N} \frac{(n+j-1)!}{j!(n-1)!}$$

weights including the basis if all of the products up to N components are considered.

The HONUs may be used in the conventional feedforward neural network structure as the hidden units to form the so-called HON. In this case, however, considerations on the higher correlation may improve the capabilities of approximation and generalization of the networks, typically only second-order networks are usually employed, in practice, to lead a tolerable number of the weights. On the other hand, if the order of the HONU is high enough, as known from the Stone-Weierstrass theorem, Eqs. (2.14), (2.15) may be considered as a network with n inputs and a single output and is capable of dealing with the problems of functional approximation and pattern recognition as seen in the following discussion.

To get a closer look on Eq. (2.14)-(2.15), we denote the higher correlation terms of an n-dimensional input $\mathbf{x} \in \Re^n$ as follows

$$\left\{ \begin{array}{ccc} u_{i_1} & = & x_{i_1} \\ u_{i_1,i_2} & = & x_{i_1} x_{i_2} \\ \vdots & & \\ u_{i_1 i_2 \ldots i_N} & = & x_{i_1} x_{i_2} \ldots x_{i_N} \end{array} \right. \tag{2.16}$$

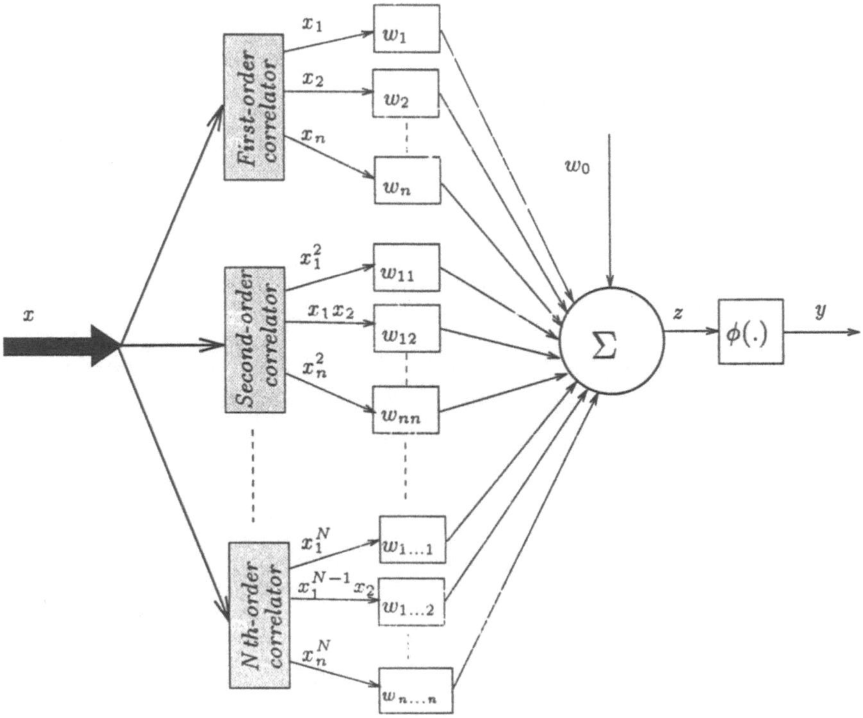

Figure 7 Block diagram of the higher-order neural unit (HONU).

then the network equations (2.14)-(2.15) may be represented as

$$y = \phi(z) \tag{2.17}$$

$$z = w_0 + \sum_{j=1}^{N} \left(\sum_{i_1 \ldots i_j} w_{i_1 \ldots i_j} u_{i_1 \ldots i_j} \right) \tag{2.18}$$

which may be treated as a two-layered neural network as shown in Figure 7. Here, $u_{i_1 \ldots i_j}$ are outputs of the hidden units being able to produce the higher order correlations between the components for each vector input pattern $\mathbf{x}$, and the output neuron is a simple linear combiner with an activation function $\phi(.)$.

To accomplish an approximation task for a given set of input-output data $\{\mathbf{x(k)}, \mathbf{y(k)}\}$, the learning algorithm for the HONU can be easily developed on the basis of the gradient descent method. Let us assume that the error function is formulated as

$$E(k) = \frac{1}{2}(d(k) - y(k))^2 = \frac{1}{2}e^2(k)$$

where $e(k) = d(k) - y(k)$ at the instant k. Minimization of the error function by a standard steepest-descent technique yields the following set of learning equations

$$w_0^{new} = w_0^{old} + \eta(d - y)\phi'(z) \tag{2.19}$$

$$w_{ij}^{new} = w_{ij}^{old} + \eta(d - y)\phi'(z)u_{i_1 \ldots i_j} \tag{2.20}$$

where $\phi'(z) = d\phi/dz$. Like the back-propagation (BP) algorithm for the MFNN (Rumelhart et al., [29]), an momentum version of the above is easily obtained.

Alternatively, since all weights appear linearly in Eq. (2.18) of the higher-order networks, one may use the method for solving linear algebraic equation to carry out the above learning task if the number of the patterns is finite. Doing so, one has to introduce the following two augmented vectors

$$\mathbf{w} = [\mathbf{w_0} \ \mathbf{w_1} \ \ldots \ \mathbf{w_n} \ \mathbf{w_{11}} \ \ldots \ \mathbf{w_{nn}} \ \ldots \ \mathbf{w_{1\ldots1}} \ \ldots \ \mathbf{w_{n\ldots n}}]^T$$

$$\mathbf{u} = [\mathbf{x_0} \ \mathbf{x_1} \ \ldots \ \mathbf{x_n} \ \mathbf{x_1^2} \ \mathbf{x_1 x_2} \ \ldots \ \mathbf{x_n^2} \ \ldots \ \mathbf{x_1^N} \ \mathbf{x_1^{N-1} x_2} \ \ldots \ \mathbf{x_n^N}]^T$$

where $x_0 = 1$, so that the network equations (2.17)-(2.18) may be rewritten into the following compact vector form

$$y = \phi\left(\mathbf{w}^{\mathbf{T}}\mathbf{u}(\mathbf{x})\right) \tag{2.21}$$

For the given p pattern pairs $\{\mathbf{x}(\mathbf{k}), \mathbf{d}(\mathbf{k})\}$ $(1 \leq k \leq p)$, define the following vectors and matrix

$$\mathbf{U} = \begin{bmatrix} \mathbf{u}^{\mathbf{T}}(\mathbf{1}) \\ \mathbf{u}^{\mathbf{T}}(\mathbf{2}) \\ \vdots \\ \mathbf{u}^{\mathbf{T}}(\mathbf{p}) \end{bmatrix}, \quad \mathbf{d} = \begin{bmatrix} \phi^{-1}(d(1)) \\ \phi^{-1}(d(2)) \\ \vdots \\ \phi^{-1}(d(p)) \end{bmatrix}$$

where $\mathbf{u}(\mathbf{k}) = \mathbf{u}(\mathbf{x}(\mathbf{k}))$, $1 \leq k \leq p$. Then, the learning problem becomes seeking a solution of the following linear algebraic equation

$$\mathbf{U}\mathbf{w} = \mathbf{d} \tag{2.22}$$

If the number of the weights is equal to the number of the data and the matrix $\mathbf{U}$ is nonsingular, Eq. (2.22) has an unique solution

$$\mathbf{w} = \mathbf{U}^{-1}\mathbf{d}$$

The more interesting case is that the dimension of the weight vector $\mathbf{w}$ is less than the number p of the data, thus, the condition for existence of the exact solution of the above linear is given by

$$rank\left[\mathbf{U} \ \vdots \ \mathbf{d}\right] = rank[\mathbf{U}] \tag{2.23}$$

In case the above condition is not satisfied the pseudo-inverse solution is usually an option and gives a best fit result. The higher-order networks are capable of dealing with the functional approximation and pattern classification problems. Also, Xu, Oja, Suen [37], Taylor and Coombes [35] demonstrated that they may be effectively applied for problem of using a model of curve, surface, and hyper-surface to fit a given data set. This problem is called the nonlinear surface fitting and is often encountered in many engineering applications. Some learning algorithms for solving such problem may be found in their papers. Moreover, if one assumes $\phi(x) = x$ in the HONU, the weights exhibit linearly in the networks and the learning algorithms for the HONs may be characterized as a linear LS procedure, the well-known local minimum problems existing in many nonlinear neural learning schemes may be avoided.

6.2 Sigma-Pi Networks

Note that the HON contains simply all linear and nonlinear correlation terms of the input components until the order n. A slight generalized structure of the HON is a polynomial network which includes weighted sums of products of selected input components with an appropriate power. Mathematically, the input-output transfer function of this network structure is given by

$$y = \phi\left(\sum_{i=1}^{N} w_i u_i\right) \tag{2.24}$$

$$u_i = \prod_{j=1}^{n} x_j^{w_{ij}} \tag{2.25}$$

where w_i, $w_{ij} \in \Re$, N is the order of the network, and u_i is the output of the ith hidden unit. As has been pointed out by Rumelhart et al. [29], this type of the feedforward networks is called the ***sigma-pi*** networks. It is easy to show that the above network satisfies the Stone-Weierstrass theorem if $\phi(x) = x$ is a linear function. From the network structure point, a sigma-pi network shown in Figure 8 may be considered as a two-layered network with a hidden layer and an output layer, where the units in the hidden layer create the products of selected input components computed with a power operation while, like the conventional weighted combiners, the output unit just makes a weighted summation of all outputs of the hidden units.

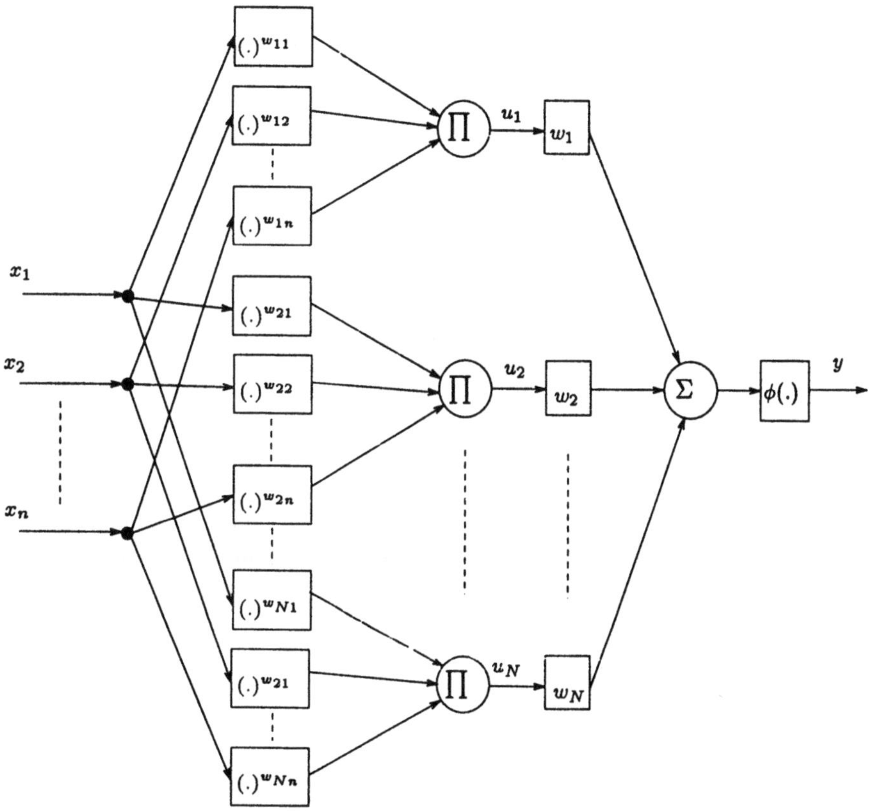

Figure 8 Block diagram of the sigma-pi network.

Moreover, a modified version of the sigma-pi networks was proposed by Hornik [14] and Cotter [4] as follows

$$y = \phi\left(\sum_{i=1}^{N} w_i u_i\right) \tag{2.26}$$

$$u_i = \prod_{j=1}^{n} \left(p(x_j) \right)^{w_{ij}} \tag{2.27}$$

where w_i, $w_{ij} \in \Re$, and $p(x_j)$ is a polynomial of x_j. It is easy to verify that the above network satisfies the Stone-Weierstrass theorem, thus, it may be a choice as an approximator for problems of functional approximation. The sigma-pi network (2.24)-(2.25) is a special case of the above network while $p(x_j)$ is assumed to be a linear function of x_j. In fact, the weights w_{ij} in both the networks (2.24)-(2.25) and (2.26)-(2.27) may be restricted to integer or nonnegative integer values.

6.3 Ridge Polynomial Networks

To maintain the fast learning and powerful mapping capabilities, and to avoid the combinatorial increase in the number of weights of the higher-order network, some modified polynomial network structures were introduced recently. One of these is the ***pi-sigma network*** (PSN) (Shin and Ghosh, [30]) which is a regular higher-order structure and involves a much smaller number of weights compared to the higher-order network (HON). The mapping equation of a pi-sigma network can be represented as

$$y = \phi \left(\prod_{i=1}^{N} \sum_{j=1}^{n} (w_{ij} x_j + \theta_i) \right) \tag{2.28}$$

The total number of weights for a Nth order pi-sigma network with n inputs is only $(n+1)N$. Compared with the sigma-pi structure the number of weights involved in this network is significantly reduced. Unfortunately, when $\phi(x) = x$ the pi-sigma network does not match the conditions provided by the Stone-Weierstrass theorem because the linear subspace condition is not satisfied. However, some studies have shown that it is a good network model for smooth functions (Shin and Ghosh, [31]).

To modify the structure of the above pi-sigma networks such that they satisfy the Stone-Weierstrass theorem, Shin and Ghosh [31] suggested to consider the so-called *ridge polynomial network* (RPN). For the vectors $\mathbf{w_{ij}} = [\mathbf{w_{ij1}}, \ldots, \mathbf{w_{ijn}}]^{\mathbf{T}}$ and $\mathbf{x} = [\mathbf{x_1}\ x_2\ \ldots\ x_n]^T$, let $< \mathbf{x}, \mathbf{w_{ij}} >= \sum_{\mathbf{k=1}}^{\mathbf{n}} \mathbf{w_{ijk}x_k}$ represent a inner product between the two vectors. A one-variable continuous function f with form $f(< \mathbf{x}, \mathbf{w_{ij}} >)$ is called ridge function. A ridge polynomial is a ridge function that can be represented as

$$\sum_{i=0}^{N}\sum_{j=0}^{M} a_{ij} < \mathbf{x}, \mathbf{w_{ij}} >^{\mathbf{i}}$$

for some $a_{ij} \in \Re$ and $\mathbf{w_{ij}} \in \Re^{\mathbf{n}}$.

The operation equation of a ridge polynomial network (RPN) is given as

$$y = \phi\left(\sum_{j=1}^{N}\prod_{i=1}^{j}\left(< \mathbf{x}, \mathbf{w_{ij}} > +\theta_{\mathbf{ji}}\right)\right) \tag{2.29}$$

when $\phi(x) = x$ the denseness of the above network is easily verified and is described in the following theorem.

Theorem 10 *Let Ω be the set of all the functions that can be represented by the ridge polynomial network on a compact set $S \supset R^n$:*

$$\Omega_N = \left\{ f(\mathbf{x}) = \sum_{\mathbf{j=1}}^{\mathbf{N}}\prod_{\mathbf{i=1}}^{\mathbf{j}}(\sum_{\mathbf{k=1}}^{\mathbf{n}} \mathbf{w_{ijk}x_k} + \theta_{\mathbf{ji}}) : \mathbf{w_{jik}}, \theta_{\mathbf{ji}} \in \Re, \mathbf{x} \in \mathbf{S} \right\}$$

$$\Omega = \bigcup_{N=1}^{\infty} \Omega_N$$

Then Ω is uniformly dense in $C[R^n]$.

The total number of weights involved in this structure is $N(N+1)(n+1)/2$. A comparison of the number of weights of the above three types of polynomial network structures show that when the networks have the same higher-order terms the weights of RPN is significantly less than that of a HON. In particular, this is a very attractive improvement offered by the RPNs.

7 GAUSSIAN NETWORKS

A typical Gaussian network is a two-layered network with an input layer, and a hidden layer of Gaussian units and an output layer of conventional summation units as shown in Figure 9. Let $\mathbf{x} = [\mathbf{x}_1, \mathbf{x}_2, ..., \mathbf{x}_l]^T$ and $\mathbf{y} = [\mathbf{y}_1, \mathbf{y}_2, ..., \mathbf{y}_m]^T$ be the input and output of the network, respectively, and $\mathbf{u} = [\mathbf{u}_1, \mathbf{u}_2, ..., \mathbf{u}_n]^T$ be the n outputs of the n hidden Gaussian neurons. A Gaussian radial basis function ϕ_i with a weighted norm is defined by

$$\phi(||\mathbf{x} - \mathbf{c}_i||_{\mathbf{K}_i}) = \mathbf{e}^{-\mathbf{d}(\mathbf{x}, \mathbf{c}_i, \mathbf{H}_i)/2} \tag{2.30}$$

where

$$d(\mathbf{x}, \mathbf{c}_i, \mathbf{H}_i) = ||\mathbf{x} - \mathbf{c}_i||_{\mathbf{K}_i} = (\mathbf{x} - \mathbf{c}_i)^T \mathbf{H}_i (\mathbf{x} - \mathbf{c}_i) \tag{2.31}$$

with

$$\mathbf{H}_i = \mathbf{K}_i^T \mathbf{K}_i$$

and $\mathbf{c}_i \in \Re^n$ and $\mathbf{H}_i \in \Re^{n \times n}$ represent, respectively the mean vector and the shape matrix defined by the inverse of the covariance matrix of the ith radial basis function.

Furthermore, $d(\mathbf{x}, \mathbf{c}_i, \mathbf{H}_i)$ can be rewritten as an expanded form

$$d(\mathbf{x}, \mathbf{c}_i, \mathbf{H}_i) = \sum_{j=1}^{n} \sum_{k=1}^{n} \mathbf{h}_{ijk} (\mathbf{x}_j - \mathbf{c}_{ij})(\mathbf{x}_k - \mathbf{c}_{ik}) \tag{2.32}$$

where c_{ij} is the jth element of $\mathbf{c_i}$, and h_{ijk} is the (j,k) the element of $\mathbf{H_i}$.

Without loss of generality, h_{ijk} can be represented based on the marginal standard deviations σ_{ij} and σ_{ik}, and the correlation coefficient k_{ijk}

$$h_{ijk} = \frac{k_{ijk}}{\sigma_{ij}\sigma_{ik}} \tag{2.33}$$

where σ_{ij} is positive real number and $k_{ijk} = 1$, if $j = k$ and $|k_{ijk}| \leq 1$ otherwise. Instead of using the general form of h_{ijk} given by Eq. (2.33), we may simply assume that the shape matrix $\mathbf{H_i}$ is positive diagonal; that is

$$h_{ijk} = \begin{cases} 1/\sigma_{ij}^2, & if \quad j = k \\ 0, & otherwise \end{cases} \tag{2.34}$$

where σ_i is so-called standard deviation or variance for controlling the width of the Gaussian function. Therefore, the input-output relationship of a Gaussian network which might have multiple outputs is described by

$$u_i = exp\left[-\frac{1}{2}\sum_{k=1}^{n}(\frac{x_k - c_{ik}}{\sigma_{ik}})^2\right], \quad 1 \leq i \leq l \tag{2.35}$$

and

$$y_j = \sum_{i=1}^{l} w_{ji}u_i = \sum_{i=1}^{l} w_{ij} exp\left[-\frac{1}{2}\sum_{k=1}^{n}(\frac{x_k - c_{ik}}{\sigma_{ik}})^2\right], \ 1 \leq j \leq m \tag{2.36}$$

where u_i is the output of the ith hidden Gaussian neuron described by a Gaussian function which forms a hyper-ellipsoid in the n-dimensional space $\Re^n$, rather than a hyper-plane, c_{ik} and σ_{ik} are said to be the center and variance parameters, respectively, of the ith Gaussian function, which determine the geometric shape and position of the hyper-ellipsoid in $\Re^n$, and l is the number of the hidden Gaussian neurons. As seen above the hidden layer in a Gaussian network consists of an array of nodes or neurons which contain some parameter vector called centers. The hidden node calculates the weighted Euclidean distance between the center and the network input vector and the result is passed through a Gaussian function. The output layer is just a set of linear combiners.

Using the notation of the variance matrices defined as follows $\mathbf{\Sigma_i} = \mathbf{diag}[\sigma_{i1}\ \sigma_{i2} \ldots \sigma_{in}]$, Eqs. (2.35) and (2.36) may be rewritten as

$$u_i = exp\left[-\frac{1}{2}(\mathbf{x} - \mathbf{c_i})^{\mathbf{T}}\mathbf{\Sigma_i^{-2}}(\mathbf{x} - \mathbf{c_i})\right] \tag{2.37}$$

and

$$y_j = \sum_{i=1}^{l} w_{ji}u_i = \sum_{i=1}^{l} w_{ij}exp\left[-\frac{1}{2}(\mathbf{x_i} - \mathbf{c_i})^{\mathbf{T}}\mathbf{\Sigma_i^{-2}}(\mathbf{x} - \mathbf{c_i})\right] \tag{2.38}$$

The approximation capability of such a Gaussian network may be addressed using the multi-point interpolation approximation technique (Poggio and Girosi, [27]). However, we may discuss again this issue using the Stone-Weierstrass theorem. Note that the exponential function can process the multiplication into addition as follows $exp(x)\ \ exp(y) = exp(x+y)$ Hence, it can be verified that the Gaussian network satisfies the Stone-Weierstrass theorem.

Theorem 11 *Let Ω be the set of all functions that can be computed by Gaussian network on a compact set $S \supset R^n$:*

$$\Omega_N = \left\{ f(\mathbf{x}) = \sum_{i=1}^{N} \mathbf{w_i exp}\left[-\frac{1}{2}\sum_{k=1}^{n}\left(\frac{\mathbf{x_k} - \mathbf{c_{ik}}}{\sigma_{\mathbf{ik}}}\right)^{2}\right] : \mathbf{w_i}, \mathbf{c_{ik}}, \sigma_{\mathbf{ik}} \in \Re, \mathbf{x} \in \mathbf{S} \right\}$$

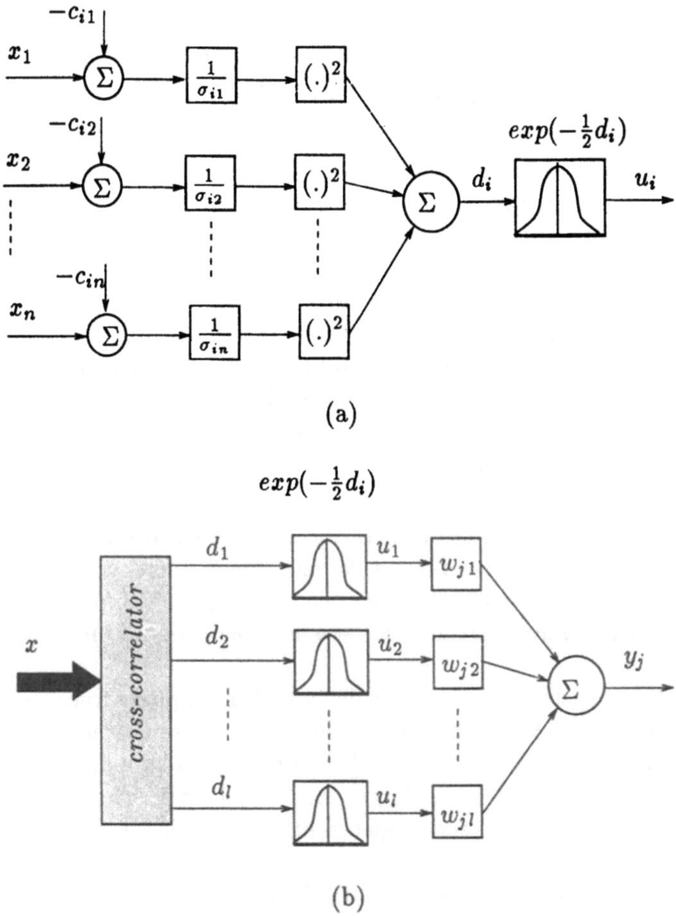

Figure 9 The schematic diagram of a Gaussian Network: (a) the connections between the inputs $x_1, x_2, \ldots, x_n$ and the ith hidden Gaussian unit, where $d_i = \sum_{k=1}^{l} (\frac{x_i - c_{ik}}{\sigma_{ik}})^2$, and $u_i = exp(-\frac{1}{2}d_i)$; (b) the connections between the output y_j and the nodes in the hidden layer.

$$\Omega = \bigcup_{N=1}^{\infty} \Omega_N$$

Then Ω is uniformly dense in $C[S]$.

8 UNIVERSAL APPROXIMATION USING FUZZY BASIS FUNCTION NETWORKS

As last topic of this chapter, the approximation capabilities of a fuzzy neural system, namely *fuzzy basis function networks* (FBFNs) (Wang and Mendel, [36]) will be now discussed. Fuzzy systems as fuzzy basis function expansions can be represented as two-layered feedforward network structures. Based on this idea, the fuzzy system may be trained to realize desired input-output relationship using various learning algorithms such as fuzzy back-propagation algorithm. As pointed out by Wang and Mendel [36], the most important advantage of using fuzzy basis functions, rather than polynomials, or radial basis functions, etc, is that a linguistic fuzzy IF-THEN rule is naturally related to a fuzzy basis function.

A fuzzy system whose basic configuration is depicted in Figure 10 has four principal elements: fuzzifier, fuzzy rule base, fuzzy inference engine, and defuzzifier. Without loss of generality, we consider multi-input and single-output fuzzy systems: $S \subset \Re^n \longrightarrow \Re$, where S is a compact set.

In such a fuzzy system, the fuzzifier deals with a mapping from the input space $S \in \Re^n$ to the fuzzy sets defined in S, which are characterized by a membership function $\mu_F : S \longrightarrow [0,1]$, and is labeled by a linguistic variables F such as "small", "medium", "large", or "very large". The most commonly-used fuzzifier is the singleton fuzzifier, which is defined as follows:

$\mathbf{x} \in \mathbf{S} \longrightarrow$ fuzzy set $A_x \subset S$ with $\mu_{A_x}(\mathbf{x}) = \mathbf{1}$ and $\mu_{A_x}(\mathbf{x}') = \mathbf{0}$ for $\mathbf{x}' \in \mathbf{S}$ and $\mathbf{x}' \neq \mathbf{x}$.

The fuzzy rule base consists of a set of linguistic rules in the form of "IF a set of conditions are satisfied, THEN a set of consequences are inferred." Moreover, we consider, in this section, the fuzzy rule base having the M rules with the following forms:

R_j $(j = 1, 2, \dots, M)$: IF x_1 is A_1^j and x_2 is A_2^j and ... and x_n is A_n^j, THEN y is B^j.

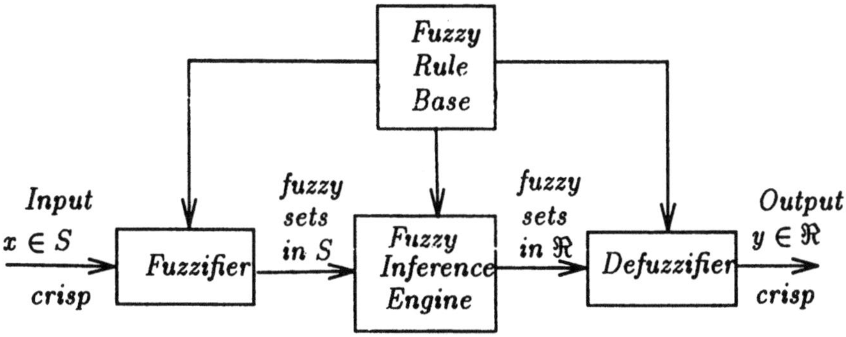

Figure 10 The schematic representation of a fuzzy system.

where x_i $(i = 1, 2, \ldots, n)$ are the input variables to the fuzzy system, y is the output variable of the fuzzy system, A_i^j and B^j are linguistic variables characterized by fuzzy membership functions $\mu_{A_i^j}$ and μ_{B^j}, respectively. A simple example is given in Figure 11. Each rule R_j can be viewed as a fuzzy implication

$$A_1^j \times \ldots \times A_n^j \longrightarrow B^j$$

which is a fuzzy set in $S \times \Re$ with

$$\mu_{A_1^j \times \ldots \times A_n^j \longrightarrow B^j}(x_1, \ldots, x_n, y) = \mu_{A_1^j}(x_1) \otimes \ldots \otimes \mu_{A_n^j} \otimes \mu_{B^j}(y)$$

for $\mathbf{x} \in \mathbf{S}$ and $y \in \Re$. The most commonly-used operations for $\otimes$ are product and min defined as

$$\mu_{A_1^j}(x_1) \otimes \mu_{A_2^j}(x_2) = \mu_{A_1^j}\mu_{A_2^j}(x_2)$$

and

$$\mu_{A_1^j}(x_1) \otimes \mu_{A_2^j}(x_2) = min[\mu_{A_1^j}(x_1), \mu_{A_2^j}(x_2)]$$

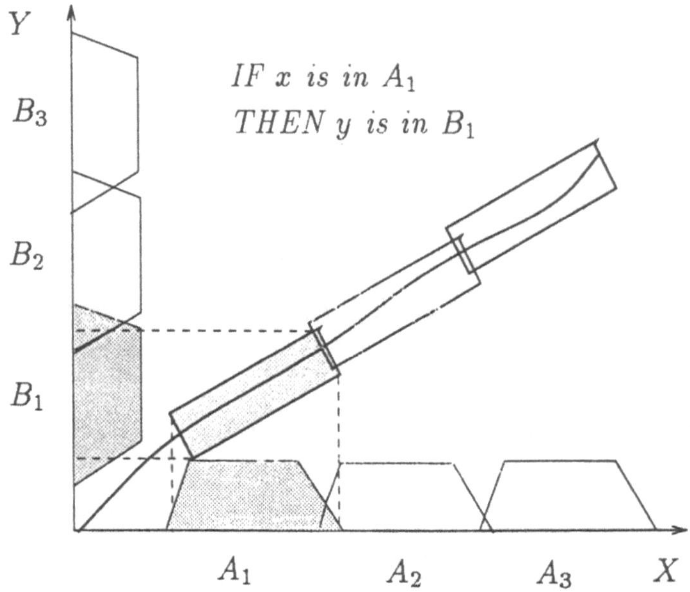

Figure 11 An example of fuzzy IF-THEN rule.

The fuzzy inference engine is a decision making logic which uses the fuzzy rules provided by the fuzzy rule base to implement a mapping from the fuzzy sets in the input space S to the fuzzy sets in the output space $\Re$. Let A_x be an arbitrary fuzzy set in S; then each R_j of the fuzzy rule base creates a fuzzy set $A_x \circ R_j$ in $\Re$ based on the sup-star composition:

$$\mu_{A_x \circ R_j} = \sup_{x' \in S} [\mu_{A_x}(\mathbf{x}') \otimes \mu_{\mathbf{A_1^j} \times \ldots \times \mathbf{A_n^j} \longrightarrow \mathbf{B_j}}(\mathbf{x}'_1, \ldots, \mathbf{x}'_n, \mathbf{z})]$$

which gives

$$\mu_{A_x \circ R_j} = \sup_{x' \in S} [\mu_{A_x}(\mathbf{x}') \otimes \mu_{\mathbf{A_1^j}}(\mathbf{x}'_1) \otimes \ldots \otimes \mu_{\mathbf{A_n^j}}(\mathbf{x}'_n) \otimes \mu_{\mathbf{B^j}}(\mathbf{z})] \qquad (2.39)$$

The defuzzifier provides a mapping from the fuzzy sets in $\Re$ to crisp points in $\Re$. The following centroid defuzzifier, which performs a mapping from the fuzzy set $A_x \circ R_j$ $(j = 1, 2, \ldots, M)$ in $\Re$ to a crisp point $y \in \Re$, is the most commonly used method

$$y = \frac{\sum_{j=1}^{M} c_j \mu_{A_x \circ R_j}(c_j)}{\sum_{j=1}^{M} \mu_{A_x \circ R_j}(c_j)} \tag{2.40}$$

where c_j is the point in $\Re$ at which $\mu_{B^j}(c_j)$ achieves maximum value, $\mu_{B^j}(c_j) = 1$.

Next, if one assumes that $\otimes$ is a product operation (product inference), note that $\mu_{A_x}(\mathbf{x}) = \mathbf{1}$ and $\mu_{A_x}(\mathbf{x}') = \mathbf{0}$ for all $\mathbf{x}' \in \mathbf{S}$ with $\mathbf{x}' \neq \mathbf{x}$, replacing $\otimes$ in Eq. (2.38) with conventional product yields

$$\mu_{A_x \circ R_j}(c_j) = \sup_{x' \in S} [\mu_{A_x}(\mathbf{x}')\mu_{\mathbf{A}_1^j}(\mathbf{x}_1') \cdots \mu_{\mathbf{A}_n^j}(\mathbf{x}_n')\mu_{\mathbf{B}^j}(\mathbf{c}_j)] = \prod_{\mathbf{i}=1}^{\mathbf{n}} \mu_{\mathbf{A}_i^j}(\mathbf{x}_i) \tag{2.41}$$

Therefore, the input-output equation of such a fuzzy system with singleton fuzzifier, product inference, and centroid defuzzifier may be expressed as

$$y = \frac{\sum_{j=1}^{M} c_j \left(\prod_{i=1}^{n} \mu_{A_i^j}(x_i)\right)}{\sum_{j=1}^{M} \left(\prod_{i=1}^{n} \mu_{A_i^j}(x_i)\right)} = \sum_{j=1}^{M} c_j \phi_j(\mathbf{x}) \tag{2.42}$$

where

$$\phi_j(\mathbf{x}) = \frac{\prod_{i=1}^{n} \mu_{A_i^j}(x_i)}{\sum_{j=1}^{M}\left(\prod_{i=1}^{n} \mu_{A_i^j}(x_i)\right)} \tag{2.43}$$

are called the *fuzzy basis functions.* Eq. (2.42) gives a fuzzy basis function expressions of the fuzzy systems.

In particular, Gaussian radial basis function, as shown in Figure 12, may be chosen as the membership function; that is

$$\mu_{A_i^j}(x_i) = exp\left[-\frac{1}{2}\left(\frac{x_i - \bar{x}_i^j}{\sigma_i^j}\right)^2\right]$$

where $\bar{x}_i^j$ and σ_i^j are real-valued parameters.

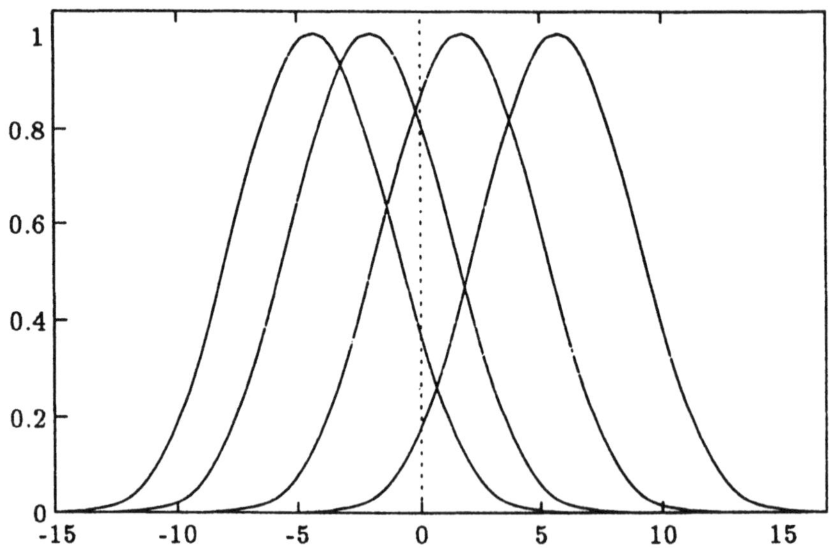

Figure 12 Gaussian membership function.

In this case, let

$$\Omega_M = \left\{ f(\mathbf{x}) = \frac{\sum_{j=1}^{M} c_j \left(\prod_{i=1}^{n} \mu_{A_i^j}(x_i) \right)}{\sum_{j=1}^{M} \left(\prod_{i=1}^{n} \mu_{A_i^j}(x_i) \right)} : c_j \in \Re, \mathbf{x} \in \mathbf{S} \right\}$$

$$\Omega = \bigcup_{M=1}^{\infty} \Omega_M$$

Then, one has the following results:

Theorem 12. *For arbitrary continuous function $g : S \subset \Re^n \longrightarrow \Re$ defined on the compact set $S \subset \Re^n$ and arbitrary $\epsilon > 0$, there exists a $f \in \Omega$ such that* $\max |g(\mathbf{x} - \mathbf{f}(\mathbf{x})| < \epsilon$ *for $x \in S$.*

A proof of the above theorem is based on the Stone-Weierstrass theorem and can be found in the Wang and Mendel's paper [36] This result shows that fuzzy basis function network with Gaussian membership function is an universal approximator. It might be noted that Eq. (2.42) defines only one type of FBFNs. Corresponding to different definitions of fuzzifier, fuzzy rule base, and defuzzifier, one may have other forms of FBFNs. In addition, to carry out such an input-output approximation process, some learning algorithms such as the analogy form of the back-propagation and Orthogonal least-squares algorithms have been proposed (Dickerson and Kosko, [6]; Wang and Mendel, [36]).

9 CONCLUSIONS

The approximation capabilities of feedforward neural networks and fuzzy systems were discussed in this chapter. Feedforward neural networks, as intelligent computing tools, contain many types of neural network structures which have various different mathematical expressions and have the different similarities to biological neural models. The studies here focus only on few commonly-used static neural network structures such as multilayered feedforward network, trigonometric networks, higher-order neural networks, and Gaussian basis function networks. On the other hand, fuzzy systems represented as fuzzy basis function networks with Gaussian membership function are also universal approximators as seen in the chapter. One may summarize that both the feedforward neural networks and fuzzy basis function networks are the universal approximators for continuous functions. However, the accuracy of the approximations not only depends on the network structures selected such as number of the layers, and the hidden units, but is also strongly related to the design of the learning phases of the network parameters.

Any lack of success in the applications of a neural network which is the universal approximator must arise from inadequate learning, an insufficient numbers of hidden units or the lack of a deterministic relationship between the input and the target. It is a fact that different neural networks result in different learning difficulties. Therefore, the choice of an appropriate approximation structure will ultimately determine the success of an application. ¿From this point of view, a successful neural approximation procedure may be divided into the following three steps: (i) determine the universal approximation structures of the neural networks; that is, ensure the inherent approximation capabilities of the neural networks through adjusting the numbers of the hidden units and layers; (ii) chose the adequate weight learning algorithms; and (iii) use the learning signals which contain sufficient information.

REFERENCES

[1] Blum, E.K. and Li, L.K. (1991). "Approximation theory and feedforward networks," Neural Networks, Vol. 4, No. 4, pp. 511-515.

[2] Cardaliaguet, P. and Euvrard, G. (1992). "Approximation of a function and its derivative with a neural network," Neural Networks, Vol. 5, No. 2, pp. 207-220.

[3] Carroll, S.M. and Dickinson, B.W. (1989). "Construction of neural nets using the Radon transform," Proceeding of International Joint Conference on Neural Networks, Vol. I, pp. 607-611, New York.

[4] Cotter, N. (1990). "The Stone-Weierstrass theorem and its application to neural networks," IEEE Trans. Neural Networks, Vol. 1, No. 4, 290-295.

[5] Cybenko, G. (1989). "Approximation by superpositions of a sigmoidal function," Math. Control Signal System, Vol. 2, No. 3, pp. 303-314.

[6] Dickerson, J. and Kosko, B. (1993). "Fuzzy function approximation with supervised ellipsoidal learning," Proc. of 1993 World Congress on Neural Networks, Vol. II, pp. 11-17, Portland, Oregon.

[7] Funahashi,K. (1989). "On the approximate realization of continuous mappings by neural networks,' vol 2, No. 3, pp. 183-192.

[8] Gallant, A.R. and White H. (1988). "There exists a neural network that does not make avoidable mistakes," Proc. of 1988 ICNN, Vol.II, pp. 657-664, San Diego.

[9] Gallant, A.R. and White H. (1992). "On learning the derivatives of an unknown mapping with multilayered feedforward networks," Neural Networks, Vol. 5, No. 1, pp. 129-138.

[10] Hecht-Nielsen, R. (1989). "Theory of the back-propagation neural network", Proc. Int'l. Joint Conf. on Neural Networks, pp. I-593-605.

[11] Hecht-Nielsen, R. (1987). "Kolmogorov's mapping neural network existence theorem," Proc. of 1987 ICNN, Vol. III, pp. 11-14.

[12] Hornik, K. (1991). "Approximation capabilities of multilayer feedforward networks," Neural Networks, Vol. 4, No. 2, pp. 251-257.

[13] Hornik, K. (1993). "Some new results on neural network approximation," *Neural Networks*, Vol. 6, No. 8, pp. 1069-1072.

[14] Hornik, K., Stinchcombe, M. and White, H. (1989). " Multilayer feedforward networks are universal approximators," Neural Networks, Vol. 2, No. 5, pp. 359-366.

[15] Hornik, K., Stinchcombe, M. and White, H. (1990). "Universal approximation of an unknown mapping and its derivatives using multilayer feedforward networks," Neural Networks, Vol. 3, No. 6, pp. 551-560.

[16] Hush, D.R. and Horne, B.G. (1993) " Progress in supervised neural networks," *IEEE* Signal Processing Magazine, No. 1, pp. 8-39.

[17] Ito, Y. (1991). "Approximation of functions on a compact set by finite sums of a sigmoid function without scaling," Neural Networks, Vol. 4, No. 1, pp. 105-115.

[18] Ito, Y. (1993). "Extension of approximation capability of three layered neural networks to derivatives," Proc. of 1993 ICNN, Vol. 1, pp. 377-381.

[19] Jin, L., Gupta, M.M. and Nikiforuk, P. N. (1994). "Approximation capabilities of feedforward and recurrent neural networks," in *Intelligent Control Systems*, Gupta, M.M. and Sinha, N.K. (Eds.), IEEE Press.

[20] Kaufmann, A. and Gupta, M.M. (1985). *Introduction to Fuzzy Arithmetic*, Van Nostrand, New York, 1985.

[21] Kolmogorov, A.N. (1957). "On the representation of continuous functions of several variables by superposition of continuous functions of one variable and addition," Dokl. Akad. Nauk USSR, Vol. 114, pp. 953-956.

[22] Kosko, B., (1992). *Neural Networks and Fuzzy Systems,* Prentice Hall, Englewood Cliffs.

[23] Kurkova, V. (1992). "Kolmogorov's theorem and multilayer neural networks," Neural Networks, Vol. 5, No. 3, pp. 501-506.

[24] Lee, S. and Kil, R. M., (1991). "A Gaussian potential function network with hierarchically self-organizing learning," *Neural Networks*, Vol. 4, pp. 207-224.

[25] Lorentz, G.G. (1986). *Approximation of Functions.* Chelsea Publishing Co., New York, 1976.

[26] Musavi, M., Ahmed, W., Chan, K. H., Faris, K. B. and Hummels, D. M., (1992). "On the training of radial basis function classifiers," *Neural Networks*, Vol. 5, pp. 595-603.

[27] Poggio, T. and Girosi, F. (1990). "Networks for approximation and learning," Proc. of IEEE, Vol. 78, pp. 1481-1497.

[28] Ray, W.O. (1988). *Real Analysis,* Prentice Hall Englewood Cliffs, New Jersey.

[29] Rumelhart, D.E., Hinton, G.E. and Williams, R.J. (1986). "Learning internal representations by error propagation," In Rumelhart, D.E. and McClelland, J. L., editors, *Parallel Distributed Processing: Explorations in the Microstructure of Cognition*, pp. 318-362, MIT Press, Cambridge, MA, 1986.

[30] Shin, Y. and Ghosh, J. (1991). "The pi-sigma network: an efficient higher-order neural network for pattern classification and function approximation," Proc. of IJCNN, Vol. 1, pp.13-18, Seattle, Washington, July.

[31] Shin, Y. and Ghosh, J. (1992). "Approximation of multivariate functions using ridge polynomial networks," Proceeding of International Joint Conference on Neural Networks, Vol. II, pp. 380-385, Baltimore, Maryland, June 1992.

[32] Sprecher, D. A. (1965). "On the structure of continuous functions of several variables," Trans. Amer. Math. Soc., 115, 340-355.

[33] Sprecher, D. A. (1993). "A universal mapping for Kolmogorov's superposition theorem," *Neural Networks*, Vol. 6, No. 8, pp. 1089-1094.

[34] Softky, R. W., and Kammen, D. M. (1991). Correlations in high dimensional or asymmetrical data sets: Hebbian neuronal processing." *Neural Networks*, Vol. 4, No. 3, pp. 337-347.

[35] Taylor, J.G. and Commbes, S. (1993). "Learning higher order correlations," *Neural Networks*, Vol. 6, No. 3, pp. 423-428.

[36] Wang, L. and Mendel, J.M. (1992). "Fuzzy basis functions, universal approximation, and orthogonal least-squares learning," IEEE Trans. On Neural Networks, Vol. 3, No. 5, pp. 807-814.

[37] Xu, L., Oja, E., and Suen, C.Y. (1992). "Modified Hebbian learning for curve and surface fitting, *Neural Networks*, Vol. 5, No. 3, pp. 441-457.

Appendix A: The Proof Of Theorem 5

We only prove the denseness of the Fourier network here. In this case, the set Ω_N is defined as

$$\Omega_N = \left\{ f(\mathbf{x}) = \sum_{\mathbf{i}=1}^{\mathbf{N}} \mathbf{u_i}\psi(\sum_{\mathbf{j}=1}^{\mathbf{n}} \mathbf{w_{ij}x_j} + \theta_\mathbf{i}) : \mathbf{u_i}, \mathbf{w_{ij}}, \theta_\mathbf{i} \in \Re, \mathbf{x} \in \mathbf{S} \right\}$$

where the function ψ is given by Eq. (2.3). Obviously, $f(\mathbf{x}) = \mathbf{1}$ is an element of the set Ω_N. Consider two arbitrary functions in Ω_N

$$f(\mathbf{x}) = \sum_{\mathbf{i_1}=1}^{\mathbf{N_1}} \mathbf{u^f_{i_1}} \psi(\sum_{\mathbf{j_1}=1}^{\mathbf{n}} \mathbf{w^f_{i_1j_1}x_{j_1}} + \theta^\mathbf{f}_\mathbf{i_1})$$

$$g(\mathbf{x}) = \sum_{\mathbf{i_2}=1}^{\mathbf{N_2}} \mathbf{u^g_{i_2}} \psi(\sum_{\mathbf{j_1}=1}^{\mathbf{n}} \mathbf{w^g_{i_2j_2}x_{j_2}} + \theta^\mathbf{g}_\mathbf{i_2})$$

Then for arbitrary constants α and $\beta \in \Re$, $\alpha f + \beta g \in \Omega$. Furthermore, note the definition of the function $\psi(x)$, we may imply

$$\psi(x)\psi(y) = \begin{cases} \psi(x), & if \;\; x, y \geq \frac{\pi}{2}, \;\; or \;\; x \leq -\frac{\pi}{2}, y \geq \frac{\pi}{2} \\ & or \;\; -\frac{\pi}{2} \leq x \leq \frac{\pi}{2}, \; y \geq \frac{\pi}{2} \\ \psi(y), & if \;\; x \geq \frac{\pi}{2}, y \leq -\frac{\pi}{2}, \\ & or \;\; -\frac{\pi}{2} \leq y \leq \frac{\pi}{2}, \; x \geq \frac{\pi}{2} \\ -\psi(x), & if \;\; -\frac{\pi}{2} \leq x \leq \frac{\pi}{2}, \; y \geq -\frac{\pi}{2} \\ -\psi(y), & if \;\; -\frac{\pi}{2} \leq y \leq \frac{\pi}{2}, \; x \leq \frac{\pi}{2} \\ \psi(z_1 - \frac{1}{2}) + \psi(z_2 + \frac{\pi}{2}), & if \;\; -\frac{\pi}{2} \leq x, y \leq \frac{\pi}{2} \\ & 0 \leq z_1 \leq \pi, \;\; -\pi \leq z_2 \leq 0 \\ -\psi(z_1 + \frac{1}{2}) + \psi(z_2 + \frac{\pi}{2}), & if \;\; -\frac{\pi}{2} \leq x, y \leq \frac{\pi}{2} \\ & -\pi \leq z_1 \leq 0, \;\; -\pi \leq z_2 \leq 0 \\ \psi(z_1 - \frac{1}{2}) - \psi(z_2 - \frac{\pi}{2}), & if \;\; -\frac{\pi}{2} \leq x, y \leq \frac{\pi}{2} \\ & 0 \leq z_1 \leq \pi, \;\; 0 \leq z_2 \leq \pi \\ -\psi(z_1 + \frac{1}{2}) - \psi(z_2 + \frac{\pi}{2}), & if \;\; -\frac{\pi}{2} \leq x, y \leq \frac{\pi}{2} \\ & -\pi \leq z_1 \leq 0, \;\; 0 \leq z_2 \leq \pi \end{cases}$$

where $z_1 = x + y$ and $z_2 = x - y$. Therefore,

$$f(\mathbf{x})\mathbf{g}(\mathbf{x}) = \sum_{\mathbf{i}=1}^{\mathbf{N}} \mathbf{u_i}\psi(\sum_{\mathbf{j}=1}^{\mathbf{n}} \mathbf{w_{ij}x_j} + \theta_\mathbf{i})$$

where the parameters u_i, w_{ij}, θ_i and N are uniquely determined by the networks of $f(\mathbf{x})$ and $g(\mathbf{x})$. From the Stone-Weierstrass theorem, the set $\Omega = \bigcup \Omega_N$ is uniformly dense in $C[S]$. □

Appendix B: The Proof Of Lemma 1

For an arbitrary $\epsilon > 0$, without loss of generality, assume $\epsilon < 1$. We will now find a finite collection of constants u_i, w_i, and θ_i such that

$$\sup_{x \in R} \left| \sum_{i=1}^{N} u_i \sigma(w_i x + \theta_i) - cosig(x) \right| < \epsilon$$

Select N such that $1/(N+1) < \epsilon/2$. For $i \in \{1, 2, \ldots, N\}$ set

$$u_i = \frac{1}{N+1}$$

Choose $M > 0$ such that $\sigma(-M) < \epsilon/2(N+1)$ and $\sigma(M) > [1 - \epsilon/2(N+1)]$. Because $\sigma(.)$ is a sigmoidal function. Furthermore, for $i \in \{1, 2, \ldots, N\}$ set

$$r_i = \sup\{\lambda : cosig(\lambda) = \frac{i}{N+1}\}$$

and

$$r_{N+1} = \sup\{\lambda : cosig(\lambda) = 1 - \frac{1}{2(N+1)}\}$$

Since $cosig(.)$ is a continuous squashing function such r_j's exist.

Next, a choice of the constants w_i and θ_i will be given. Let

$$w_i r_i + \theta_i = M \tag{2.44}$$

and

$$w_i r_{i+1} + \theta_i = -M \tag{2.45}$$

Then, a unique set of w_i and θ_i may be determined by the above two-equations as follows

$$w_i = \frac{2M}{r_i - r_{i+1}} \tag{2.46}$$

$$\theta_i = \frac{-M(r_i + r_{i+1})}{r_i - r_{i+1}} \tag{2.47}$$

It is easy to verify that for u_i, w_i and θ_i given by Eqs. (1.44)-(1.46)

$$\left| \sum_{i=1}^{N} u_i \sigma(w_i x + \theta_i) - cosig(x) \right| < \epsilon$$

on each of the intervals $(-\infty, r_1]$, $(r_1, r_2]$, ... , $(r_N, r_{N+1}]$, $(r_{N+1}, +\infty)$. □

Appendix C: The Proof Of Lemma 2

Since S is a compact set and N is finite, there is a $M > 0$ such that for $i \in \{1, 2, \ldots, N\}$

$$-M \leq \sum_{j=1}^{n} w_{ij}x_j + \theta_i \leq M, \qquad \mathbf{x} \in \mathbf{S}$$

From Lemma 6.1, for every $\epsilon > 0$, there is a set of constants $\bar{u}_l$, $\bar{w}_l$, and $\bar{\theta}_l$ such that

$$\sup_{\lambda \in R} \left| \sum_{l=1}^{Q} \bar{u}_l \sigma(\bar{w}_l \lambda + \bar{\theta}_l) - cosig(\lambda) \right| < \epsilon / N \sum_{i=1}^{N} |u_i|$$

Hence,

$$\sup_{x \in S} \left| \sum_{l=1}^{Q} \bar{u}_l \sigma \left[\bar{w}_l (\sum_{j=1}^{n} w_{ij}x_j + \theta_i) + \bar{\theta}_l \right] - cosig(\sum_{j=1}^{n} w_{ij}x_j + \theta_i) \right| < \epsilon / N \sum_{i=1}^{N} |u_i|$$

that is,

$$\sup_{x \in S} \left| \sum_{i=1}^{N} u_i \sum_{l=1}^{Q} \bar{u}_l \sigma \left[\bar{w}_l (\sum_{j=1}^{n} w_{ij}x_j + \theta_i) + \bar{\theta}_l \right] - \sum_{i=1}^{N} u_i cosig(\sum_{j=1}^{n} w_{ij}x_j + \theta_i) \right| < \epsilon$$

Let

$$f(x) = \sum_{i=1}^{N} \sum_{l=1}^{Q} u_i \bar{u}_l \sigma \left[\bar{w}_l (\sum_{j=1}^{n} w_{ij}x_j + \theta_i) + \bar{\theta}_l \right]$$

Then

$$\sup_{x \in S} |f(x) - g(x)| < \epsilon.$$

□

Appendix D: The Proof Of Corollary 2

We need only to prove that the three-layered network with two hidden neural layers described by

$$f(\mathbf{x}) = \sum_{i=1}^{N_1} u_i \sigma \left[\sum_{j=1}^{N_2} v_{ij} \sigma(\sum_{k=1}^{n} w_{ijk} x_k + \theta_j) + l_i \right]$$

can approximate g on S with error $< \epsilon$.

For every $\epsilon > 0$, using Theorem 6.8, there is a three-layered network

$$\bar{f}(\mathbf{x}) = \sum_{i=1}^{N} u_i \sigma(\sum_{j=1}^{n} w_{ij} x_j + l_i)$$

such that

$$|\bar{f}(\mathbf{x}) - g(\mathbf{x})| < \frac{\epsilon}{2}$$

for all $\mathbf{x} \in \mathbf{S}$. On the other hand, the sigmoidal function $\sigma(x)$ is uniformly continuous on the compact set S, then for a given set of constants

$$\epsilon_i' = \epsilon/(N_1 |u_i|), \qquad i = 1, \ldots, N_1$$

there are the constants δ_i, we may find a set of the three-layered networks

$$\sum_{j=1}^{N_2} v_{ij} \sigma(\sum_{k=1}^{n} w_{ijk} x_k + \theta_j)$$

such that

$$\left| \sum_{j=1}^{N_2} v_{ij}\sigma(\sum_{k=1}^{n} w_{ijk}x_k + \theta_j) - \sum_{j=1}^{n} w_{ij}x_j \right| < \delta_i$$

and

$$\left| \sigma\left[\sum_{j=1}^{N_2} v_{ij}\sigma(\sum_{k=1}^{n} w_{ijk}x_k + \theta_j) + l_i \right] - \sigma(\sum_{j=1}^{n} w_{ij}x_j + l_i) \right| < \epsilon_i'$$

Hence

$$|f(\mathbf{x}) - \bar{\mathbf{f}}(\mathbf{x})| < \sum_{\mathbf{i}=\mathbf{1}}^{\mathbf{N_1}} \frac{\epsilon_{\mathrm{i}}'}{|\mathbf{u_i}|} = \epsilon$$

Finally,

$$|f(\mathbf{x}) - \mathbf{g}(\mathbf{x})| \leq |\mathbf{f}(\mathbf{x}) - \bar{\mathbf{f}}(\mathbf{x})| + |\bar{\mathbf{f}}(\mathbf{x}) - \mathbf{g}(\mathbf{x})| < \epsilon$$

□

3

ORDERING FUZZY REAL QUANTITIES

Robert Lowen

Department Wiskunde en Information
Universiteit Antwerpen
Froenenborgerlaan 171, 2020 Antwerpen, Belgium

1 INTRODUCTION

Decision making, at one stage or another, usually involves ranking or ordering "things," [1], [2] and in applications of fuzzy logic these "things" are often modeled or described by fuzzy real quantities. Fuzzy real quanities themselves have been described in different ways in the literature [3], [4]. The model which we shall be considering in this paper involves probability measures on the real line [4], [5]. We shall, as is usual in this context, call these "stochastic fuzzy real numbers."

It is our purpose in this chapter to give some canonical definitions of ordering of stochastic fuzzy real numbers, and we shall prove some basic properties of these orderings. The interesting aspect of these orders is that they are themselves fuzzy, in the sense that degrees are given by which one stochastic fuzzy real number is smaller or larger than another one. We give several examples which shall demonstrate the naturality and applicability of our concepts.

2 STOCHASTIC FUZZY REAL NUMBERS

Given a set X we denote the characteristic function of a subset $A \subset X$ by 1_A, and if $A = \{x\}$ then we denote this characteristic function simply by 1_x.

We shall also require t-norms and t-conorms, so we briefly recall the definitions. A function $T : I \times I \to I$ is called *t-norm* if it fulfills the following properties

1. T is non-decreasing, i.e.

$$\forall x, y, x', y' \in I : x \leq x' \text{ and } y \leq y' \;\Rightarrow\; T(x, y) \leq T(x', y'),$$

2. T is commutative, i.e.

$$\forall x, y \in I : T(x, y) = T(y, x),$$

3. T is associative, i.e.

$$\forall x, y, z \in I : T(x, T(y, z)) = T(T(x, y), z),$$

4. T has 1 as a unit, i.e.

$$\forall x \in I : T(1, x) = T(x, 1).$$

A function $S : I \times I \to I$ is called a *t-conorm* if it fulfills the following properties

1. S is non-decreasing, i.e.

$$\forall x, y, x', y' \in I : x \leq x' \text{ and } y \leq y' \;\Rightarrow\; S(x, y) \leq S(x', y'),$$

2. S is commutative, i.e.

$$\forall x, y \in I : S(x, y) = S(y, x),$$

3. S is associative, i.e.

$$\forall x, y, z \in I : S(x, S(y, z)) = S(S(x, y), z),$$

4. S has 0 as a unit, i.e.

$$\forall x \in I : S(0, x) = S(x, 0).$$

We shall in particular be needing the so-called bounded product and bounded sum, also called the Lukasiewicz-connectives. Together with many other t-norms and t-conorms they are interesting because unlike minimum and maximum, they take the additive structure of the numbers in the unit interval into consideration. Bounded product is the t-norm defined as

$$T_m(x, y) \doteq \max(x + y - 1, 0),$$

and bounded sum is the t-conorm defined as

$$S_m(x, y) \doteq \min(x + y, 1).$$

The probabilistic view of the fuzzy real numbers is that they constitute stochastic quanities, which can be described by real numbers, but not necessarily by one real number in particular. In order to explain this view exactly we need some preliminary concepts from probability theory.

If X is an arbitrary measurable space, then $\mathcal{M}(X)$ stands for the set of all probability measures on X. Given a point $x \in X$ we shall denote P_x the probability measure defined by

$$P_x(A) \doteq \begin{cases} 0 & x \ni A \\ 1 & x \in A \end{cases}$$

for all $A \subset X$ which are measurable. Such a measure is called a Dirac-measure in x.

If Y is another measurable space and

$$f : X \longrightarrow Y$$

is a measurable map, then it is well-known that we can "extend" f to a map between $\mathcal{M}(X)$ and $\mathcal{M}(Y)$ in the following way. Let $\mathcal{B}_X$ and $\mathcal{B}_Y$ stand for the measurable sets in X and Y respectively. Then we define

$$f^* : \mathcal{M}(X) \longrightarrow \mathcal{M}(Y) : P \longrightarrow f^*(P)$$

by setting

$$\forall P \in \mathcal{M}(X), \forall B \in \mathcal{B}_Y : f^*(P)(B) \doteq P\left(f^{-1}(B)\right).$$

2.1 Definition

A *stochastic fuzzy real number* is a probability measure on $\Re$. The set of all stochastic fuzzy real numbers therefore is $\mathcal{M}(\Re)$.

Stochastic fuzzy real numbers can also be introduced in a different way. We put $\Re_l(I)$, (resp. $\Re_r(I)$) the set of all left-continuous, (resp. right-continuous) non-decreasing fuzzy sets μ on $\Re$, for which

$$\inf_{x\in\Re} \mu(x) = 0,$$
$$\sup_{x\in\Re} \mu(x) = 1.$$

The proof of the following theorem can be found in any standard textbook on probability theory, and we therefore refrain from repeating it.

2.2 Theorem

The sets $\mathcal{M}(\Re), \Re_l(I)$, and $\Re_r(I)$ are canonically isomorphic. The following are descriptions of canonical isomorphisms.

1. *The map*

$$\mathcal{M}(\Re) \longrightarrow \Re_l(I) : P \longrightarrow F_P^l$$

where for all $x \in \Re, F_P^l(x) \doteq P(]-\infty, x[)$, is a bijection.

2. *The map*

$$\mathcal{M}(\Re) \longrightarrow \Re_r(I) : P \longrightarrow F_P^r$$

where for all $x \in \Re, F_P^r(x) \doteq P(]-\infty, x])$, is a bijection. □

In the sequel we shall always use that form of the stochastic fuzzy real line which is best adapted to our purpose. The ordinary real numbers are of course nicely embedded into the stochastic fuzzy real line.

2.3 Theorem

The following are embeddings.

1.

$$\Re \longrightarrow \Re_l(I) : x \longrightarrow F_x^l$$

where $F_x^l \doteq 1_{]x,\infty[}$.

2.

$$\Re \longrightarrow \Re_r(I) : x \longrightarrow F_x^r$$

where $F_x^r \doteq 1_{[x,\infty[}$.

3.

$$\Re \longrightarrow \mathcal{M}(\Re) : x \longrightarrow P_x$$

where for any Borel set B

$$P_x(B) \doteq \begin{cases} 1 & if \;\; x \in B, \\ 0 & if \;\; x \ni B. \square \end{cases}$$

Thus we see that we can interpret $\mathcal{M}(\Re)$, as an extension of $\Re$. The points of $\Re$, or the Dirac measures, are deterministic points, and the points of $\mathcal{M}(\Re)/\Re$ are points with uncertainty. For instance, the height of man is not a real number. However when using this expression, one thinks and resons with real numbers. Mathematically the height of man is represented as a real random variable defined on the set (or on some representative subset) of all men. If one then makes a histogram of this varible one sees that it is distributed approximately normally. Thus the "real quantity" which we call the height of man can represented by a normal probability measure. Now suppose we ask the question: Is the height of man between 1 and 2 meters? Most people will be inclined to answer, yes. However strictly speaking, the question is absurd and the answer is incorrect. The question should have been: What is the probability that a randomly chosen man has a height between 1 and 2 meters? We can use fuzzy sets to model this in a perfectly simple and beautiful way which is much closer to our intuitive feeling. Rather than saying that the height of a man lies in the interval between 1 and 2 meters we shall give a "degree" that it belongs to that interval. We reason as follows. Always thinking of $\mathcal{M}(\Re)$ as an extension of $\Re$, we also extend measurable subsets from $\Re$ to $\mathcal{M}(\Re)$. This is where fuzzy sets are really required. Suppose $A \subset \Re$ is measurable, there is no canonical set-extension to $\mathcal{M}(\Re)$, but there is a uniquely defined and canomical fuzzy set-extension defined by

$$\delta_A : \mathcal{M}(\Re) \longrightarrow I : P \longrightarrow P(A).$$

Now if A is the interval [1, 2] and P stands for the normal probability representing the height of man, then

$$\delta_A(P) = P(A),$$

the degree that P belongs to δ_A is precisely the probability of having the height of a randomly chosen man lying in the interval A.

3 ORDERING OF THE STOCHASTIC FUZZY REAL LINE

3.1 Definition

Given two stochastic fuzzy real numbers $P, Q \in \mathcal{M}(\Re)$, we define

$$\rho(P,Q) \doteq \sup_{x\in\Re} P\left(]-\infty, x[\right) \wedge Q\left(]x, \infty[\right).$$

$\rho(P,Q)$ should be interpreted as the degree that P is strictly smaller than Q.

3.2 Proposition

The fuzzy relation

$$\rho : \mathcal{M}(\Re) \times \mathcal{M}(\Re) \longrightarrow I$$

is an extension of the strict order relation on $\Re$, in the sense that if $x, y \in \Re$ then

$$x < y \Leftrightarrow \rho\left(P_x, P_y\right) = 1.$$

Proof

It suffices to note that

$$\begin{aligned} \rho\left(P_x, P_y\right) &= \sup_{t\in\Re} P_x(]-\infty, t[) \wedge P_y(]t, \infty[) \\ &= \sup_{t>x} P_y(]t, \infty[). \square \end{aligned}$$

3.3 Examples

1. If $P = P_x$ for some $x \in \Re$, then

$$\rho(P_x, Q) = Q(]x, \infty[),$$

and

$$\rho(P, P_x) = P(]-\infty, x[).$$

This nicely illustrates the intuitive correctness of the fuzzy order relation which we have defined. The degree that the "deterministic real number" x is strictly smaller than the "stochastic fuzzy real number" Q equals the Q-probability of finding points strictly larger than x. Analogously, the degree that the "deterministic real number" x is strictly larger than the "stochastic fuzzy real number" P equals to the P-probability of finding points strictly smaller than x.

2. let us suppose that P is a normal probability measure with mean and standard deviation respectively, m_1 and s_1. If we put

$$F(x) \doteq \frac{1}{\sqrt{2\pi}} \int_{-\infty}^{x} \exp^{\frac{-t^2}{2}} dt$$

the distribution function of the normal probability measure with mean and standard deviation respectively 0 and 1, then since

$$P(]-\infty, x[) = F\left(\frac{x - m_1}{s_1}\right),$$

and

$$Q(]x, \infty[) = F\left(\frac{x - m_2}{s_2}\right),$$

it follows that

$$\rho(P, Q) = \sup_{x \in \Re} P(]-\infty, x[) \wedge Q(]x, \infty[).$$

is attained in the point $x \in \Re$ where

$$F\left(\frac{x - m_1}{s_1}\right) = 1 - F\left(\frac{x - m_2}{s_2}\right)$$

Since for any $x \in \Re, F(x) = 1 - F(-x)$, it follows that

$$\frac{x - m_1}{s_1} = -\frac{x - m_2}{s_2},$$

i.e.

$$x = \frac{m_1 s_2 + m_2 s_1}{s_1 + s_2}.$$

Consequently

$$\rho(P,Q) = F\left(\frac{m_2 - m_1}{s_1 + s_2}\right) = \frac{1}{\sqrt{2\pi}} \int_{-\infty}^{\frac{m_2 - m_1}{s_1 + s_2}} \exp^{-\frac{t^2}{2}} dt$$

Again this example nicely illustrates the meaning and naturality of the fuzzy order relation ρ. Thus we have e.g. that

$$\lim_{m_2 - m_1 \to \infty} \rho(P,Q) = \lim_{m_2 - m_1 \to \infty} \frac{1}{\sqrt{2\pi}} \int_{-\infty}^{\frac{m_2 - m_1}{s_1 + s_2}} \exp^{-\frac{t^2}{2}} dt = 1,$$

that

$$\lim_{m_2 - m_1 \to -\infty} \rho(P,Q) = \lim_{m_2 - m_1 \to -\infty} \frac{1}{\sqrt{2\pi}} \int_{-\infty}^{\frac{m_2 - m_1}{s_1 + s_2}} \exp^{-\frac{t^2}{2}} dt = 0,$$

and that

$$\rho(P,Q) = \frac{1}{\sqrt{2\pi}} \int_{-\infty}^{0} \exp^{-\frac{t^2}{2}} dt = \frac{1}{2} \text{ if } m_1 = m_2.$$

3. lets us suppose that P is a uniform probability measure on an interval I_1 with length 1 and with a mean m_1, and that Q is a uniform probability measure on an interval I_2 with length 1 and with a mean m_2.

   Then one can easily verify that

$$\rho(P,Q) = \begin{cases} 1 & \text{if } m_2 \geq m_1 + 1, \\ \frac{1}{2}(m_2 - m_1 - 1) & \text{if } m_1 - 1 \leq m_2 \leq m_1 + 1, \\ 0 & \text{if } m_2 \leq m_1 - 1. \end{cases}$$

   This implies that $\rho(P,Q) = 1$ as long as the entire interval I_1 lies on the left of the entire interval I_2. The value of $\rho(P,Q)$ then gradually decreases as I_1 moves over I_2 towards the right, until it reaches 0 when I_2 lies at the left of I_1.

4. As a last example let us suppose that P and Q have triangular density functions symmetric round the points $p \in \Re$ and $q \in \Re$ respectively

$$f_P(x) \doteq \begin{cases} x - p + 1 & x \in [p-1, p] \\ -x + p + 1 & x \in [p, p+1] \\ 0 & \text{elsewhere.} \end{cases}$$

and

$$f_Q(x) \doteq \begin{cases} x-q+1 & x \in [q-1,q] \\ -x+q+1 & x \in [q,q+1] \\ 0 & \text{elsewhere.} \end{cases}$$

Then we find that

$$\rho(P,Q) = \begin{cases} 0 & q-p \leq -2 \\ \frac{1}{2}\left(1+q-p+\frac{(q-p)^2}{4}\right) & -2 \leq q-p \leq 0 \\ \frac{1}{2}\left(1+q-p-\frac{(q-p)^2}{4}\right) & 0 \leq q-p \leq 2 \\ 1 & 2 \leq q-p \end{cases}$$

Again we have that $\rho(P,Q) = 1$ as long as the entire triangular region of P lies on the left of the triangular region of Q. The value of $\rho(P,Q)$ then gradually decreases as the regions move over one another, until it reaches 0 when the triangular region Q lies completely on the left of the triangular region of P. When $P = Q$ we have $\rho(P,Q) = \frac{1}{2}$ as it should be.

We shall now prove the basic theorem concerning the properties of the fuzzy order relation ρ, which shows that it has good properties.

3.4 Therom

The following properties are fulfilled for all $P, Q \in \mathcal{M}(\Re)$:

1. *Antireflexivity*

$$T_m\left(\rho\left(P,Q\right),\rho\left(Q,P\right)\right) = 0,$$

2. *Transitivity*

$$\sup_{R\in\mathcal{M}(\Re)} T_m\left(\rho\left(P,R\right),\rho\left(R,Q\right)\right) \leq \rho\left(P,Q\right),$$

3. *Linearity*

$$P \neq Q \Rightarrow S_m\left(\rho\left(P,Q\right),\rho\left(Q,P\right)\right) > 0.$$

Proof

1. It is sufficient to note that if $x \leq y$ then

$$\begin{aligned} P(]-\infty,x[) \quad &\wedge \quad Q(]x,\infty[) + Q(]-\infty,y[) \wedge P(]y,\infty[) \\ &\leq \quad P(]-\infty,x[) + P(]x,\infty[) \\ &\leq \quad 1, \end{aligned}$$

and analogously, if $y \leq x$ then

$$\begin{aligned} P(]-\infty,x[) \quad &\wedge \quad Q(]x,\infty[) + Q(]-\infty,y[) \wedge P(]y,\infty[) \\ &\leq \quad Q(]y,\infty[) + P(]-\infty,y[) \\ &\leq \quad 1. \end{aligned}$$

Consequently

$$\rho(P,Q) + \rho(Q,P) \leq 1.$$

2. Let $P, Q, R \in \mathcal{M}(\Re)$, then it is sufficient to show that for any $x, y \in R$, there exists $z \in \Re$, such that

$$\begin{aligned} P(]-\infty,x[) \quad &\wedge \quad R(]x,\infty[) + R(]-\infty,y[) \wedge Q(]y,\infty[) \\ &\leq \quad P(]-\infty,z[) \wedge Q(]z,\infty[) + 1. \end{aligned}$$

Now if $x \leq y$ then

$$\begin{aligned} P(]-\infty,x[) \quad &\wedge \quad R(]x,\infty[) + R(]-\infty,y[) \wedge Q(]y,\infty[) \\ &\leq \quad P(]-\infty,y[) + Q(]y,\infty[) \\ &\leq \quad P(]-\infty,y[) + Q(]y,\infty[) + 1. \end{aligned}$$

and thus we can put $z \doteq y$ if $y \leq x$,then

$$\begin{aligned} P(]-\infty,x[) \quad &\wedge \quad R(]x,\infty[) + R(]-\infty,y[) \wedge Q(]y,\infty[) \\ &\leq \quad R(]y,\infty[) + R(]-\infty,y[) \\ &\leq \quad 1, \end{aligned}$$

and thus we can let z be any point in $\Re$.

3. Let $P \neq Q \in \mathcal{M}(\Re)$, and suppose that $S_m(\rho(P,Q), \rho(Q,P)) = 0$, i.e. suppose that $\rho(P,Q) = \rho(Q,P) = 0$. This implies that for all $x \in \Re$:

$$P(]-\infty,x[) \quad \wedge \quad Q(]x,\infty[) = 0,$$

and

$$Q(]-\infty,x[) \quad \wedge \quad P(]x,\infty[) = 0.$$

If we now put

$$\begin{aligned} A_-(P) &\doteq \{x \in |P(]-\infty, x[) = 0\}, \\ A_+(P) &\doteq \{x \in |P(]x, \infty[) = 0\}, \\ A_-(Q) &\doteq \{x \in |Q(]-\infty, x[) = 0\}, \\ A_+(Q) &\doteq \{x \in |Q(]x, \infty[) = 0\}, \end{aligned}$$

The $A_-(P)$ and $A_-(Q)$ are intervals extending to $-\infty$, and $A_+(P)$ and $A_+(Q)$ are intervals extending to ∞. Furthermore we obviously have that

$$A_-(P) \cup A_+(Q) = A_-(Q) \cup A_+(P) = \Re,$$

and that none of the sets $A_-(P), A_+(Q), A_-(Q)$ or $A_+(P)$ can be empty. Consequently we have

$$-\infty < \inf A_+(Q) \leq \sup A_-(P) < \infty.$$

Now we have to consider cases.

Case 1. $\inf A_+(Q) < \sup A_-(P)$.. Then there exists $a < b$ such that $P(]-\infty, b[) = Q(]a, \infty[) = 0$ for any $x \in]a, b[$ we consequently have

$$P(]-\infty, x[) = Q(]x, \infty[) = 1$$

i.e. $\rho(P, Q) = 1 > 0$, which is in contradiction with our supposition.

Case 2. $\inf A_+(Q) = \sup A_-(P)$.. Then obviously

$$P(]a, \infty[) = Q(]-\infty, a[) = 1.$$

Now if $P(]a, \infty[) = 0$ then $P = P_a$ and it follows from the fact that $P \neq Q$, that there exists $b < a$ such that $Q(]-\infty, b[) > 0$ Consequently

$$Q(]-\infty, b[) \wedge P(]b, \infty[) > 0$$

i.e. $\rho(Q, P) > 0$, which is in contradiction with our supposition. If on the other hand $P(]a, \infty[) > 0$, then some $b > a$ we also have $P(]b, \infty[) > 0$, and it again follows that

$$Q(]-\infty, b[) \wedge P(]b, \infty[) > 0$$

i.e. $\rho(Q, P) > 0$, which once again is in contradiction with our supposition. Consequently

$$S_m\left(\rho\left(P, Q\right), \rho\left(Q, P\right)\right) > 0. \quad \square$$

3.5 Remark

In the foregoing theorem we have used the t-norm T_m and the t-conorm S_m as logical connectives. The interpretation of antireflexivity is that one cannot have at the same time a high degree by which P is strictly smaller than Q, and a high degree by which Q is strictly smaller than P. The interpretation of transitivity is that the degree that P is strictly smaller than Q increases as the degrees that P is strictly smaller than R, and that R is strictly smaller than Q increase, more in particular that

$$\rho(P,R)+\rho(R,Q)\leq\rho(P,Q)+1.$$

The interpretation of linearity finally is that either the degree by which P is strictly smaller than Q has to be strictly positive, or the degree by which Q is strictly smaller than P has to be positive. In other words, at least to some degree, P has to be strictly smaller than Q or Q has to be strictly smaller than P.

We now look at the same properties which we described in the foregoing theorem, but using a minimum and maximum as logical connectives.

3.6 Theorem

The following properties are fulfilled for all $P, Q \in \mathcal{M}(\Re)$:

1. *Antireflexivity*

$$\min\left(\rho(P,Q),\rho(Q,P)\right)\leq\frac{1}{2},$$

2. *Transitivity*

$$\sup_{R\in\mathcal{M}(\Re)}\min\left(\rho(P,R),\rho(R,Q)\right)\leq\max\left(\rho(P,Q),\frac{1}{2}\right),$$

3. *Linearity*

$$P\neq Q\Rightarrow\max\left(\rho(P,Q),\rho(Q,P)\right)>0.$$

Proof

1. This is an immediate consequence of the foregoing theorem since for any $a, b \in I$ we have

$$T_m(a,b) = 0 \Rightarrow \min(a,b) \leq \frac{1}{2}.$$

2. Let $P, Q, R \in \mathcal{M}(\Re)$ be such that

$$\min\left(\rho(P,R), \rho(R,Q)\right) > \alpha > \frac{1}{2}.$$

Then there exists $x, y \in \Re$ such that

$$P(]-\infty,x[) \wedge R(]x,\infty[) \wedge R(]-\infty,y[) \wedge Q(]y,\infty[) > \alpha.$$

Now if $y < x$ then $R(]-\infty,y[) \wedge R(]x,\infty[) \leq \frac{1}{2}$ and thus $\alpha < \frac{1}{2}$, a contradiction. Consequently $x \leq y$ and it follows that

$$P(]-\infty,x[) \wedge Q(]x,\infty[) > \alpha$$

which implies that $\rho(P,Q) > \alpha$.

3. This is again an immediate consequence of the foregoing theorem since for any $a, b \in I$ we have

$$S_m(a,b) > 0 \Leftrightarrow \max(a,b) > 0. \quad \square$$

3.7 Examples

We shall now show by means of examples that all inequalities given in the foregoing theorems are best possible.

1. Let P be a normal probability measure with a mean of 0 and standard deviation 1. It follows from 3.3 that

$$\rho(P,P) = F\left(\frac{0-0}{2}\right) = F(0) = \frac{1}{2},$$

and consequently

$$\rho(P,P) + \rho(P,P) = 1.$$

This shows that the inequality hidden in the formula fo antireflexivity in 3.4 is the best possible.

2. Let $x < y < z \in \Re$, and put

$$P \doteq P_x, R \doteq P_y \text{ and } Q \doteq P_z$$

then it follows that

$$T_m\left(\rho(P,R), \rho(R,Q)\right) = 1 = \rho(P,Q).$$

this shows that the transitivity formula in 3.4 is the best possible.

3. Let $x < y \in \Re$, let $\alpha, \beta \in I$, and put

$$\begin{aligned} P &\doteq \alpha P_x + (1-\alpha) P_y, \\ &\text{and} \\ Q &\doteq \beta P_x + (1-\beta) P_y, \end{aligned}$$

Then it follows that

$$\begin{aligned} &\rho(P,Q) \\ &= \sup_{t\in\Re} \left(\alpha P_x(]-\infty,t[) + (1-\alpha)P_y(]-\infty,t[)\right) \\ &\quad \wedge \left(\beta P_x(]t,\infty[) + (1-\beta)P_y(]t,\infty[)\right) \\ &= \sup_{t\le x}(0) \vee \sup_{x<t<y} \left(\alpha \wedge (1-\beta)\right) \vee \sup_{y\le t}(0) \\ &= \alpha \wedge (1-\beta). \end{aligned}$$

Analogously we find that

$$\rho(Q,P) = \beta \wedge (1-\alpha).$$

Consequently, given $0 < \epsilon < \frac{1}{2}$, if we put

$$\alpha \doteq \frac{\epsilon}{3} \text{ and } \beta \doteq \frac{2\epsilon}{3},$$

then it follows that

$$\begin{aligned} S_m\left(\rho(P,Q), \rho(Q,P)\right) &= \min\left(\alpha \wedge (1-\beta) + \beta \wedge (1-\alpha), 1\right) \\ &= \min\left(\frac{\epsilon}{3} \wedge \left(1 - \frac{2\epsilon}{3}\right) + \frac{2\epsilon}{3} \wedge \left(1 - \frac{\epsilon}{3}\right), 1\right) \\ &= \frac{\epsilon}{3} + \frac{2\epsilon}{3} = \epsilon. \end{aligned}$$

This shows that the formula concerning linearity in 3.4 is best possible.

4. The same example as under 1 shows that the formula for antireflexivity in 3.6 is the best possible.

5. The same example as under 2 shows that the formula for transitivity in 3.6 is the best possible.

6. The same example as under 3 shows that the formula for linearity in 3.6 is the best possible.

REFERENCES

[1] J. C. Fodor, "Strict Preference Relations Based on Weak t-norms," Fuzzy Sets and Systems, vol. 43, pp. 327-336, 1991.

[2] S. Ovchinnikov and M. Roubens, "On Strict Preference Relations," Fuzzy Sets and Systems, vol. 43, pp. 319-326, 1991.

[3] D. Dubois and H. Prade, "Operations on Fuzzy Numbers," International Journal on System Science, vol. 9, pp. 613-626, 1978.

[4] R. Lowen, "On (R(L),+)," Journal of Fuzzy Sets and Systems, vol. 9, pp. 203-209, 1983.

[5] R. Lowen, "The Order Aspect of the Fuzzy Real Line," Manuscripta Mathematics, vol. 39, pp. 293-309, 1985.

4

FUZZY LOGIC CONTROLLERS FOR AIRCRAFT FLIGHT CONTROL

Jia Luo and Edward Lan

Department of Aerospace Engineering, University of Kansas
Lawrence, KA 66045, USA

ABSTRACT

Fuzzy logic proportional-integral-differential (PID) controllers are developed to perform both stability augmentation and automatic flight control functions. It operates for controlling both longitudinal and lateral- directional motions for an example aircraft, the X-29. The controllers for pitch, roll and yaw control are generated by analyzing a mathematical model describing aircraft static and dynamic characteristics. Each set of fuzzy rules consists of coarse and fine rules. The coarse rules are designed to supply fast response with large control input; while the fine rules are designed to make fine adjustments to improve dynamic stability. Nonlinear numerical simulations are performed to verify the controller performance at the design and off-design conditions for angle of attack hold, bank angle hold and sideslip angle hold. The results indicate that fuzzy PID controllers can provide fast, well damped response to pilot commands and thus improve flight performance, and increase agility, and also are robust.

1 INTRODUCTION

Fuzzy set was introduced by Zadeh to describe complex or ill-defined systems [1]. Soon after, it was applied to process control of systems for which the state variables could not be precisely measured [2]. Since then, fuzzy logic-based controllers have found many successful industrial applications and have demonstrated significant improvements in performance and robustness. However, they have not been utilized in flight control of operational aircraft in the past. Only recently has fuzzy logic been investigated for aerospace applications. For example, Larkin considered a fuzzy logic controller to control an aircraft's final-approach flight path [3]. Chiu, et al. investigated active control of a flexible wing aircraft [4]. Application to helicopter design with multiple objectives has also been studied [5]. Application to space operations was discussed in [6]. However, the "fuzzy" concept has not been embraced by control engineers for high performance aircraft, such as fighters. It seems natural that fuzzy logic can be effectively implemented in a fighter control system since the aircraft characteristics have wide variation over a large flight envelope and some of these characteristics are not accurately known.

Flight control systems of high performance aircraft generally consist of two subsystems, one being the stability augmentation system (SAS) and the other being the automatic flight control system (AFCS)[7]. The former is to provide an aircraft with both statically and dynamically stable behaviors. The latter is mainly designed to avoid an unnecessary or dangerously high workload and to carry out all basic piloting functions. For the longitudinal motion, a typical SAS includes a pitch damper and a longitudinal stability SAS (α - SAS). The pitch damper is used to improve the short period handling quality by increasing damping, and the longitudinal stability SAS is to provide longitudinal static stability or stiffness. For the lateral-directional motion, a typical SAS consists of a yaw damper, a roll damper and a directional stability SAS (β -SAS). These are used to improve the Dutch-roll damping, decrease the roll mode time constant and increase the directional stiffness, respectively. Basic AFCS's have pitch attitude hold, altitude hold, bank angle hold, heading angle hold and other modes. Static and dynamic stabilities are very important properties an aircraft must possess. If the inherent stability is not available or not enough, stability augmentation should be employed in the flight control systems.

Agility is another important contributor to success in the modern air combat arena. A fighter aircraft with good longitudinal and lateral agility is capable of generating large instantaneous pitch, roll and yaw rates, without sacrificing controllability [8]. However, a fighter using a linear controller, even with gain scheduling, does not always have good agility. Large instantaneous pitch, roll or yaw rate may induce an unacceptable overshoot or total loss of controllability.

Fuzzy logic-based controllers for industrial applications are usually designed based on heuristic grounds to model a human operator's behavior. When a dynamic model describing a system's characteristics is available, albeit it is only approximate, it is advantageous to make use of such model to develop the fuzzy logic controller to create a truly robust control system [9]. In this chapter a fuzzy proportional-integral-differential (PID) flight control system with some SAS and AFCS functions is developed for a sample aircraft, the X-29, to demonstrate this application. Fuzzy flight control system to provide fast, well damped responses to pilot pitch, yaw and roll commands will be presented.

2 FUZZY PID CONTROLLER

For a process described by linear mathematical models (transfer functions or state space equations), the linear proportional-integral-differential (PID) controllers can be employed to satisfy certain control specifications. In many engineering tasks, a linear model is obtained by linearizing a complicated nonlinear model at some operating condition to characterize the dynamics of the process. A PID controller designed based on this model is effective only in a small region around this operating condition. To describe the process dynamics quantitatively at other operating conditions, this model may not be precise enough. However, it may offer some qualitative information and physical insight, which can be used to develop a PID-like fuzzy controller. The objective is to develop a fuzzy controller which performs well in a wide range of conditions.

In the conventional linear control theory, the PID controller takes the form of

$$u = K_P(x - x_s) + K_I \int (x - x_s)dt + K_D \frac{dx}{dt} \tag{4.1}$$

where K_P, K_I and K_D are the proportional, integral and differential gains respectively. The integral part is included to drive x to its steady state, x_s, and the proportional and differential parts are used to modify the transient response of the closed loop system.

The incremental form for the PID controller, Eq. (4.1), from $t = (n-1)T$ to nT is

$$\begin{aligned} \Delta u(n) &= u(n) - u(n-1) \\ &= K_P\,[x(n) - x(n-1)] + (K_I T)\,[x(n) - x_s] + \\ &\quad (\frac{K_D}{T})\,[x(n) - 2x(n-1) + x(n-2)] \end{aligned} \tag{4.2}$$

$$\approx \quad (K_P T)\dot{x}(n) + (K_I T)\left[x(n) - x_s\right] + K_D T\ddot{x}(n)$$

Fuzzy PID controllers [10] have a similar structure to the linear PID controller and can be described in the following conditional statements:

If $\dot{x}$ is VP_j, $x - x_s$ is VI_j, and $\ddot{x}$ is VD_j,
then Δu is VU_j, $(j = 1, \ldots, n)$

where VP_j, VI_j, VD_j and VU_j represent suitable linguistic fuzzy values for the jth fuzzy rule.

The input of a fuzzy PID controller normally includes the first derivative of the state variable, $\dot{x}$, the error between the state variable and its set point, $x - x_s$, and the second derivative of the state variable, $\ddot{x}$. The controller output is the incremental control Δu. Given $\dot{x}$, $x - x_s$ and $\ddot{x}$, the value of Δu is generated through the fuzzification, fuzzy logic inference and defuzzification stages.

In the fuzzification stage, $\dot{x}$, $x - x_s$ and $\ddot{x}$ are scaled to the universe of discourse by suitable scaling factors G_p, G_i and G_d. The scaled variables or measurements are therefore

$$\bar{x_p} = Q\left[\frac{\dot{x}}{G_p}\right] \quad \bar{x_i} = Q\left[\frac{x - x_s}{G_i}\right] \quad \bar{x_d} = Q\left[\frac{\ddot{x}}{G_d}\right] \tag{4.3}$$

where Q represents quantifying a real number to its nearest integer.

In the fuzzy inference stage, by evaluating all n fuzzy rules in parallel the value of the degree of fulfillment of fuzzy incremental control at each point k on the universe of discourse is obtained as

$$\mu_U(k) = \max_{j=1}^{n}\{\min\left[VP_j(\bar{x_p}), VI_j(\bar{x_d}), VU_j(\nu_k)\right]\} \tag{4.4}$$

where ν_k represents the magnitude of the kth point on the universe of discourse. The action taken in Eq. (4.4) can be interpreted as follows [9].

For given $\bar{x_p}, \bar{x_i}$ and $\bar{x_d}$, and at the kth point for Δu, the degree of satisfaction or the compatibility of each fuzzy rule can be evaluated by the intersection or min-operation:

$$\omega_j = \min\left[VP_j(\bar{x_p}), VI_j(\bar{x_i}), VD_j(\bar{x}_d)VU_j(\nu_k)\right]$$

For example, VP_j may be positive medium, VI_j may be positive small, etc., and they are related by the connective operator AND so that the min-operation is needed. The degree of fulfillment involving all of these rules ($j = 1, \ldots, n$) is obtained by taking the union (or the max-operation) of the degrees of satisfaction of each rule. In other words, the rule that is most compatible with the values of $\bar{x_p}, \bar{x_i}, \bar{x_d}$, and ν_k is picked up for the kth point:

$$\mu_U(\nu_k) = \max_{j=1}^{n} \omega_j$$

With the membership function, $\mu_U(\nu_k)$, the fuzzy incremental control can be converted into a deterministic control Δu by two possible defuzzification methods: the mean-of-maximum and the center-of-gravity procedures.

The mean-of-maximum procedure generates a value of Δu corresponding to the maximum grade of membership, $\mu_U(\nu_k)$. If there is more than one maximum value with the same magnitude, a value representing the mean of all local maxima is generated. This procedure does not consider the continuous variation in $\mu_U(\nu_k)$.

In the center-of-gravity procedure, the value of Δu representing the center of gravity of the membership function is generated. In this procedure the fuzzy incremental control at each point k on the universe of discourse has a contribution to Δu with a degree of $\mu_U(\nu_k)$ as follows:

$$\Delta u = G_u \frac{\sum_{k=1}^{m} \mu_U(\nu_k)\nu_k w_k}{\sum_{k=1}^{m} \mu_U(\nu_k)} \tag{4.5}$$

where G_u is the scaling factor for incremental control, ν_k represents the magnitude of the kth point in the universe of discourse, and w_k is the weight of the used fuzzy rule, which is employed to make further fine adjustments. In the present application, the center-of-gravity method will be used.

In the following a fuzzy PID flight control system is developed for the X-29 aircraft.

3 FUZZY FLIGHT CONTROL SYSTEM FOR THE X-29 AIRCRAFT

3.1 Conventional Stability Augmentation and Automatic Flight Control

To make an aircraft acceptable in flying qualities, certain levels of static and dynamic stability are required. Pitch, yaw and roll dampers are used to augment damping or dynamic stability by sensing angular rates and moving control surfaces to oppose the angular motion. To avoid serious divergence problems caused by a lack of inherent longitudinal or directional static stability, the α - SAS or β - SAS is employed to artificially increase the level of static stability or stiffness. Stability Augmentation Systems are normally in the inner loop of flight control systems. In the outer loop automatic flight control systems are employed to carry out certain tasks.

It should be noted that pitch control surface (elevator, or canard, symmetric and strake flaps for the X-29), yaw control surface (rudder), and roll control surface (aileron, or differential flap for the X-29) are functionally distinct in that the aileron is for the angular rate (p) control while the elevator and rudder are for the angular displacement (α or β) control. Therefore, from an analytical viewpoint, a PID angle of attack hold controller can be simply written as

$$\delta_c = KP_\alpha(\alpha - \alpha_s) + KI_\alpha \int (\alpha - \alpha_s)dt + KD_\alpha q \tag{4.6}$$

A PID sideslip hold controller can be represented by

$$\delta_r = KP_\beta(\beta - \beta_s) + KI_\beta \int (\beta - \beta_s)dt + KD_\beta r \tag{4.7}$$

and a PI bank angle hold controller can be

$$\begin{aligned}\delta_{df} &= KP_p(p-p_s)+KI_p\int(p-p_s)dt \\ &\approx KP_p(p-p_s)+KI_p(\phi-\phi_s)\end{aligned} \tag{4.8}$$

In the above equations, δ_c, δ_r and δ_{df} represent the deflections of canard, rudder and differential flap. For the X-29 aircraft, the other two longitudinal control surfaces, symmetric and aftbody strake flaps should have similar control functions to Eq. (4.6).

The differential parts in Eqs. (4.6) and (4.7), and the proportional part in Eq. (1.8) function as pitch, yaw and roll dampers and therefore improve dynamic stability. The proportional parts in Eqs. (4.6) and (4.7) are used to increase longitudinal and directional static stability, respectively. The integral parts in the above equations perform the outer loop tasks of command tracking.

In a digital flight control system the incremental forms of control functions shown in Eqs. (4.6), (4.7) and (4.8) can be implemented as follows:

$$\begin{aligned}\Delta\delta_c(n) &= KP_\alpha\,[\alpha(n)-\alpha(n-1)]+(KI_\alpha T)\,[\alpha(n)-\alpha_s] \\ &\quad +KD_\alpha\,[q(n)-q(n-1)] \\ &\approx (KP_\alpha T)\dot{\alpha}(n)+(KI_\alpha T)\,[\alpha(n)-\alpha_s]+(KD_\alpha T)\dot{q}(n)\end{aligned} \tag{4.9}$$

$$\begin{aligned}\Delta\delta_r(n) &= KP_\beta\,[\beta(n)-\beta(n-1)]+(KI_\beta T)\,[\beta(n)-\beta_s] \\ &\quad +KD_\beta\,[r(n)-r(n-1)] \\ &\approx (KP_\beta T)\dot{\beta}(n)+(KI_\beta T)\,[\beta(n)-\beta_s]+(KD_\beta T)\dot{r}(n)\end{aligned} \tag{4.10}$$

$$\begin{aligned}\Delta\delta_{df}(n) &= KP_p\,[p(n)-p(n-1)]+(KI_pT)\,[p(n)-p_s] \\ &\approx (KP_pT)\dot{p}(n)+(KI_pT)\,[p(n)-p_s]\end{aligned} \tag{4.11}$$

The above simplified PID control functions can be used for the analysis and development of corresponding fuzzy control rules and fuzzy control structure.

Variable	Unit	Trim Value
Altitude, h	ft	10,000
Mach Number, Mach		1.5
Angle of Attack,α	deg	0.86
Velocity, V	ft/sec	1616.0
Pitch Angle, θ	deg	0.86
Canard Deflection, δ_c	deg	0.59
Symmetric Flap, δ_{sf}	deg	-6.00
Strake Flap, δ_{st}	deg	-0.38
Throttle Position, δ_{th}		0.88

Table 1 Trim condition of the xxample aircraft.

3.2 Dynamic Characteristics Of The X-29 Aircraft

The X-29 aircraft has a forward-swept wing, and a high level of static instability (i.e. negative stiffness). The aircraft incorporates closed-coupled canards to provide a low-drag configuration. Longitudinal stability and control of the aircraft is obtained with canard, symmetric flap, and aftbody strake surfaces. Lateral-directional motion is controlled by the conventional rudder and differential flap deflection.

To develop a fuzzy logic based flight system for the X-29, the aircraft dynamic characteristics are examined first. From a nonlinear aerodynamic database, the X-29 linear model and trim condition can be generated [11]. In Table 1 a trim condition at a Mach number of 1.5 and h = 10,000 ft is listed,

and the locally linearized equations of motion are,

$$\begin{bmatrix} \dot{V} \\ \dot{\alpha} \\ \dot{q} \\ \dot{\theta} \end{bmatrix} = \begin{bmatrix} -0.07292 & 69.66020 & 0.00000 & -32.14209 \\ -0.00002 & -2.30449 & 1.00000 & 0.00000 \\ 0.00052 & 71.96335 & -3.56228 & 0.00000 \\ 0.00000 & 0.00000 & 1.00000 & 0.00000 \end{bmatrix} \begin{bmatrix} V \\ \alpha \\ q \\ \theta \end{bmatrix}$$

$$+\begin{bmatrix} -1.03806 & 1.10598 \\ -0.00088 & -0.00301 \\ 0.92282 & -0.58371 \\ 0.00000 & 0.00000 \end{bmatrix} \begin{bmatrix} \delta_c \\ \delta_{st} \end{bmatrix} \tag{4.12}$$

$$\begin{bmatrix} \dot{\beta} \\ \dot{p} \\ \dot{r} \\ \dot{\phi} \end{bmatrix} = \begin{bmatrix} -0.49268 & 0.01524 & -0.99988 & 0.01989 \\ -61.17613 & -7.83522 & 4.99085 & 0.00000 \\ 31.80377 & -0.23455 & -0.99414 & 0.00000 \\ 0.00000 & 1.00000 & -0.01524 & 0.00000 \end{bmatrix} \begin{bmatrix} \beta \\ p \\ r \\ \phi \end{bmatrix}$$

$$+\begin{bmatrix} -0.00197 & 0.00200 \\ 8.24620 & 1.84901 \\ 0.24926 & -0.43630 \\ 0.00000 & 0.00000 \end{bmatrix} \begin{bmatrix} \delta_{df} \\ \delta_r \end{bmatrix} \tag{4.13}$$

It should be noted that the state and control variables in the above linear equations are the perturbation variables from the trim conditions. From the above two equations it follows that the longitudinal static stability derivative is found to be

$$M_\alpha \approx 71.96 I_y > 0$$

which represents a high level of longitudinal static instability. The directional static stability derivative

$$N_\beta \approx 31.08 I_z > 0$$

however, indicates that directional static stability (or weathercock stability) is available. Because of the longitudinal static instability, no short period or phugoid mode exists, and any disturbance will cause divergence in the longitudinal motion.

The open-loop eigenvalues are listed in Table 2. The dynamic stability can be determined with the open-loop eigenvalues.

	Eigenvalues
Longitudinal	-11.4396, 5.5731, -0.0715, -0.0017
Lateral-Directional	-7.8164 (Roll), 0.0076 (Spiral), -0.7566±5.8062i (Dutch Roll)

Table 2 Open loop eigenvalues of the example aircraft.

Based on these stability analyses, an α - SAS and a pitch damper are required in the flight control systems to make the aircraft acceptable in flying qualities. However, excessive static stability can degrade overall performance. Since the directional static stability is inherently available, an β - SAS is not needed in the fuzzy flight control system; but a yaw damper is needed to increase dutch roll damping. The lateral- directional dynamic stability also needs to be improved. Optimal control theory can be applied to the above linear models to determine gains of the linear control functions given by Eqs. (4.6)-(4.8), which can be further used to generate the corresponding fuzzy control rules.

An integrator equation

$$\dot{\xi}_\alpha = \alpha - \alpha_s \tag{4.14}$$

will be added to Eq. (4.12) to find the control functions (Eq. 4.6) for the angle of attack hold. If the specified closed loop eigenvalues are -5.6±4.2i (short period), -0.25 and -0.015 (non-oscillatory phugoid), and -0.5 (integrator), then the resulting incremental control functions [12] can be obtained through a linear optimal control theory to be:

$$\begin{aligned}\Delta\delta_c \approx\ & [-(90.98T)\dot{\alpha} - (0.968T)(\alpha - \alpha_s) - (11.449T)\dot{q}] \\ & + (0.0416T)\dot{V} - (17.092T)\dot{\theta},\end{aligned} \tag{4.15}$$

$$\Delta\delta_{st} \approx [(19.638T)\dot{\alpha} - (10.152T)(\alpha - \alpha_s) - (1.908T)\dot{q}]$$
$$- (0.007\ T)\ \dot{V} - (1.673T)\dot{\theta}.$$

For the lateral-directional motion, an integrator equation

$$\dot{\xi}_\beta = \beta - \beta_s \tag{4.16}$$

will be added to Eq. (4.13) to find the control functions (Eqs. 4.7 and 4.8). If the specified closed loop eigenvalues are -8.00 (roll), -0.05 (spiral), -4.883.66i (dutch roll), and -0.7 (integrator), then the resulting incremental control function 12 is determined, again by a linear optimal control theory, to be:

$$\begin{aligned} \Delta\delta_r \quad \approx \quad & \left[(3.41T)\dot{\beta} - (3.536T)(\beta - \beta_s) + (15.947T)\dot{r}\right] \\ & - (0.4126T)\dot{p} - (0.865T)(p - p_s) \end{aligned} \tag{4.17}$$

$$\begin{aligned} \Delta\delta_{df} \quad \approx \quad & [-(0.1576T)\dot{p} - (0.667T)(p - p_s)] \\ & + (1.01T)r - (0.8166T)\dot{r} + (0.2904T)(\beta - \beta_s) \end{aligned} \tag{4.18}$$

The sampling time T is taken to be 0.025 sec. It should be noted that in eqs. (4.15), (4.17) and (4.18) the terms included in the brackets have dominant roles in augmenting dynamic stability and keeping zero steady state errors. The relationships among fuzzy variables and the control structure in the fuzzy flight control system can be formulated based on these control functions.

3.3 Architecture Of Fuzzy Flight Control System

Fuzzy PID longitudinal and lateral-directional controllers for the X-29A Advanced Technology Demonstrator are set up and illustrated in Figs. 1 and 2.

Three sets of fuzzy control rules are generated for the angle of attack, sideslip angle and bank angle controls. These fuzzy PID control rules have similar structures to their conventional counterparts. The variables on the right hand side of Eqs. (4.6), (4.7) and (4.8) form the premises of fuzzy control rules, and the variables on the left hand side of these equations form the conclusions of the fuzzy control rules. In addition, fuzzy PID pitch attitude hold and heading

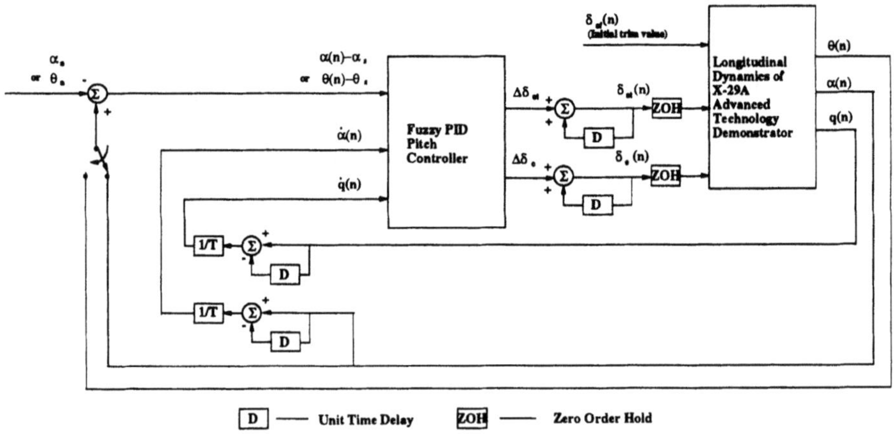

Figure 1 Block diagram of longitudinal fuzzy control for X-29.

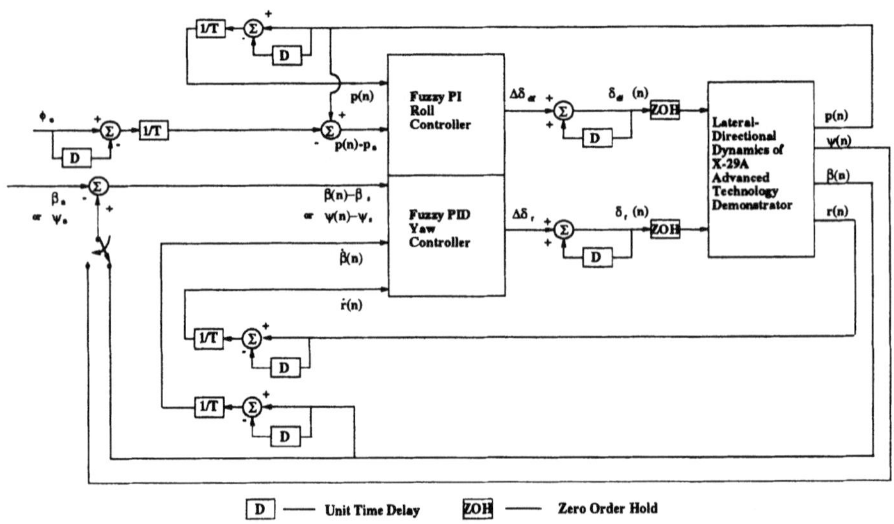

Figure 2 Block diagram of lateral-directional fuzzy control for X-29.

hold modes are also included in the present fuzzy flight control system. These fuzzy rules can be described by conditional statements as listed in Table 3.

Fuzzy PID α hold rules:	Fuzzy PID θ hold rules:
If $\alpha - \alpha_s$ is VI_j, $\dot{\alpha}$ is VP_j, and $\dot{q}$ is VD_j, then $\Delta\delta_c$ is VU_j, $\Delta\delta_{st}$ is VV_j, $(j = 1, \ldots, m_\alpha)$	If $\theta - \theta_s$ is VI_j, $\dot{\alpha}$ is VP_j, and $\dot{q}$ is VD_j, then $\Delta\delta_c$ is VU_j, $\Delta\delta_{st}$ is VV_j, $(j = 1, \ldots, m_\theta)$
Fuzzy PID β hold rules:	Fuzzy PID Ψ hold rules:
If $\beta - \beta_s$ is VI_j, $\dot{\beta}$ is VP_j, and $\dot{r}$ is VD_j, then $\Delta\delta_r$ is VU_j, $(j = 1, \ldots, m_\beta)$	If $\Psi - \Psi_s$ is VI_j, β is VP_j, and $\dot{r}$ is VD_j, then $\Delta\delta_r$ is VU_j, $(j = 1, \ldots, m_\Psi)$
Fuzzy PI ϕ hold rules:	
If $p - p_s$ is VI_j, and $\dot{p}$ is VP_j, then $\Delta\delta_{df}$ is VU_j, $(j = 1, \ldots, m_p)$	

Table 3 Forms of fuzzy control rules.

In the table VP_j, VI_j, VD_j and $VU_j (or VV_j)$ represent suitable linguistic fuzzy values for the jth fuzzy control rule. These fuzzy values are chosen as: Negative Big(NB), Negative Medium(NM), Negative Small(NS), Around Zero(AZ), Positive Big(PB), Positive Medium(PM), Positive Small(PS). They are defined in the universe of discourse [-20, 20]. The corresponding membership functions are shown in Fig. 3.

3.4 Fuzzy Inference Rules

One of the most difficult and important problems in the design of fuzzy logic controllers is the generation of linguistic fuzzy control rules. Human operators' experience has played an important role in forming rules in many successful industrial applications. However, the empirical design approach is usually less efficient, time-consuming and even frustrating because some unclear, irrelevant and even incorrect information could be generated by human operators. For aircraft fuzzy controller design, the analysis of dynamic characteristics along with optimal control applications can make the generation of fuzzy rules more efficient, as will be presented in the following.

As described in the preceding section, the present fuzzy PID flight control system for the X-29 aircraft consists of three sets of rules to perform both stability augmentation and command tracking functions for pitch, yaw and

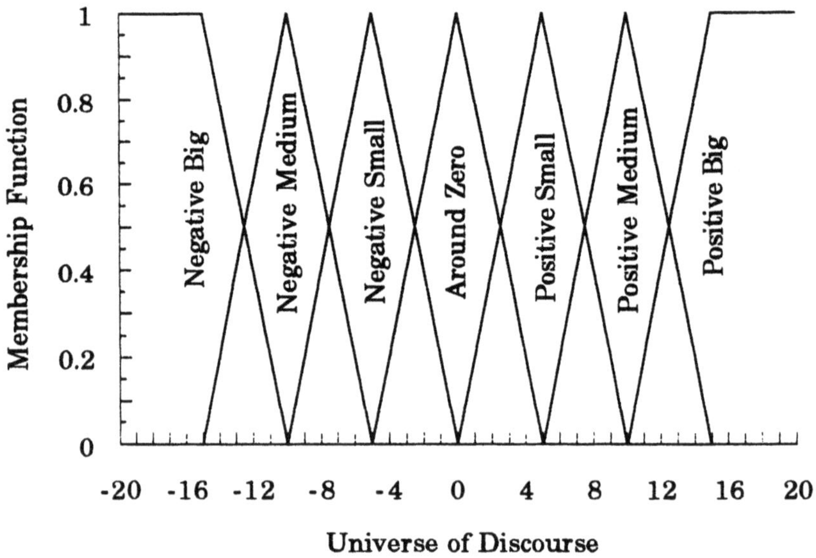

Figure 3 Membership functions.

roll motions. The number of fuzzy inference rules for each controller is basically determined by the combination of linguistic fuzzy values for each of premise variables in a fuzzy rule. For example, the pitch PID controller could have $7^3 = 343$ rules since all three premise variables, $\alpha - \alpha_s, \dot{\alpha}, \dot{q}$ as, , andq , are described by seven fuzzy values. Likewise, the yaw PID controller and the roll PI controller could have $7^3 = 343$ and $7^2 = 49$ rules, respectively. A large collection of fuzzy rules, however, may result in a relatively slow procedure for the fuzzy inference and evaluation, and thus makes the real-time implementation difficult. In the present fuzzy flight control system, the fuzzy inference rules are simplified by considering possible human operators' strategies which are quite different from those of linear PID controllers with constant gains. The present fuzzy control system includes coarse and fine rules generated for each set of rules [13]. The coarse rules are formed to use a large control action to quickly drive the controlled variables to set points or equilibrium points and also provide necessary static stability. Dynamic stability is augmented through the fine rules only when the variables come close to the set points or equilibrium points. The pitch, yaw and roll control rules for the X-29 are discussed below.

		$\dot{\alpha}$						
		NB	NM	NS	AZ	PS	PM	PB
	NB	PB	PB	PB	PB	PB	PB	PS
	NM	PB	PB	PB	PB	PB	PS	NB
	NS	PB	PB	PB	PB	AZ	NB	NB
$\alpha - \alpha_s$	AZ	PB	PB	PB		NB	NB	NB
	PS	PB	PB	AZ	NB	NB	NB	NB
	PM	PB	NS	NB	NB	NB	NB	NB
	PB	NS	NB	NB	NB	NB	NB	NB

Table 4 Coarse rules for a hold with canard (weight = 1) .

Pitch PID Control Rules

The X-29 aircraft has a high level of longitudinal instability as indicated earlier . Static stability should be provided with active canard, symmetric flap and aftbody strake flap. In the present study the deflection of the symmetric flap is kept at the initial trim value. Only canard and strake flap are employed in the fuzzy control. Referring to the linear PID control given by Eq. (4.15), the proportional gains (coefficients of $\dot{\alpha}$) are obviously larger than the integral and differential gains (coefficients of $\alpha - \alpha_s$ and $\dot{q}$). A small and positive $\dot{\alpha}$ could result in relatively large incremental changes of canard (negative) and strake flap (positive) to offer artificial static stability. Also, deviation of the angle of attack from its set point, $\alpha - \alpha_s$, has a minor contribution to the incremental changes of canard and strake flap. The fuzzy pitch control strategy can then be summarized as: as long as $\alpha - \alpha_s$ and $\dot{\alpha}$ are not close to zero, a larger control action should be taken to increase static stability and also quickly drive α to α_s with coarse rules no matter how large $\dot{q}$ is. When the angle of attack is close to the specified set point and $\dot{\alpha}$ is around zero, pitch damping is strongly imposed through fine rules to reduce oscillation around the set point. As a result, $7^2 - 1 = 48$ coarse rules are formed and listed in Tables 4 and 5 for both canard and strake flap control, respectively.

Notice that the upper left half of Table 4 and the lower right half of Table 5 represent for positive incremental changes in canard and strake flap, and the lower right half of Table 4 and upper left half of Table 5 represent negative incremental changes of these two control surfaces, respectively. The shaded areas include 7 fine rules which are listed in Tables 6 and 7.

		$\dot{\alpha}$						
		NB	NM	NS	AZ	PS	PM	PB
$\alpha - \alpha_s$	NB	NB	NB	NB	NB	NB	NB	PM
	NM	NB	NB	NB	NB	NB	PS	PB
	NS	NB	NB	NB	NB	AZ	PB	PB
	AZ	NB	NB	NB		PB	PB	PB
	PS	NB	NB	AZ	PB	PB	PB	PB
	PM	NB	NS	PB	PB	PB	PB	PB
	PB	NM	PB	PB	PB	PB	PB	PB

Table 5 Coarse rules for a hold with strake flap (weight = 1) .

For both *alpha* $- \alpha_s$ and $\dot{\alpha}$ are AZ,
if $\dot{q}$ is NB, then $\Delta\delta_c$ is PB,
else if $\dot{q}$ is NM, then $\Delta\delta_c$ is PM,
else if $\dot{q}$ is NS, then $\Delta\delta_c$ is PS,
else if $\dot{q}$ is AZ, then $\Delta\delta_c$ is AZ,
else if $\dot{q}$ is PS, then $\Delta\delta_c$ is NS,
else if $\dot{q}$ is PM, then $\Delta\delta_c$ is NM,
else if $\dot{q}$ is PB, then $\Delta\delta_c$ is NB.

Table 6 Fine rules for α hold with canard (weight < 1).

For both *alpha* $- \alpha_s$ and $\dot{\alpha}$ are AZ,
if $\dot{q}$ is NB, then $\Delta\delta_{st}$ is NB,
else if $\dot{q}$ is NM, then $\Delta\delta_{st}$ is NM,
else if $\dot{q}$ is NS, then $\Delta\delta_{st}$ is NS,
else if $\dot{q}$ is AZ, then $\Delta\delta_{st}$ is AZ,
else if $\dot{q}$ is PS, then $\Delta\delta_{st}$ is PS,
else if $\dot{q}$ is PM, then $\Delta\delta_{st}$ is PM,
else if $\dot{q}$ is PB, then $\Delta\delta_{st}$ is PB.

Table 7 Fine rules for α hold with strake flap (weight < 1) .

The fine rules are employed to reduce pitch oscillations only in regions around the set point α_s and therefore the values are reduced by weights being less than one, and actually function as a fuzzy pitch damper. The total number of coarse and fine rules for each control surface turns out to be $48 + 7 = 55$, which is much less than $7^3 = 343$. The selected 2×55 rules greatly speed up the process of fuzzy inference and evaluation.

For a small sideslip, the kinematic equation

$$\alpha \approx \theta - \theta_W \tag{4.19}$$

holds, where θ_W is the flight path angle. From Eq. (4.19) it follows that a positive $\Delta\theta$ normally corresponds to a positive $\Delta\alpha$. The fuzzy PID control rules for the angle of attack hold can be simply taken for the fuzzy pitch attitude hold with the premise variables $\alpha - \alpha_s$ replaced by $\theta - \theta_s$.

Yaw PID Control Rules

As far as the fuzzy yaw PID control is concerned , the fuzzy rules are supposed to take a similar form to the pitch control rules based on the same control strategy. The actual yaw control rules, however, are chosen slightly differently because of the particular directional dynamic characteristics. As analyzed in Section 3.2, the inherent directional static stability is available. Excessive static stability offered by a set of linguistic rules would degrade the performance and result in sluggish response. In the present yaw fuzzy controller the function performed by the proportional part ($\dot{\beta}$) is simply removed from coarse rules. If the angular displacement error $\beta - \beta_s$ is not close to zero, the incremental change of rudder deflection is heavily dependent on the error regardless of the other variables. Therefore, a large and positive $\beta - \beta_s$ results in a large but negative $\Delta\delta_r$ based on the linear PID control given by Eq. (4.17), and so forth. When the sideslip is close to its set point β_s, the yaw rate may be very large due to the rapid decrease in $\beta - \beta_s$. The differential part ($\dot{r}$) will play an important role in reducing the yaw rate and also the roll rate caused by cross-coupling. Table 8 presents the coarse rules for yaw control.

In Table 8 the upper part represents the area for a positive rudder incremental change and the lower part is for a negative change. It should be noted that only 6 coarse rules are formed since the rudder incremental change

		$\dot{\beta}$						
		NB	NM	NS	AZ	PS	PM	PB
$\beta - \beta_s$	NB	PB	PB	PB	PB	PB	PB	PB
	NM	PM	PM	PM	PM	PM	PM	PM
	NS	PS	PS	PS	PS	PS	PS	PS
	AZ							
	PS	NS	NS	NS	NS	NS	NS	NS
	PM	NM	NM	NM	NM	NM	NM	NM
	PB	NB	NB	NB	NB	NB	NB	NB

Table 8 Coarse rules for β hold with rudder (weight = 1) .

For $\beta - \beta_s$ is AZ,
if $\dot{r}$ is NB, then $\Delta\delta_r$ is NB,
else if $\dot{r}$ is NM, then $\Delta\delta_r$ is NM,
else if $\dot{r}$ is NS, then $\Delta\delta_r$ is NS,
else if $\dot{r}$ is AZ, then $\Delta\delta_r$ is AZ,
else if $\dot{r}$ is PS, then $\Delta\delta_r$ is PS,
else if $\dot{r}$ is PM, then $\Delta\delta_r$ is PM,
else if $\dot{r}$ is PB, then $\Delta\delta_r$ is PB.

Table 9 Fine rules for β hold with rudder (weight < 1).

is independent of $\dot{\beta}$. In addition, the shaded area represents 7 fine rules which are listed in Table 9.

Similar to the pitch controller, the fine rules are used to reduce yaw oscillations only in regions around the set point β_s, and the control action is relatively small compared with that of the coarse rules. According to Eq. (4.17), this set of fine rules can be thought of as a fuzzy yaw damper. The total number of directional control rules is $6 + 7 = 13$, which is even smaller than that of pitch control rules.

At low angles of attack, the following kinematic equation holds according to the definition in flight dynamics,

$$\beta \approx \Psi_W - \Psi \tag{4.20}$$

Coarse Rules (Weight = 1)	Fine Rules (Weight ¡ 1)
if $p - p_s$ is NB, then $\Delta\delta_{df}$ is PB, else if $p - p_s$ is NM, then $\Delta\delta_{df}$ is PM, else if $p - p_s$ is NS, then $\Delta\delta_{df}$ is PS, else if $p - p_s$ is PS, then $\Delta\delta_{df}$ is NS, else if $p - p_s$ is PM, then $\Delta\delta_{df}$ is NM, else if $p - p_s$ is PB, then $\Delta\delta_{df}$ is NB.	For $p - p_s$ is AZ, if $\dot{p}$ is NB, then $\Delta\delta_{df}$ is PB, else if $\dot{p}$ is NM, then $\Delta\delta_{df}$ is PM, else if $\dot{p}$ is NS, then $\Delta\delta_{df}$ is PS, else if $\dot{p}$ is AZ, then $\Delta\delta_{df}$ is AZ, else if $\dot{p}$ is PS, then $\Delta\delta_{df}$ is NS, else if $\dot{p}$ is PM, then $\Delta\delta_{df}$ is NM, else if $\dot{p}$ is PB, then $\Delta\delta_{df}$ is NB.

Table 10 Roll PI control rules with differential flap.

in which Ψ_W is the azimuth angle of flight path. A positive $\Delta\Psi$ normally corresponds to a negative $\Delta\beta$. Accordingly the fuzzy PID rules for heading hold are similar to those for β hold except that the positive and negative areas of coarse rules in Table 8 are interchanged.

Roll PI Control Rules

The fuzzy PI roll control is intended to provide bank angle hold and also to improve dynamic lateral stability around the set point or the desired roll rate p_s through the differential flap . The fuzzy controller is also designed to model a human operator's strategy: No matter how large the roll acceleration $\dot{p}$ is, as long as the roll rate p is not close to its set point p_s, relatively large incremental changes in the differential flap should be used to quickly drive roll rate to its set point, the integral part being dominant. Once the roll rate comes close to its set points rapidly, the proportional part will play a more important role to impose damping and keep the roll rate p as close to p_s as possible with little or no oscillation. As a result, this fuzzy PI controller is highly nonlinear and time-varying. Based on the above control strategies and the linear control function given by Eq. (4.18), the roll control rules are formed and listed in Table 10.

The coarse rules correspond to the integral fuzzy roll control to reduce the roll rate error with relatively large incremental controls. Fine rules are employed to reduce or eliminate oscillations only in regions around the set point. Thus, incremental controls for the fine rules are weighted smaller than those used for the coarse rules.

From a dynamic point of view, the fine rule can be thought of as a fuzzy roll damper according to Eq. (4.18).

The fuzzy PID control algorithm described in Section 2 can now be adopted to compute the non-fuzzy control adjustments $\Delta\delta c, \Delta\delta st, \Delta\delta df$ and $\Delta\delta_r$ at each time step.

Care should be taken when choosing scaling factors because they have significant effects on the performance of the system being controlled. The determination of scaling factors used in the present application is discussed in following section.

3.5 Scaling Factors

The scaling factors have important effect on the system response [14]. They can take many possible values which can give just as good but probably not better response. The scaling factors of incremental control $\Delta\delta c, \Delta\delta st, \Delta\delta df$ and $\Delta\delta_r$ can be determined by using the given maximum deflection rates for each control surfaces. For the premise variables of control rules, the scaling factors are first estimated by evaluating possible large value of each variable, or using linear control gains. These estimated scaling factors are then adjusted to get better performance in terms of quick response and small tracking errors.

Firstly, the scaling factors of $\Delta\delta c, \Delta\delta st, \Delta\delta df$ and $\Delta\delta_r$ are determined based on an assumption that the maximum possible values of these incremental controls correspond to the maximum value on the universe of discourse, which is 20 in the present controller. For the X-29, the position and rate limits of deflection for these control surfaces are listed in Table 11 [15].

The scaling factors are then chosen as

$$\begin{aligned} G_c &= \tfrac{100T}{20}\deg = \frac{100T}{20 \times 57.3} rad, \\ G_{st} &= \tfrac{30T}{20}\deg = \frac{30T}{20 \times 57.3} rad, \\ G_{df} &= \tfrac{70T}{20}\deg = \frac{70T}{20 \times 57.3} rad, \end{aligned}$$

Control surface	Position limit (deg)	Rate limit (deg/sec)
Canard	-60, 30	±100
Symmetric flap	-10, 25	±70
Strake flap	-30, 30	±30
Differential flap	-17.5, 17.5	±70
Rudder	-30, 30	±125

Table 11 Position and rate limits of control surfaces of the X-29.

$$G_r \ = \tfrac{100T}{20}\,\text{deg} = \ \frac{100T}{20 \times 57.3} rad,$$

where the sampling time T is taken to be 0.025 sec. In the linear control given by Eq. (4.18), the value of $p - p_s$ alone, corresponding to the maximum $\Delta\delta_{df}$, is $\frac{70T}{0.667T} = 105\,\text{deg}$. Therefore the initial estimate of the scaling factor for $p - p_s$ is $G_p = \frac{105}{20}\frac{deg}{sec} = \frac{105}{(20\times 57.3)}\frac{rad}{sec}$. Similarly, the initial estimates of scaling factors for $\dot{p}$ is $G_{\dot{p}} = \frac{100T}{(57.3\times 0.1567T)} = \frac{11}{20}\frac{rad}{sec^2}$. The initial estimates of G_p and $G_{\dot{p}}$ can also be found by evaluating p_{max} and $\dot{p}_{max}$ from simulation or flight test data. Because the roll moment of inertia is much smaller than pitch and yaw moments of inertia, $\dot{q}_{max}$ and $\dot{r}_{max}$ are less than $\dot{p}_{max}$. Hence, the estimates of $G_{\dot{q}}$ and $G_{\dot{r}}$ may be reduced to be less than $\frac{11}{20}\frac{rad}{sec^2}$. Also based on simulation and flight test data the reasonably large values for $|\alpha - \alpha_s|, |\beta - \beta_s|$ and $|\dot{\alpha}|$ can be set to $30\,\text{deg}, 30\,\text{deg}$ and $50\frac{deg}{sec}$, respectively. Their initial estimates could be $G_\alpha = G_{beta} = \frac{30}{20}$ deg, and $G_{\dot{\alpha}} = \frac{50}{20}\frac{deg}{sec}$. With the above estimates, extensive simulations are then made to update and modify these scaling factors to obtain good response to various commands. For this purpose, the scaling factors for each control system are systematically incremented from the initial estimates and tested for different types of command input (such as step, sinusoidal and ramp) and initial disturbances. The values giving response with small overshoots and tracking errors are those to be used. In the present fuzzy control system, the scaling factors for the angular acceleration are determined to be

$$G_{\dot{p}} = \frac{14}{20}\frac{rad}{sec^2}, \quad G_{\dot{q}} = \tfrac{6}{20}\tfrac{rad}{sec^2}, \quad G_{\dot{r}} = \frac{2}{20}\frac{rad}{sec^2}$$

and the scaling factor for $\dot{\alpha}$ is

$$G_{\dot{\alpha}} = \frac{40}{20}\frac{\text{deg}}{sec} = \frac{40}{20 \times 57.3}\frac{rad}{sec}$$

Slightly larger G_α, G_β and G_p offer better response to larger commands. In the present control system, they have been chosen as linear functions of $|\alpha - \alpha_s|, |\beta - \beta_s|$ and $|p - p_s|$, respectively to achieve better performance:

$$\begin{aligned}
G_\alpha(\alpha) &= \tfrac{25+0.5|\alpha-\alpha_s|}{20}\,\text{deg} &= \frac{25+0.5|\alpha-\alpha_s|}{20 \times 57.3} rad \\
G_\beta(\beta) &= \tfrac{32+0.5|\beta-\beta_s|}{20}\,\text{deg} &= \frac{32+0.5|\beta-\beta_s|}{20 \times 57.3} rad \\
G_p(p) &= \tfrac{30+2|p-p_s|}{20}\,\tfrac{\text{deg}}{sec} &= \frac{30+2|p-p_s|}{20 \times 57.3}\frac{rad}{sec}
\end{aligned}$$

It should be pointed out that if the fuzzy control rules are generated based on dynamic analysis, the fuzzy controller performance is robust to a slight change of scaling factors.

For the pitch attitude and heading hold, the scaling factors of $\theta - \theta_s$ and $\Psi - \Psi_s$ can be simply chosen as

$$G_\theta(\theta) = G_\alpha(\theta), \quad G_\Psi(\Psi) = G_\beta(\Psi)$$

to provide good performance.

4 VALIDATION THROUGH NONLINEAR SIMULATION

To validate the fuzzy PID controllers developed above, nonlinear simulations are performed. The six degree-of-freedom nonlinear simulation program, SIMX29 [11] was modified by implementing the fuzzy PID control system. Nonlinear aerodynamic data from wing tunnel tests and thrust data are incorporated in the simulation. A 4th order Runge-Kutta scheme is used to numerically integrate the equations of motion. Also, a set of lateral-directional linear conventional PIF-CGT controllers [16] is compared with the present fuzzy controller in performance.

The deflection and rate limits of each control surfaces listed in Table 11 are used in the present nonlinear simulations. Aircraft tracking performance, stability, controllability and robustness with the fuzzy flight control system will be analyzed and discussed.

4.1 Fuzzy Pitch Control

To test the X-29 longitudinal performance with the fuzzy PID control system, closed loop response under angle of attack, pitch angle step commands, and initial disturbances are examined .

Figs. 4 and 5 present the time histories of angle of attack and pitch rate under a 10° angle of attack step command. The corresponding deflection angles of canard and strake flap are given in Figs. 6 and 7. As shown in Fig. 4, the angle of attack can quickly reach the command value without overshoot. Canard and strake flap approach constant deflections to accomplish this maneuver.

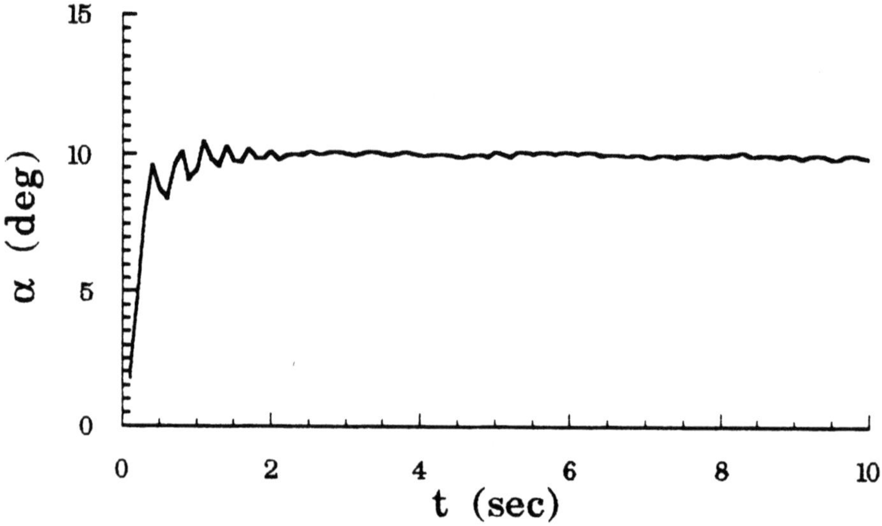

Figure 4 Angle of attack at step command $\alpha_s = 10^o$.

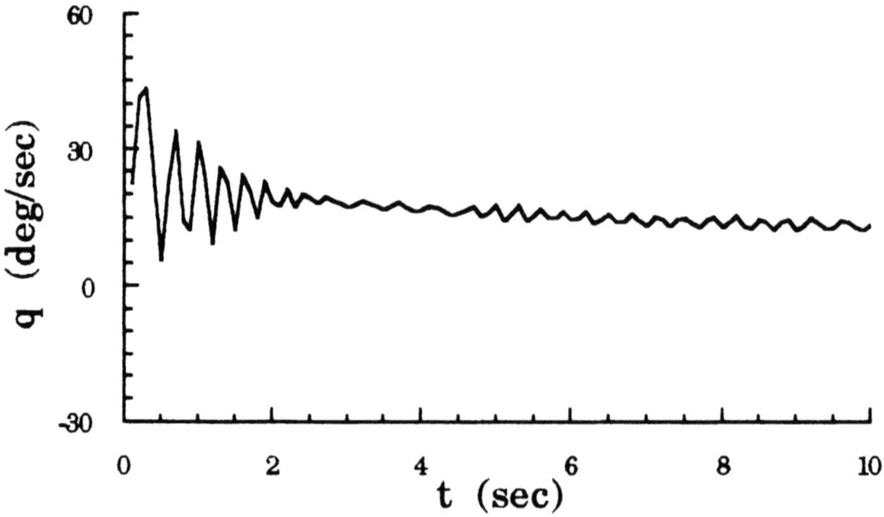

Figure 5 Pitch rate at step command $\alpha_s = 10^o$.

The same set of fuzzy control rules are employed to perform a 10° pitch angle step command. The simulation results are given in Figs. 8 and 9. There is an

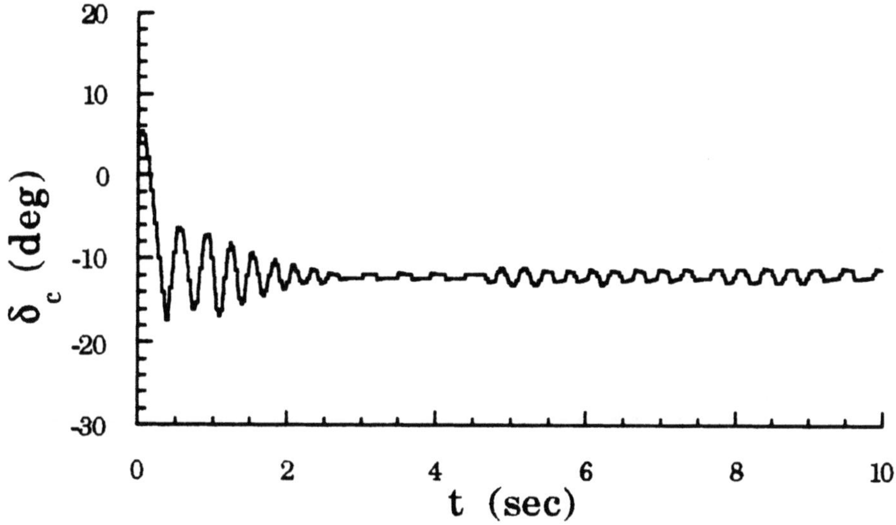

Figure 6 Canard deflection at $\alpha_s = 10^o$.

overshoot in the pitch angle response as shown in Fig. 8. This is perhaps caused by using the same control rules as the α-hold indicated earlier. When the pitch angle reaches the commanded value $\theta_s = 10^o$, the pitch rate is stabilized to zero. As a result, a new trim flight condition is reached through the fuzzy controller.

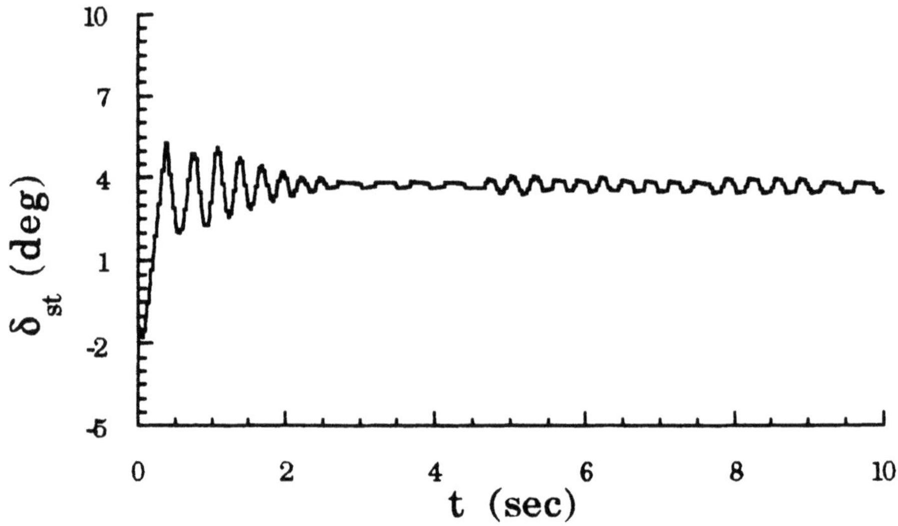

Figure 7 Strake flap deflection at $\alpha_s = 10^o$.

Fig. 10 shows the aircraft trajectories in the vertical plane at $\alpha_s = 10^o$ and $\theta_s = 10^o$. At the command of $\theta_s = 10^o$ the aircraft keeps a steady climb flight, while at the command of $\alpha_s = 10^o$ the aircraft makes a loop maneuver.

In Figs. 11 and 12, responses of angle of attack and pitch rate at a command of $\alpha_s = 15^o$ (an off- design condition) are given. The well-behaved responses illustrate the robustness of the fuzzy controller.

4.2 Fuzzy Yaw Control

The directional fuzzy PID controller is developed to perform yaw dynamic stability augmentation and automatic flight control functions such as sideslip hold and heading hold. This fuzzy controller can be evaluated by examining the aircraft response to step commands of sideslip and heading angle, and initial disturbances. Some of the responses by the fuzzy controller are also compared with those by a linear PIF-CGT controller .

Figures 13 and 14 present the time histories of sideslip angle and yaw rate under a 10° sideslip step command. The sideslip response with a rise time of

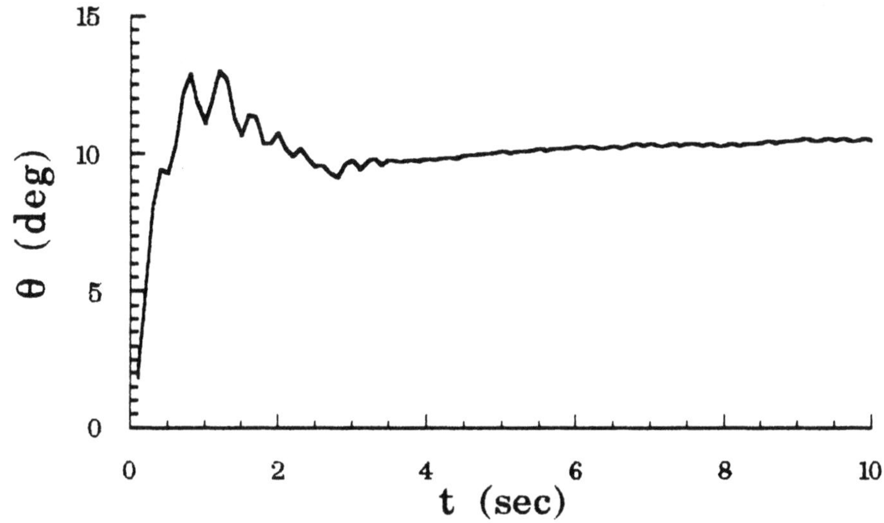

Figure 8 Pitch angle at step command $\theta_s = 10^o$.

0.55 sec by the fuzzy controller was much faster than that with a rise time of 2.10 sec by a linear controller in Fig. 13. The yaw rate generated by the fuzzy controller was larger compared with that of the linear controller as shown in Fig. 14. Short transition times and engagement which are needed by modern fighter aircraft can be obtained with the fuzzy controller, and thus good yaw agility can be achieved. The corresponding deflection angles and rates of differential flap and rudder are given in Figs. 15 and 16. Control of this maneuver mainly comes from the rudder deflection. The differential flap deflection also makes a small contribution to this maneuver in addition to keeping the roll rate close to zero at a steady state sideslip.

The responses of heading angle and yaw rate to a heading step command of $\Psi = 10^o$ are shown in Figs. 17 and 18. With the fuzzy yaw PID controller, the heading angle is quickly driven to the command value. The yaw rate is also quickly damped to zero with the fine rules in the fuzzy controller. Consequently, the aircraft is switched to a new trim condition that is the same as the initial flight condition except the change in heading angle.

Figure 19 shows the aircraft trajectories in the horizontal plane at $\beta_s = 10^o$ and $\Psi_s = 10^o$. At the command of $\Psi_s = 10^o$ the aircraft keeps a straight and

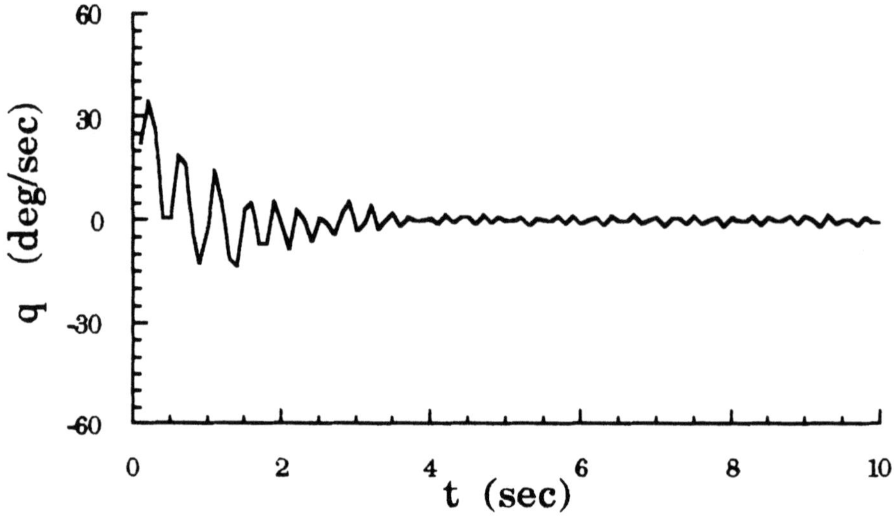

Figure 9 Pitch rate at step command $\theta_s = 10^o$.

level flight with a heading angle of 10°, while at the command of $\beta_s = 10^o$ the aircraft makes a slight left turn with a 10° sideslip and zero bank angle.

In Figures 20 and 21, under a 15° sideslip step command (an off-design condition), the fuzzy controller still performed well, but the linear controller lost the controllability of both surfaces with divergent sideslip and yaw rate as a result.

4.3 Fuzzy Roll Control

The X-29 roll performance with the fuzzy flight control system is tested through bank-to-bank maneuvers . Figs. 22 and 23 show the time histories of bank angle and roll rate during a 40° bank-to-bank maneuver. In Fig. 22 the bank angle response by the fuzzy controller is shown to almost exactly follow the command, but the response of the linear PIF-CGT controller has large overshoots at turning points. The roll rate by the fuzzy controller is much faster than that of the linear controller as shown in Fig. 23. This indicates that the fuzzy controller can significantly improve the roll performance and

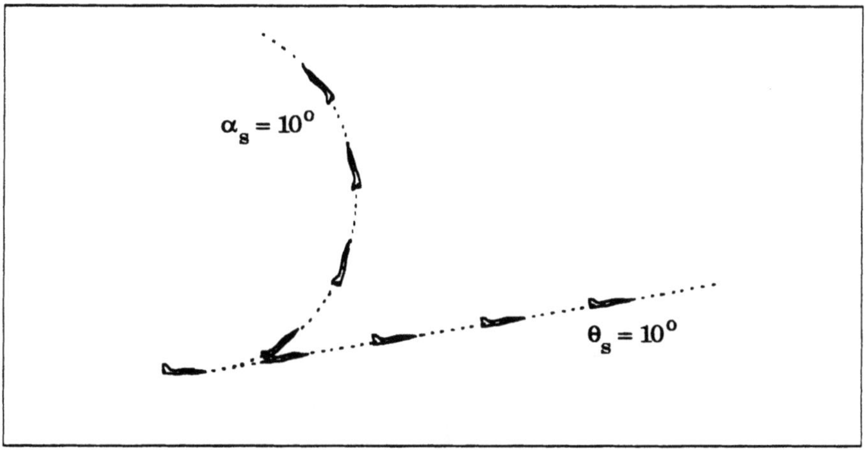

Figure 10 Trajectory in vertical plane for $\alpha_s = 10^o$ hold and $\theta_s = 10^o$.

increase its agility. Control surface deflections of differential flap and rudder in Figs. 24-25 are within their limits. This maneuver is accomplished mainly by the differential flap control as shown in Fig. 24. Rudder deflection is mainly used to reduce sideslip excursion during the bank-to-bank maneuver.

In Figure 26, the bank angle responses by the same fuzzy controller and a linear controller during a 70° bank-to-bank maneuver are given. An excellent response is obtained by the fuzzy controller, but the linear controller generates a divergent response. The roll rate response by the fuzzy controller during this maneuver has slight overshoots but is damped quickly to the commanded value while the linear controller induces an unacceptable roll rate response as shown in Fig. 27.

For additional simulation results, readers should consult Reference 12.

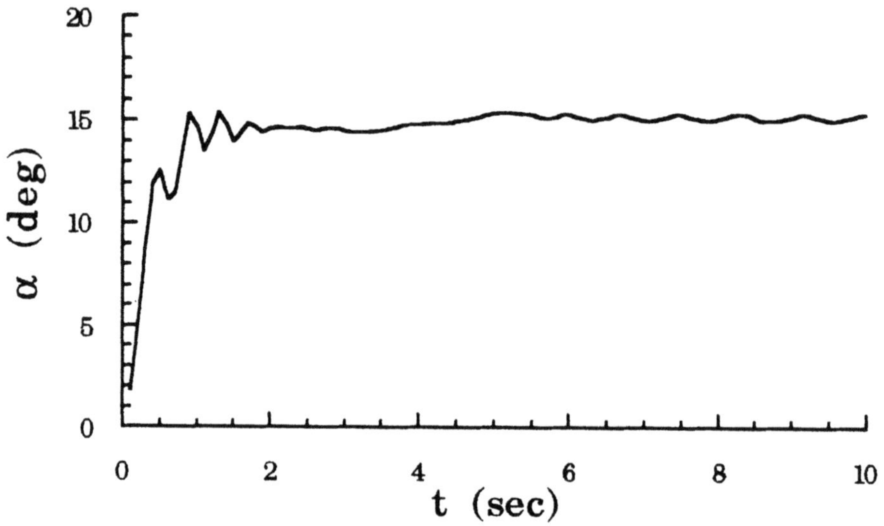

Figure 11 Angle of attack at step command $\alpha_s = 15^o$.

5 CONCLUSIONS

Fuzzy logic technology has been applied to flight control to enhance aircraft performance. The design procedure of a fuzzy flight control system was studied and presented with a sample aircraft, the X- 29. Through nonlinear simulations of the system performance, it could be concluded that:

1. A pure fuzzy logic controller could perform both stability augmentation and automatic flight control functions with coarse and fine control rules.
2. A fuzzy flight control system could be developed based on aircraft dynamic analysis and optimal control of a linearized model along with possible human operators' strategy.
3. Aircraft with a fuzzy control system behaved nonlinearly even with a linear open loop system. A fuzzy controller could offer excellent performance in terms of quick response and small tracking errors to achieve highly nonlinear control objectives.

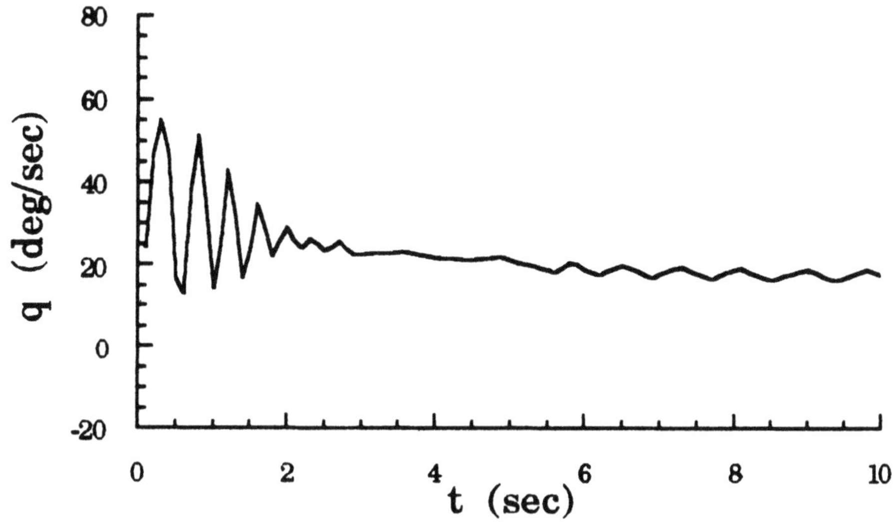

Figure 12 Pitch rate at step command $\alpha_s = 15^o$.

4. A fuzzy flight control system was capable of utilizing the potential of control surfaces to generate large angular rates without sacrificing controllability, and thus to greatly improve aircraft agility.

5. A fuzzy flight control system was robust compared with a conventional control because all the control rules were evaluated for an existing flight condition before any control action was taken.

REFERENCES

[1] Zadeh, L. A., "Outline of a New Approach to the Analysis of Complex Systems and Decision Processes," IEEE Trans. Syst., Man, Cybern., Vol. SMC-3, No. 1, pp. 28-44, 1973.

[2] Mamdani, E. H. and Assilian, S., "An Experiment in Linguistic Synthesis with a Fuzzy Logic Controller," Int. J. Man-Machine Studies, Vol. 7, pp. 1-13, 1975.

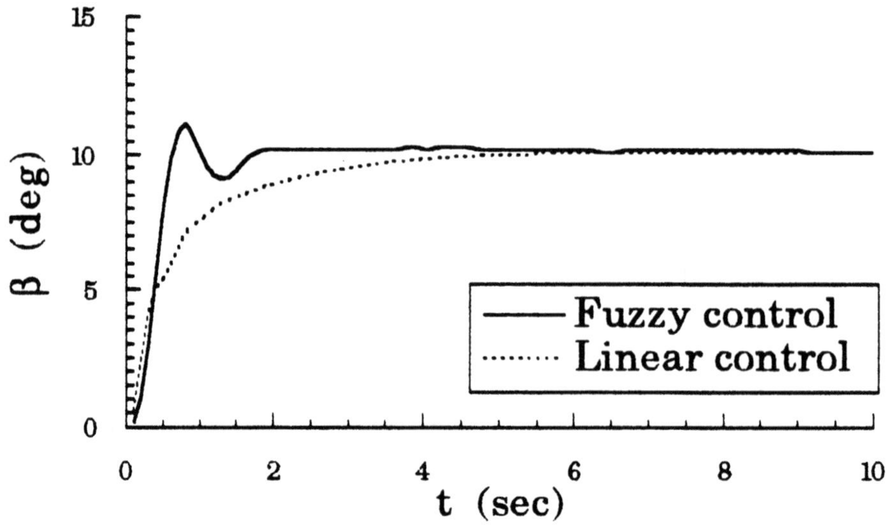

Figure 13 Sideslip angle at step command $\beta_s = 10^o$.

[3] Larkin, L. I., "A Fuzzy Logic Controller for Aircraft Flight Control," Industrial Applications of Fuzzy Control, edited by M. Sugeno, Elsevier Science Publishers B. V. (North-Holland), 1985.

[4] Chiu, S., Chand, S., Moore, D., and Chaudhary, A., "Fuzzy Logic for Control of Roll and Moment for a Flexible Wing Aircraft," IEEE Control Systems Magazine, June 1991, pp.42-48.

[5] Rao, S. S. and Dhingra, A. K., "Applications of Fuzzy Theories to Multi-Objective System Optimization," NASA CR-177573, Jan. 1991.

[6] Villarreal, J. A.; Lea, R. N.; and Savely, R. T., "Fuzzy Logic and Neural Network Technologies," AIAA Paper 92-0868, Jan. 1992.

[7] Roskam, J., Airplane Flight Dynamics and Automatic Flight Controls, Part II, Roskam Aviation and Engineering Corporation, Ottawa, KS, 1982.

[8] Valasek, J., Eggold, D., and Downing, D., "A Study of a Proposed Modified Torsional Agility Metric," AIAA Paper 91-2883-CP, August 1991.

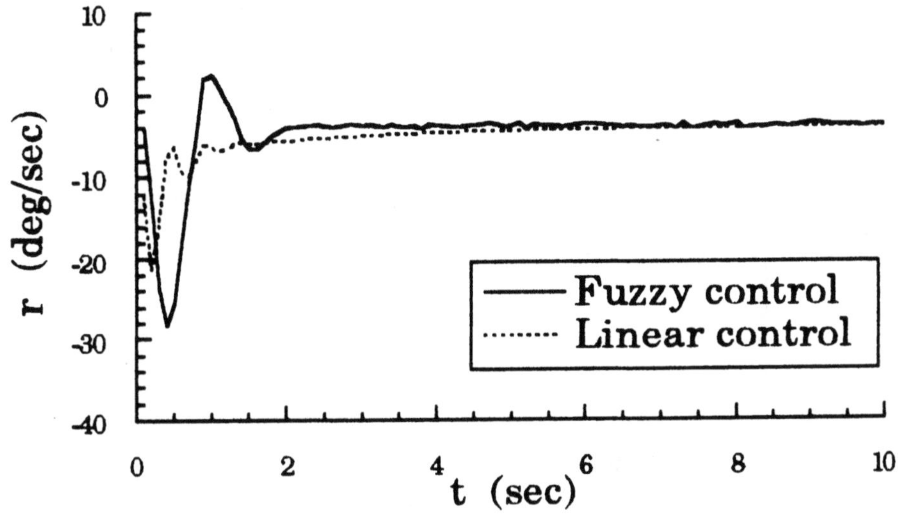

Figure 14 Yaw rate at step command $\beta_s = 10^o$.

[9] Bernard, J. A., "Use of a Rule-Based System for Process Control," IEEE Control Systems Magazine, Oct. 1988, p.3.

[10] Peng, X., Liu, S., Yamakawa, T., Wang, P., and Liu, X., "Self-regulating PID Controller and its Applications to a Temperature Controlling Process," Fuzzy Computing-Theory, Hardware, and Applications, edited by Gupta, M., and Yamakawa, T., Elsevier Science Publishers, 1988, pp.355- 364.

[11] Linse, D., "Design and Analysis of a High Angle of Attack Flight Controls System," MS thesis, the University of Kansas, 1987.

[12] Luo, J., "Aircraft Control Based on Fuzzy Logic," Ph.D Dissertation, the University of Kansas, 1994.

[13] Luo, J., and Lan, C. Edward, "Development and Performance of a Fuzzy Logic Lateral Controller for X-29 Aircraft," International Fuzzy Systems and Intelligent Control Conference, March 1993, Louisville, KY.

[14] Procyk, T., and Mamdani, E., "A Linguistic Self-Organizing Process Controller," Automatica, Vol. 15, 1979, pp. 15-30.

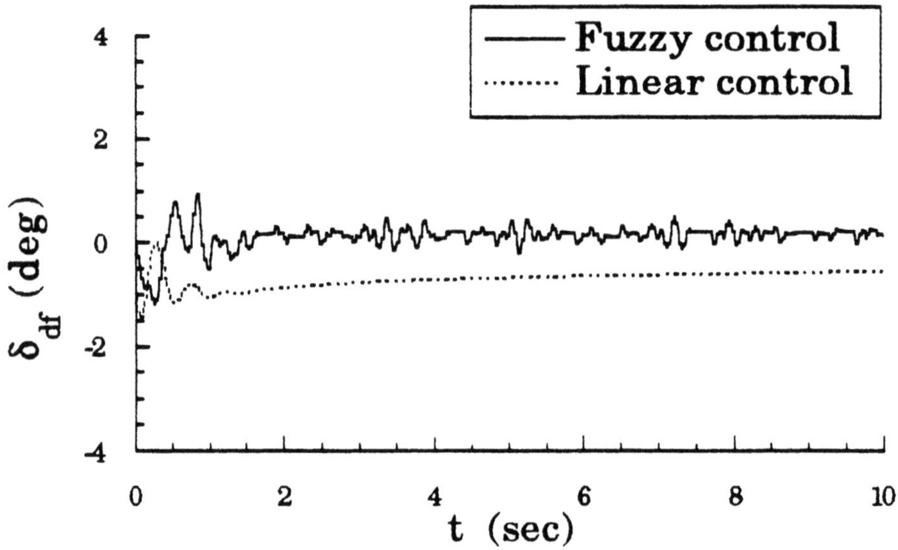

Figure 15 Differential flap deflection at step command $\beta_s = 10^o$.

[15] Bosworth, John T., "Linearized Aerodynamic and Control Law Models of the X-29A airplane and Comparison with Flight Data," NASA Technical Memorandum 4356, February 1992.

[16] Suikat, R., "An Optimal Pole Placement Gain Scheduling Algorithm Using Output Feedback," Ph.D Dissertation, the University of Kansas, 1987.

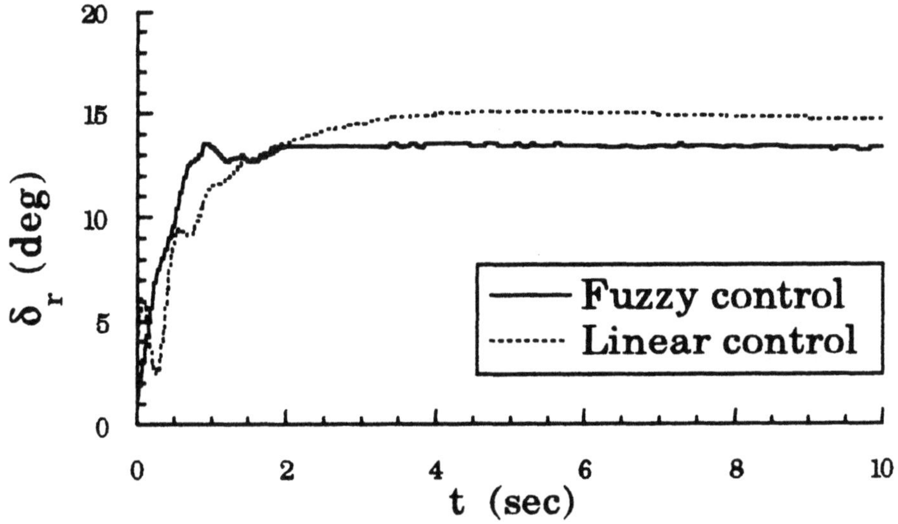

Figure 16 Rudder deflection at step command $\beta_s = 10^o$.

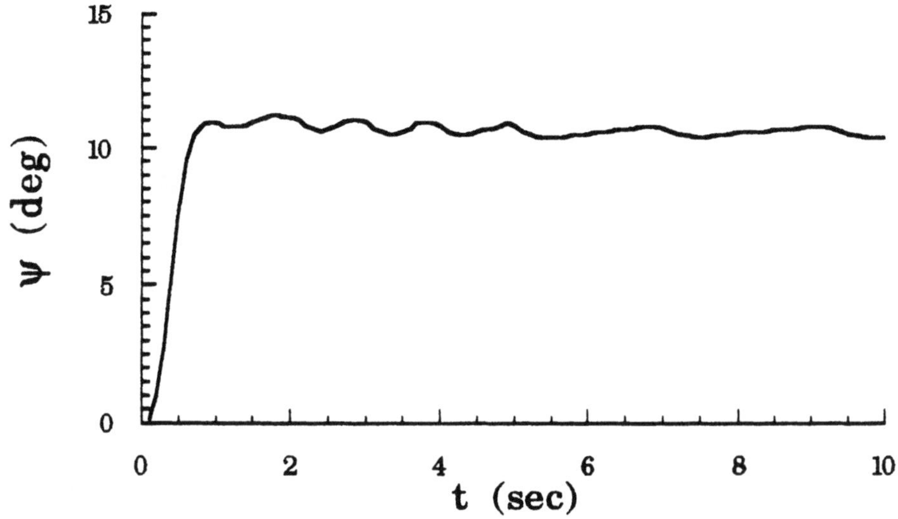

Figure 17 Heading angle at step command $\Psi_s = 10^o$.

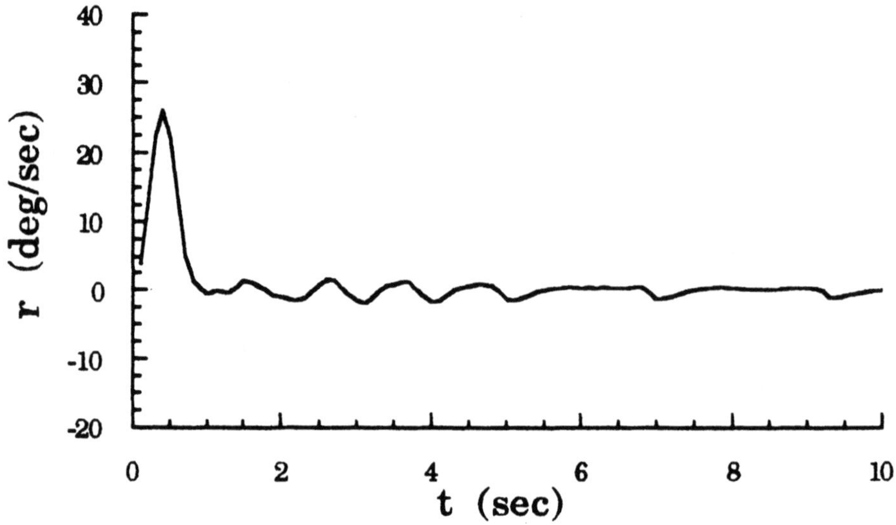

Figure 18 Yaw rate at step command $\Psi_s = 10^o$.

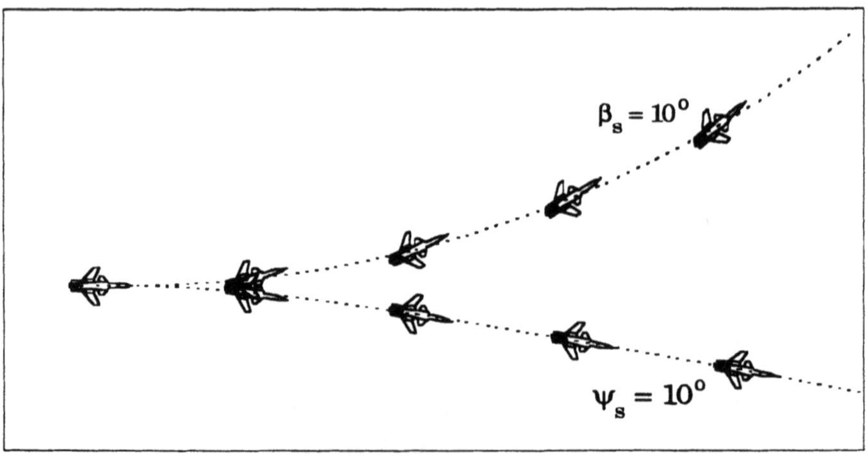

Figure 19 Trajectories in horizontal plane for $\beta_s = 10^o$ hold and $\Psi_s = 10^o$.

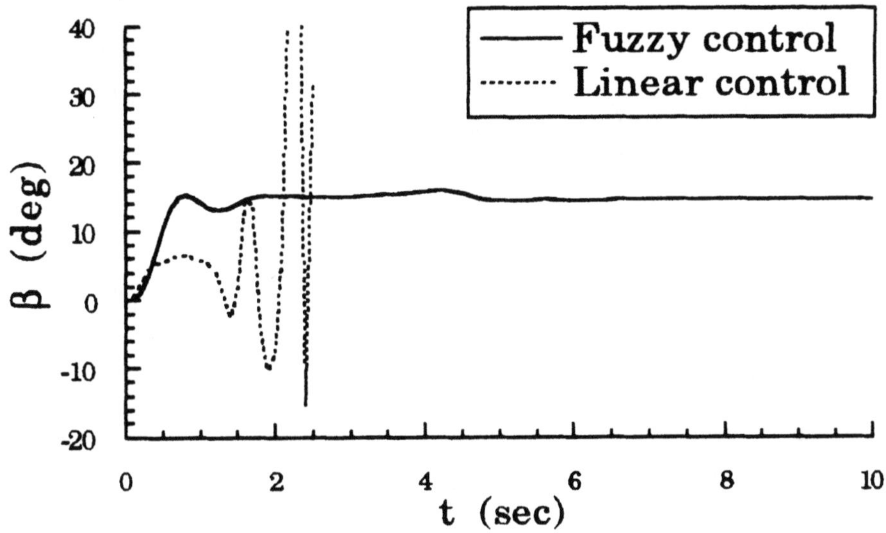

Figure 20 Sideslip angle at step command $\beta_s = 15^o$.

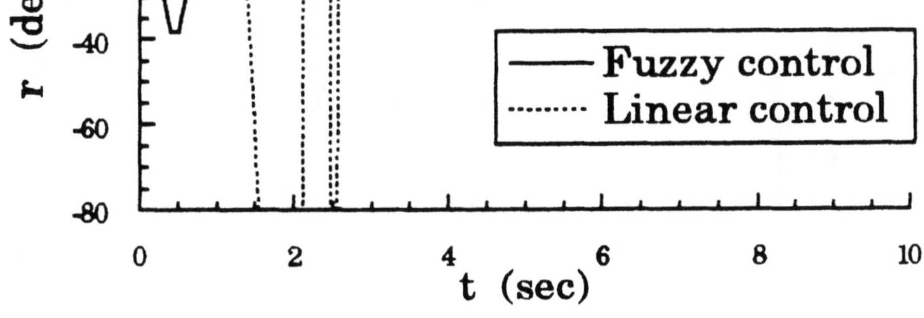

Figure 21 Yaw rate at step command $\beta_s = 15^o$.

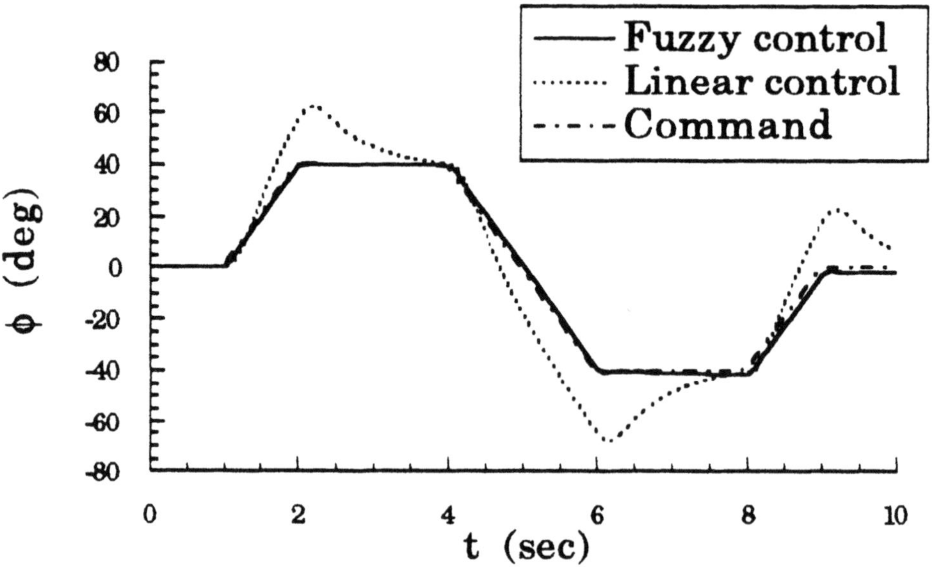

Figure 22 Bank angle at a 40° bank-to-bank command.

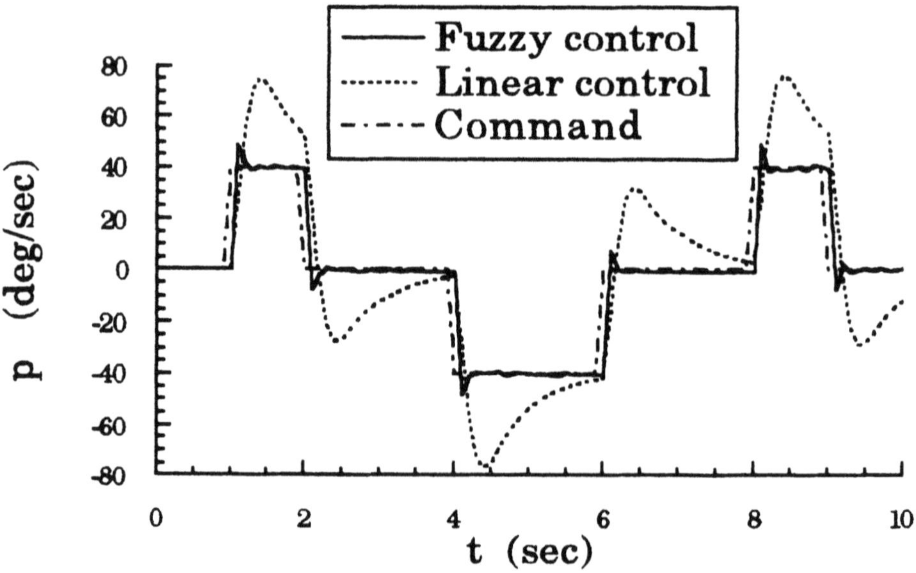

Figure 23 Roll rate at a 40° bank-to-bank command.

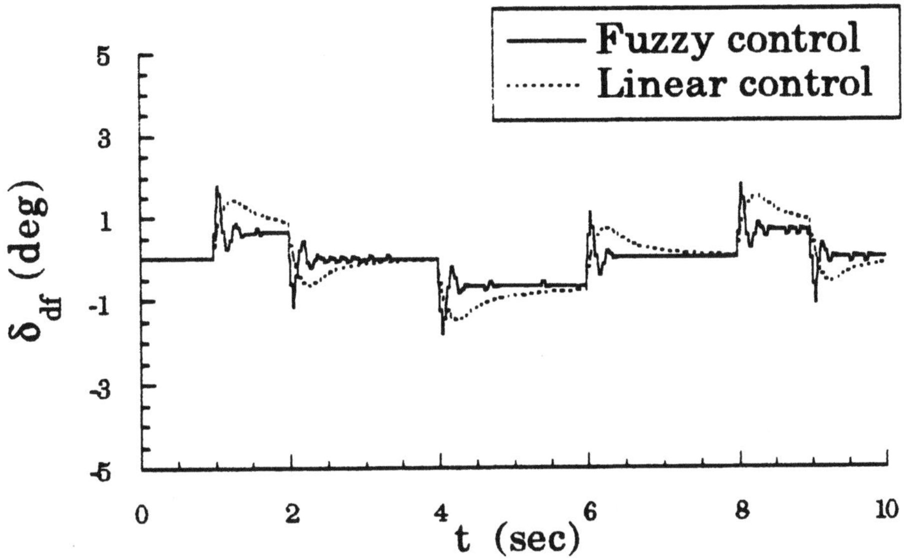

Figure 24 Differential flap deflection at a 40° bank-to-bank command.

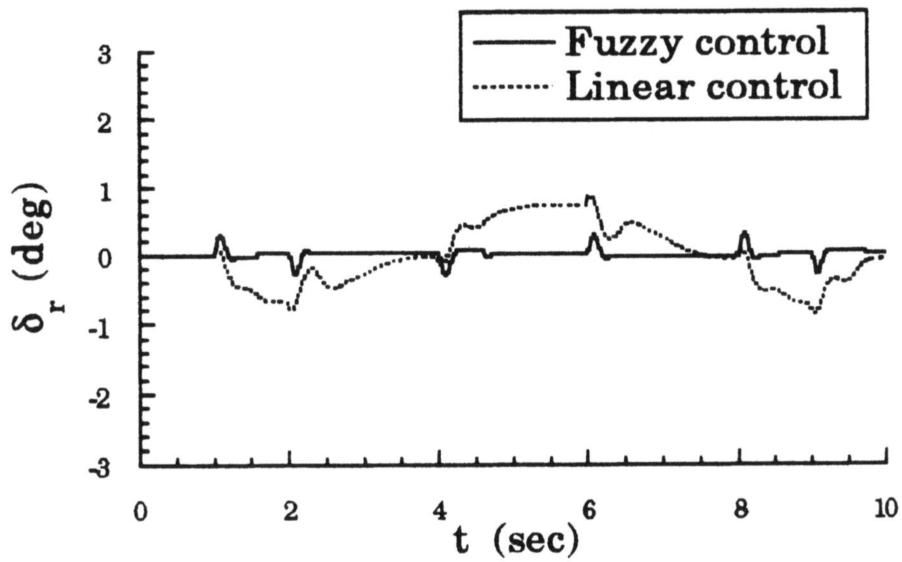

Figure 25 Rudder deflection at a 40° bank-to-bank command.

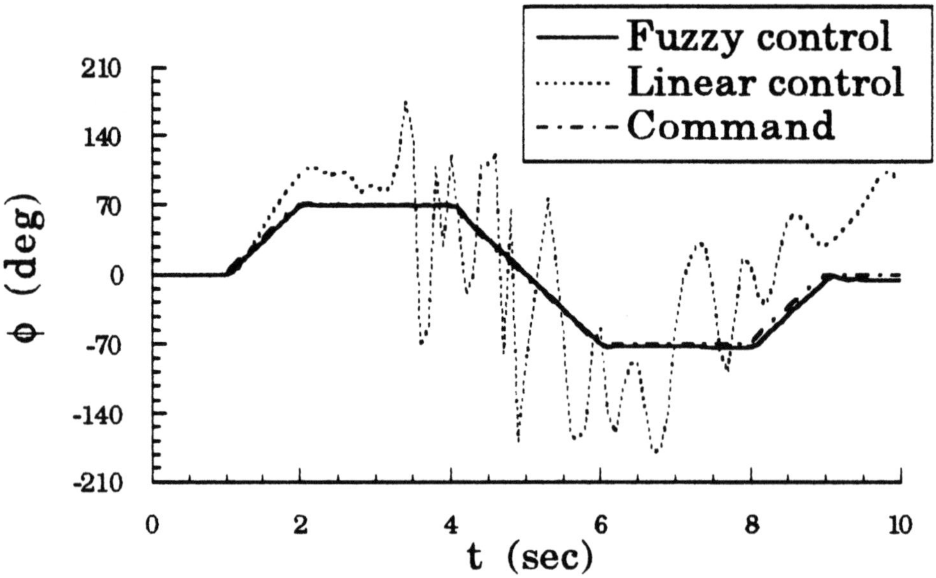

Figure 26 Bank angle at a 70° bank-to-bank command.

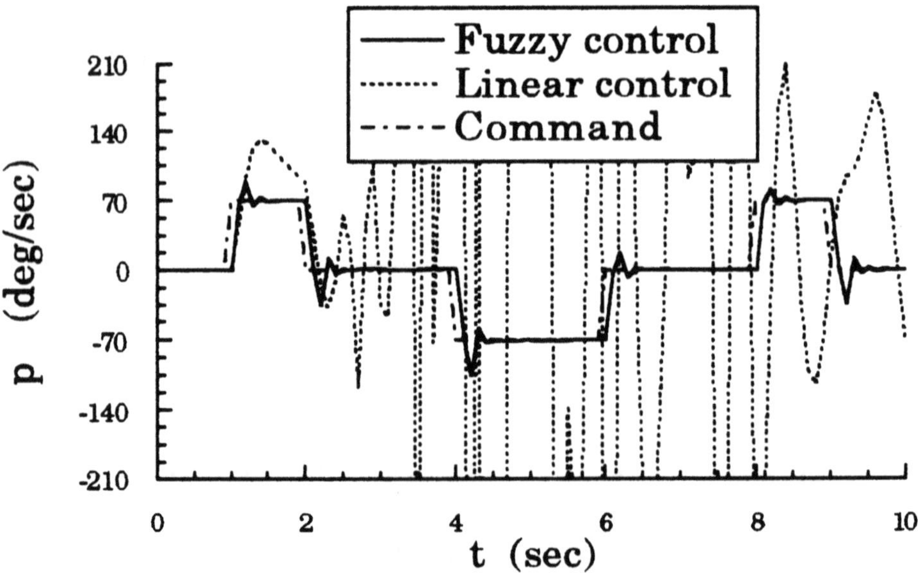

Figure 27 Roll rate at a 70° bank-to-bank command.

5

A CLASSICAL CONTROLLER: A SPECIAL CASE OF THE FUZZY LOGIC CONTROLLER

Thomas Brehm and Kuldip Rattan

Department of Computer Science and Engineering
**Department of Electrical Engineering, Wright State University*
Dayton, OH 45435, USA

ABSTRACT

The objective of this work is to demonstrate that a classical controller is equivalent to a special case of the Fuzzy Logic Controller (FLC). The FLC is basically a piecewise linear controller. Analysis of a mathematical equation for an FLC shows that the resultant equation is a linear combination of the inputs over different ranges of operations. If the linear combination is the same over all ranges of operation, then the FLC is equivalent to the classic controller. A control surface analysis gives a clear visual interpretation of the results. Simulations of a control system produce the same results for the FLC and the classic controller.

1 INTRODUCTION

Fuzzy logic controllers (FLCs) demonstrate excellent performance in numerous applications such as industrial processes [10] and flexible arm control [8]. Mamdami introduced this control technology that Zadeh pioneered with his work in fuzzy sets [11]. Unlike "two valued" logic, fuzzy set theory allows the degree of truth for a variable to exist somewhere in the range of [0,1]. For example, if pressure is a linquistic variable that describes an input, then the terms low, medium, high and dangerously high describe the fuzzy sets for the pressure variable. If the universe of discourse for pressure is [0, 100], then low could be defined as "close to 10", "medium" could be "around 40", and so on. For control applications, linguistic variables describe the inputs of the dynamic plant and the rules define the relationships between the inputs and outputs. Thus, precise knowledge of a plant's transfer function is not necessary for design and implementation of the FLC. The thrust of earlier efforts involved replacing humans in the control loop by describing the operations' actions in terms of linguistic rules.

The classical controller is mathematical based and requires knowledge of the transfer function of the plant for its design. A proportional (P) controller uses a fixed gain to scale the error to produce an output. A proportional-derivative (PD) controller has an additional gain that scales the change in error input. The classical controller is a linear controller with a fixed operating point. However, the output of the FLC is dependent on the current state of its input(s). Therefore, the output of the FLC is not necessarily a constant linear combination of its input(s). Recent research into fuzzy control has applied classical techniques to stability analysis [5] and design [7, 12].

The goal of this chapter is to present a derivation of an output equation of a proportional FLC (PFLC) and a proportional-derivative FLC (PDFLC). Analysis of an FLC output equation developed by Sabharwal [10] shows that the PDFLC is a piecewise linear controller with many similarities to the classical proportional-derivative(PD) controller [1,2,5]. This chapter verifies this hypothesis and further shows that the classical controller is a special case of the FLC.

2 FLC ARCHITECTURE AND TERMINOLOGY

Figure 1 shows the components of a feedback control system that has an FLC in place of a classical controller. An FLC can be divided into three components; the fuzzification process, inference, and the defuzzification process. The fuzzification process interprets the inputs as linguistic values. Inference uses a knowledge base of rules to determine the output sets for the input linguistic values. Finally, the defuzzification process uses the output of the inference process to derive a single "crisp" output value.

Fuzzification involves dividing each input variables' unverse of discourse into ranges called fuzzy sets. A function applied across each range determines the membership of the variable's current value in the fuzzy sets. The value at which the membership is maximum is called the peak value. Width of a fuzzy set is the distance from the peak value to the point where the membership is zero.

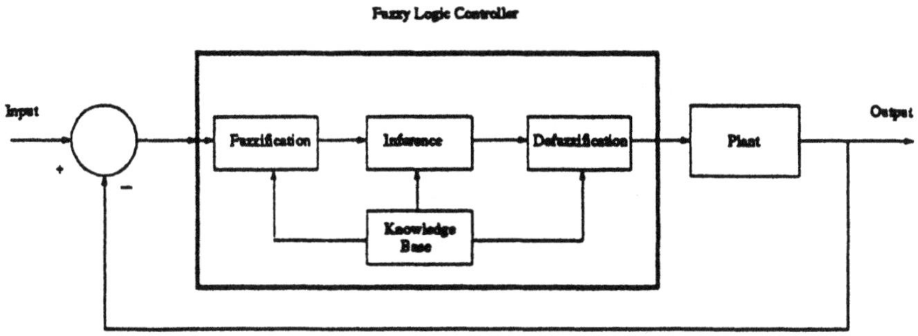

Figure 1 FLC in feedback control loop.

Linguistic rules express the relationship between input variables. Table I is an example of matrix of rules that covers all possible combinations of fuzzy sets for two input variables. The rules describe a PDFLC. The rule matrix is just a convenient way of representing all rules in "English" of the form:

R_N: If error is E_i and change in error is ΔE_j,
then output is U_{ij} .

where 1≤ i ≤ number of sets for error, $1 \leq j \leq$ number of sets for change in error and $1 \leq N \leq$ (number of sets for error multiplied by the number of set for change in error). E_i and ΔE_j are fuzzy sets for error and change in error, respectively and U_{ij} are the output fuzzy sets. In this case, each variable has seven fuzzy sets that gives a totle of 49 rules. The notation PB means positive big; PM means positive medium; PS means positive small; ZO means zero; NS means negative small; NM means negative medium; and NB means negative big.

The defuzzification process determines the "crisp output" by resolving the applicable rules into a single output value. One method of defuzzification is the **simplified reasoning method** (also commonly referred to as the modified centroid of area method) [3, 4]. This method uses weighted averaging of the input membership values and the center points of the output fuzzy sets to determine the crisp output. For this method, the output membership functions must be symmetric around the center value. Input

		Error						
		NB	NM	NS	ZO	PS	PM	PB
	NB	NB	NB	NB	NB	NM	NB	ZO
Change	NM	NB	NB	NB	NM	NS	ZO	PS
in	NS	NB	NB	NM	NS	ZO	PS	PM
Error	ZO	NB	NM	NS	ZO	PS	PM	PB
	PS	NM	NS	ZO	PS	PM	PB	PB
	PM	NS	ZO	PS	PM	PB	PB	PB
	PB	ZO	PS	PM	PB	PB	PB	PB

Table 1 Rule matrix for PDFLC.

membership values are determines on a per rule basis and product of these input membership values are used to determine the crisp output. For the rule:

If x is A_1 and y is B_1 then z is U_1,

the strength of the antecedent for this rule is given by:

$$\mu_{Rule1} = \mu_{A_1 \cdot B_1} = \mu_{A_1}(x) \cdot \mu_{B_1}(y).$$

The weighted average using the output values and the products input membership gives the output expression of the general form:

$$\begin{aligned} u &= \frac{\mu_{rule1}U_{rule1} + \mu_{rule2}U_{rule2} + \ldots + \mu_{rulen}U_{rulen}}{\mu_{rule1} + \mu_{rule2} + \ldots + \mu_{rulen}} \\ &= \frac{\sum_{n=1}^{\#Rules} \mu_{rulen}U_{rulen}}{\sum_{n=1}^{\#Rules} \mu_{rulen}}, \end{aligned}$$

where μ_{rulen} is the product of the membership and U_{rulen} is the output value for rule n.
There are several type of fuzzification and defuzzification schemes. Analysis of all fuzzification and defuzzification schemes is beyond the scope of this work. Therefore, the fuzzification and defuzzification processes, and the knowledge base are constrained as follows:

CONSTRAINT 1: The fuzzufication process uses the triangular membership function. The triangular membership function has only one point where the membership is one and to each side of that point the membership decreases linearly to zero. Since the goal is to provide a piecewise linear like PD controller, the linear nature of the function is required.

CONSTRAINT 2: The width of a fuzzy set extends to the peak value of each ajacent fuzzy set and vice versa as shown in Figure 2. The sum of the membership values over the interval between two adjacent sets will be one. Therefore, the sum of all membership over the universe of discourse at any instant for a control variable will always be equal to one. This constraint is aslo commonly referred to a fuzzy partitioning [3, 4].

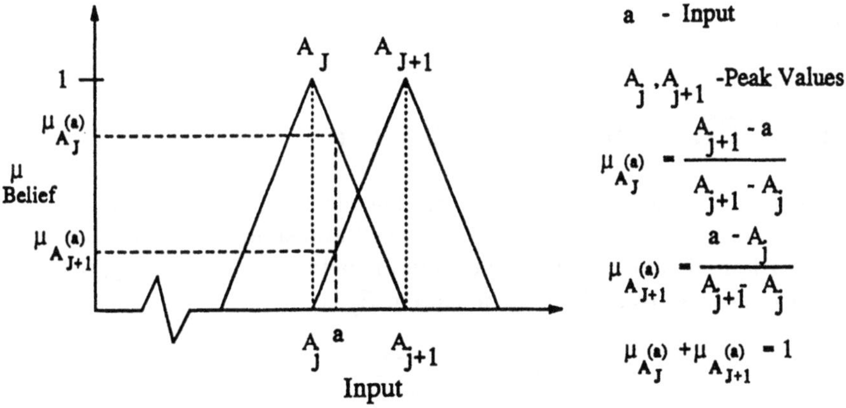

Figure 2 Two membership functions in the universe of discourse for the variable a_0.

CONSTRAINT 3: The defuzzification method used is the modified center of area method. This method is similar to obtaining a weighted average of all possible output values. Therefore, this component is also linear.

CONSTRAINT 4: The rules in the knowledge base will cover all possible memberships of the input variables. As shown in Table I, each element of the control matrix has a value.

CONSTRAINT 5: The controllers in this chapter are for a normalized input. A non-normalized step input requires a gain to normalize the input and a gain to scale the output.

3 FLC OUTPUT EQUATION DERIVATION

FLC is based on linguistic expression of the desired control action. Therefore, to derive an output equation for the FLC, the numerical expressions of the fuzzification and defuzzification processes are used to translate the "English" like terms to a mathematic form. The fuzzification process uses functions to return membership values for the input. These membership functions are substituted into the equation for defuzzification to give the output expression of the FLC.

3.1 PFLC Output Equation

For the PFLC , there is one control variable which has membership in exactly two fuzzy sets as shown in Figure 3. Figure 3 also shows that if the error value, e (desired value minus actual value) is between E_j and E_{j+1}, fuzzy sets E_j and E_{j+1} is active. The membership in E_j is:

$$\mu_j = \frac{E_{j+1} - e}{E_{j+1} - E_j}, \tag{5.1}$$

and E_{j+1} is:

$$\mu_{j+1} = \frac{e - E_j}{E_{j+1} - E_j}. \tag{5.2}$$

As required, the sum of the memberships given by (1) and (2) is one.

For the control variable, e, that has membership in two fuzzy sets, there will be two applicable rules expressed as:

R_j: if e is E_j then u is U_j,
R_{j+1}: if e is E_{j+1} then u is U_{j+1}.

where e and u are the error and output. These two rules form a 2 element sub vector from the fuzzy rule vector are shown in Table 2.

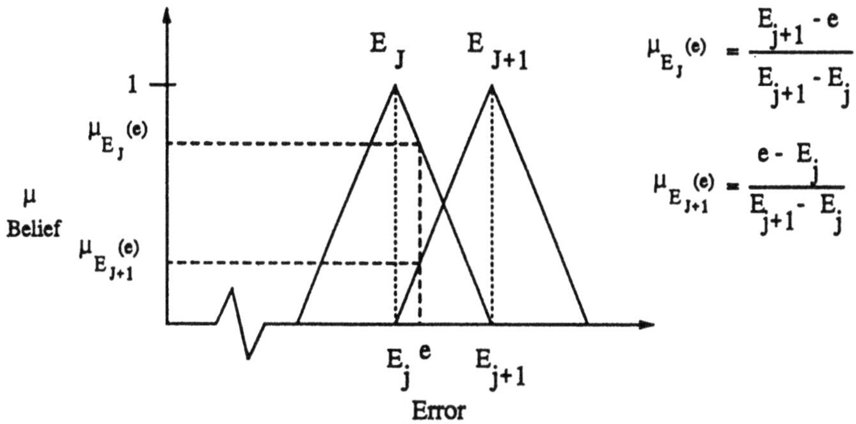

Figure 3 Example membership functions for error input.

Error	NB	E_j	$Ej+1$	PS	PB
Output	NB	U_j	U_{j+1}	PS	PB

Table 2 PFLC control rule vector.

The crisp output action is determined by applying the modified centroid of area defuzzification scheme to the two control rules and is given by:

$$u = \frac{\mu_j U_j + \mu_{j+1} U_{j+1}}{\mu_j + \mu_{j+1}}. \tag{5.3}$$

The expressions for μ and the values U are substituted into equation (5.3) giving:

$$\frac{\left[\frac{E_{j+1}-e}{E_{j+1}-E_j}\right] U_j + \left[\frac{e-E_j}{E_{j+1}-E_j}\right] U_{j+1}}{\left[\frac{E_{j+1}-e}{E_{j+1}-E_j}\right] + \left[\frac{e-E_j}{E_{j+1}-E_j}\right]}.$$

Removing the common denominator $(E_{j+1} - E_j)$ gives:

$$\frac{(E_{j+1} - e)U_j + (e - E_j)U_{j+1}}{(E_{j+1} - e) + (e - E_j)}.$$

Expanding the terms in the numerator and denominator and grouping like terms yields the final expression for the PFLC output:

$$u = e\frac{U_{j+1} - U_j}{E_{j+1} - E_j} + \frac{E_{j+1}U_j - E_jU_{j+1}}{E_{j+1} - E_j}. \tag{5.4}$$

3.2 PDFLC Output Equation

Derivation for the PDFLC output equation follows the same procedures as the PFLC. However, the PFLC has two input control variable; error (desired value minus actual value) and change in error(current error minus previous error divided by the time interval). Like the PFLC, each input variable has membership in exactly two fuzzy sets. The membership functions for error are the same as the PFLC and are expressed in equation (5.1) and (5.2). Figure 4 shows the second input control variable, change in error. For the change in error input, if the value Δe is between ΔE_k and ΔE_{k+1}, then the membership for ΔE_k is:

$$\mu_k = \frac{\Delta E_{k+1} - \Delta e}{\Delta E_{k+1} - \Delta E_k}, \tag{5.5}$$

and the membership for ΔE_{k+1} is

$$\mu_{k+1} = \frac{\Delta e - \Delta E_k}{\Delta E_{k+1} - \Delta E_k}. \tag{5.6}$$

For the two control variable, error and change in error, with two active sets, there will be four applicable rules expressed as:

$R_{j,k}$: if e is E_j and Δe is ΔE_k then u is $U_{j,k}$,
$R_{j+1,k}$: if e is E_{j+1} and Δe is ΔE_k then u is $U_{(j+1),k}$,
$R_{j,k+1}$: if e is E_j and Δe is ΔE_{k+1} then u is $U_{j,(k+1)}$,
$R_{j+1,k+1}$: if e is E_{j+1} and Δe is ΔE_{k+1} then u is $U_{(j+1),(k+1)}$.

where e, Δe, and u are the error, change in error and output, respectively. These four rules form a 2x2 sub matrix from the fuzzy rule matrix shown in Table 3.

The crisp output control action is determined by applying the modified centroid of area defuzzification scheme to the four control rules and is given by:

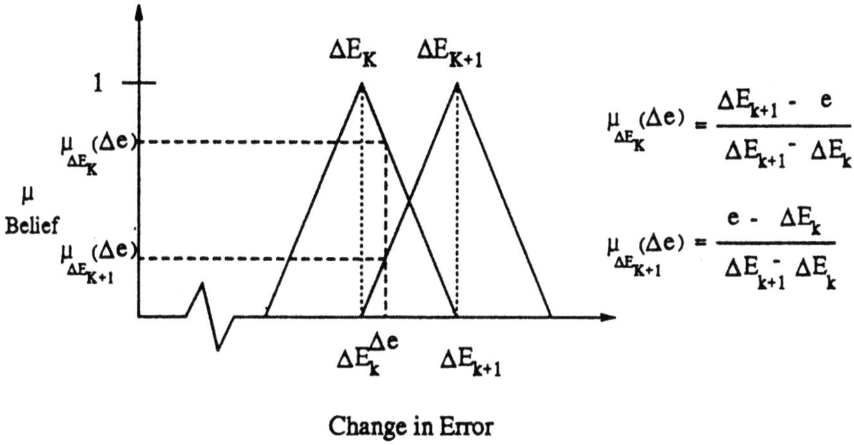

Figure 4 Example membership functions for change in error.

Error

		NB	E_j	E_{j+1}	PS	PB
	NB	NB	NB	NB	NS	ZO
Change of Error	ΔE_k	NB	$U_{j,k}$	$U_{(j+1),k}$	ZO	PB
	ΔE_{k+1}	NB	$U_{j,(k+1)}$	$U_{(j+1),(k+1)}$	PS	PB
	PS	NS	ZO	PS	PB	PB
	PB	ZO	PS	PB	PB	PB

Table 3 PD control rule matrix for FLC.

$$u = \frac{\sum_{i=1}^{4} \mu_i U_i}{\sum_{i=1}^{4} \mu_i}, \tag{5.7}$$

where μ_i is calculated by the product rule applied to the antecent of the fuzzy rule and U_i is the output set for the ith rule. The product rule is defined as the product of each membership value. Therefore, for a given rule, μ_i is calculated by multiplying the value of membership of the error input for the given error fuzzy subset and the value of membership of the change in error input for the change in error fuzzy subset as given in (5.8).

$$\mu_i = [\text{membership of } e \text{ in } E] \times [\text{membership of } \Delta e \text{ in } \Delta E]. \tag{5.8}$$

The product rule is necessary to obtain an expression for the PDFLC output, u [5, 7, 9]. For the four applicable rules, membership functions (5.1), (5.2), (5.5) and (5.6) are substituted for membership values in the product rule to obtain the expression for each μ_i. The expression for μ_i and the corresponding values of U_i are:

$$\begin{array}{lll}
\text{Rule } R_{j,k} & : \mu_1 = \frac{E_{j+1}-e}{E_{j+1}-E_j} \cdot \frac{\Delta E_{k+1}-\Delta e}{\Delta E_{k+1}-\Delta E_k}, & U_1 = U_{j,k} \,, \\
\text{Rule } R_{(j+1),k} & : \mu_2 = \frac{e-E_j}{E_{j+1}-E_j} \cdot \frac{\Delta E_{k+1}-\Delta e}{\Delta E_{k+1}-\Delta E_k}, & U_2 = U_{(j+1),k} \,, \\
\text{Rule } R_{j,(k+1)} & : \mu_3 = \frac{E_{j+1}-e}{E_{j+1}-E_j} \cdot \frac{\Delta e-\Delta E_k}{\Delta E_{k+1}-\Delta E_k}, & U_3 = U_{j,(k+1)} \,, \\
\text{Rule } R_{(j+1),(k+1)} & : \mu_4 = \frac{e-E_j}{E_{j+1}-E_j} \cdot \frac{\Delta e-E_k}{\Delta E_{k+1}-\Delta E_k}, & U_4 = U_{(j+1),(k+1)} \,.
\end{array}$$

The expression for μ_i and the output values U_i are substituted into equation (5.7) giving:

$$\frac{A_1 U_{j,k} + A_2 U_{j+1,k} + A_3 U_{j,k+1} + A_4 U_{j+1,k+1}}{A_1 + A_2 + A_3 + A_4},$$

where

$$\begin{aligned}
A_1 &= \left[\frac{E_{j+1}-e}{E_{j+1}-E_j} \cdot \frac{\Delta E_k - \Delta e}{\Delta E_{k+1} - \Delta E_k}\right], \\
A_2 &= \left[\frac{e-E_j}{E_{j+1}-E_j} \cdot \frac{\Delta E_k - \Delta e}{\Delta E_{k+1} - \Delta E_k}\right], \\
A_3 &= \left[\frac{E_{j+1}-e}{E_{j+1}-E_j} \cdot \frac{\Delta e - \Delta E_{k+1}}{\Delta E_{k+1} - \Delta E_k}\right], \\
A_4 &= \left[\frac{e-E_j}{E_{j+1}-E_j} \cdot \frac{\Delta e - \Delta E_k}{\Delta E_{k+1} - \Delta E_k}\right].
\end{aligned}$$

Removing the common denominator $[(E_{j+1} - E_j)(\Delta E_{K+1} - \Delta E_k)]$ gives:

$$\frac{B_1 U_{j,k} + B_2 U_{j+1,k} + B_3 U_{j,k+1} + B_4 U_{j+1,k+1}}{B_1 + B_2 + B_3 + B_4},$$

where

$$\begin{aligned}
B_1 &= (E_j - e) \cdot (\Delta E_k - \Delta e)\,, \\
B_2 &= (e - E_j) \cdot (\Delta E_k - \Delta e)\,, \\
B_3 &= (E_{j+1} - e) \cdot (\Delta e - \Delta E_{k+1})\,, \\
B_4 &= (e - E_j) \cdot (\Delta e - \Delta E_k)\,.
\end{aligned}$$

Expanding the terms in the numerator and denominator and grouping like terms yields the final expression for the PDFLC output:

$$\begin{aligned} u = {} & e\frac{\Delta E_k(U_{j,k+1} - U_{j+1,k+1}) + \Delta E_{k+1}(U_{j+1,k} - U_{j,k})}{(E_{j+1} - E_j)(\Delta E_{k+1} - \Delta E_k)} + \\ & \Delta e\frac{E_{j+1}(U_{j,k+1} - U_{j,k}) + E_j(U_{j+1,k} - U_{j+1,k+1})}{(E_{j+1} - E_j)(\Delta E_{k+1} - \Delta E_k)} + \\ & (e^{\star}\Delta e)\frac{(U_{j,k} - U_{j,k+1}) + (U_{j+1,k+1} - U_{j+1,k})}{(E_{j+1} - E_j)(\Delta E_{k+1} - \Delta E_k)} + \\ & \frac{E_{j+1}(\Delta E_{j+1}U_{j,k} - \Delta E_k U_{j,k+1}) + E_j(\Delta E_k U_{j+1,k+1} - \Delta E_{k+1}U_{j+1,k})}{(E_{j+1} - E_j)(\Delta E_{k+1} - \Delta E_k)}. \end{aligned} \tag{5.9}$$

4 FLC AS A PIECEWISE LINEAR CONTROLLER

The output equation for the PFLC is a function of the input error and for the PDFLC, the output is a function of error and change in error. However, for both FLC types, the equations are dependent on the fuzzy sets for the current range of operations . Therefore, as the values of the input control variables change , the controller output equation changes.

4.1 PFLC as a Piecewise Linear Classical P Controller

As demonstrated in equation (5.4), the output of the PFLC is similar to the classical proportional controller. Like the equation for a classical proportional controller, equation (5.4) has error multiplied by a gain term but there is an additional constant term. Equation (5.4) can be written as:

$$u = K_{p-eff} \cdot e + Const,$$

where the effective proportional gain is given by:

$$K_{p-eff} = \frac{U_{j+1} - U_j}{E_{j+1} - E_j}, \tag{5.10}$$

and the constant controller output is given by:

$$Const = \frac{E_{j+1}U_j - E_jU_{j+1}}{E_{j+1} - E_j}. \tag{5.11}$$

If the effects of the constant term are negligible, then the form of the PFLC is identical to the classical proportional controller.

K_{p-eff} and *Const* given by equations (5.10) and (5.11) have a common denominator the width between the adjacent fuzzy sets. The numerator of the effective gain is the difference between the adjacent output values. The numerator of the constant term is dependent on the value of the peak values of the error fuzzy sets and the value of the outputs. Therefore, a change in either the error fuzzy sets or the output values will change the effective gain, K_{p-eff} and the constant term.

Equations (5.10) and (5.11) are valid for error in the range E_j to E_{j+1}. If the value of error were to fall in another range (i.e. E_{j+1} and E_{j+2}), then the value of the effective gain and the constant term would be:

$$K'_{p-eff} = \frac{U_{j+1} - U_{j+1}}{E_{j+2} - E_{j+1}}, \tag{5.12}$$

$$Const' = \frac{E_{j+2}U_{j+1} - E_{j+1}U_{j+2}}{E_{j+2} - E_{j+1}}. \tag{5.13}$$

The effective gin and the constant term are dependent on the difference between the peak values of the error fuzzy sets and the output values. Therefore, for the new ranges, if the difference $(E_{j+2} - E_{j+1})$ is not the sane as $(E_{j+1} - E_j)$, the effective gain and the constant term given in equation (5.12) and (5.13) will change. The same is true for the output values (i.e. $(U_{j+2} - U_{j+1})$ is not the same as $(U_{j+1} - U_j)$. For the constant term, unless $(E_{j+2}U_{j+1} - E_{j+1}U_{j+2})$ and $(E_{j+1}U_j - E_jU_{j+1})$ are both zero, a change in either the output or error peak values change the constant for that range.

4.2 PDFLC as a Piecewise Linear Classical PD Controller

Equation (5.9) demonstrates that output of the PDFLC is similar to a classical PD controller output. The controller equation consists of four terms: error multiplied by a gain, change in error multiplied by a gain, a nonlinear term ($e\Delta e$) multiplied by a gain, and a constant term. The effects of the nonlinear term are minimal and can be ignored. Equation (5.9) can thus be written in the form:

$$u = K_{p-eff}e + K_{d-eff}\Delta e + Const,$$

where

$$K_{p-eff} = \frac{\Delta E_k[U_{j,k+1} - U_{j+1,k+1}] + \Delta E_{k+1}[U_{j+1,k} - U_{j,k}]}{(E_{j+1} - E_j)(\Delta E_{k+1} - \Delta E_k)}, \quad (5.14)$$

$$K_{d-eff} = \frac{E_{j+1}(U_{j,k+1} - U_{j,k}) + E_j(U_{j+1,k} - U_{j+1,k+1})}{(E_{j+1} - E_j)(\Delta E_{k+1} - \Delta E_k)}, \quad (5.15)$$

$$Const = \frac{E_{j+1}(\Delta E_{k+1} U_{j,k} - \Delta E_k U_{j,k+1})}{(E_{j+1} - E_j)(\Delta E_{k+1} - \Delta E_k)} + \frac{E_j(\Delta E_k U_{j+1,k+1} - \Delta E_{k+1} U_{j+1,k})}{(E_{j+1} - E_j)(\Delta E_{k+1} - \Delta E_k)}. \quad (5.16)$$

K_{p-eff}, K_{d-eff} and Const given by equations (5.14)-(5.16) have a common denominator whose value is determined by the product of the width of the error and the width of the change in error sets. The numerator for each term uses the output values from the 2x2 rule sub matrix. The error gain, K_{p-eff}, uses the difference between the values in the rows (i.e.$U_{j,(k+1)} - U_{(j+1),(k+1)}$ and $U_{(j+1),k} - U_{j,k}$) times change in error sets peak values. The change in error gain, K_{d-eff}, uses the differences between the columns (i.e. $U_{j,(k+1)} - U_{j,k}$ and $U_{(j+1),k} - U_{(j+1),(k+1)}$) times the error sets peak values. The constant term uses the peak values for both input fuzzy sets and the output values. Therefore, since both effective gain terms and the constant term are made up of error and change in error peak values, changes in any one of the peak values will affect all terms. The same is also true with the output values. Changes in any one of the four output values affects both gain terms and the constant term.

The terms in (5.14) through (5.16) only apply to the range of operation between the peak values E_j and E_{j+1} for error and between ΔE_k and E_{j+1} for change in error. The width between the peak values for the next fuzzy sets may not be the same as the previous sets. Also, the difference in output values may not be the same for the next range of operation. Therefore, the effective gain values and the constant term may be different. As an example, the terms for the next range operation could be:

$$K'_{p-eff} = \frac{\Delta E_{k+1}(U_{j+1,k+2} - U_{j+2,k+2})}{(E_{j+2} - E_{j+1})(\Delta E_{k+2} - \Delta E_{k+1})} + \frac{\Delta E_{k+2}(U_{j+2,k+1} - U_{j+1,k+1})}{(E_{j+2} - E_{j+1})(\Delta E_{k+2} - \Delta E_{k+1})}, \quad (5.17)$$

$$K'_{d-eff} = \frac{E_{j+2}(U_{j+1,k+2} - U_{j+1,k+1})}{(E_{j+2} - E_{j+1})(\Delta E_{k+2} - \Delta E_{k+1})} + \frac{E_{j+1}(U_{j+2,k+1} - U_{j+2,k+2})}{(E_{j+2} - E_{j+1})(\Delta E_{k+2} - \Delta E_{k+1})}, \quad (5.18)$$

$$Const' = \frac{E_{j+2}(\Delta E_{k+2}U_{j+1,k+1} - \Delta E_{k+1}U_{j+1,k+2})}{(E_{j+2} - E_{j+1})(\Delta E_{k+2} - \Delta E_{k+1})} + \frac{E_{j+1}(\Delta E_{k+1}U_{j+2,k+2} - \Delta E_{k+2}U_{j+2,k+1})}{(E_{j+2} - E_{j+1})(\Delta E_{k+2} - \Delta E_{k+1})}. \quad (5.19)$$

As demonstrated in (5.17)-(5.19), if the difference between the set peak values is different as compared to the previous range, the product term $[(E_{j+2} - E_{j+1})(\Delta E_{k+2} - \Delta E_{k+1})]$ for the gains and constant term are different. The difference between the output terms along the rows (i.e. $U_{(j+1),(k+2)} - U_{(j+2),(k+2)}$) and $U_{(j+2),(k+1)} - U_{(j+1),(k+1)}$ and along the columns (i.e. $U_{(j+1),(k+2)} - U_{(j+1),(k+1)}$ and $U_{(j+2),(k+1)} - U_{(j+2),(k+2)}$) for 2x2 rule sub matrix may also be different. Thus, K'_{p-eff}, K'_{d-eff} and $Const'$ for this range of operation are not the same as K_{p-eff}, K_{d-eff} and the $Const$ for the previous range. This example demonstrates the piecewise linear nature of the PDFLC.

Equations (5.14) and (5.15) indicate that the effective gains of each input control variable are mutually dependent. Changing a fuzzy set of either input control variable will affect the magnitude of both effective gains. However, the piecewise linearity of either effective gain is still dependent on the corresponding input control variable. For example, to calculate the piecewise linear error gain K_p, the value of change in error is held constant which gives $K_{p-eff} = \frac{C_0(U_{j,k+1} - U_{j+1,k+1}) + C_1(U_{j+1,k} - U_{j,k})}{(E_{j+1} - E_j)C_2}$, where C_0, C_1, and C_2 are fixed values determined by the change in error sets. As the error input changes for the fixed value of change in error. the K_{p-eff} will depend on the error and output fuzzy sets. The same analysis can be applied to determine the effective piecewise linear K_{d-eff} for a fixed value of error.

5 CLASSICAL CONTROLLER: A SPECIAL CASE OF FLC

Development of a mathematical equation for a FLC based on triangular membership functions demonstrates that the FLC approximates a piecewise linear classical controller. However, the PFLC or the PDFLC can attain a constant gain linear control action if the gain terms can be made equivalent over all ranges of operation. This special case of the PFLC or PDFLC is equivalent to the classical controller counterpart whose gain is constant over its entire range of operation. This hypothesis is verified by comparing controller output equations for the FLC and the classical controller and using graphical analysis techniques.

5.1 Classical P Controller: A Special Case of PFLC

The classical P controller output is the input error multiplied by a gain. As shown previously, the output of a PFLC is the error multiplied by an effective gain plus a constant term. The values of the gain and the constant term are dependent on the current operating range. To make the PFLC equivalent to the classical P controller, the gain needs to be constant and the constant term should be zero over all the ranges. Analysis of the PFLC output equation gives the conditions for an equivalent classical P controller. A graphical study of the input-output relationship verifies the output equation analysis.

PFLC Output Equation Analysis

The PFLC equation is expressed as

$$u = e\frac{U_{j+1} - U_j}{E_{j+1} - E_j} + \frac{E_{j+1}U_j - E_jU_{j+1}}{E_{j+1} - E_j},$$

where e is the input error, U is the peak value of the output fuzzy set, E is the peak value of a fuzzy set for error and u is the output value. The effective gain is the difference between adjacent fuzzy set peak values for error.

To make the effective gain constant over all intervals of operation, the numerator and denominator must be constant. Therefore, the difference between adjacent fuzzy set peak values must be constant over all intervals (i.e. $(E_{j+1} - E_J)$ must equal $(E_{j+2} - E_{j+1})$ etc.). For the numerator, the difference between output values for adjacent rules must also be equivalent (i.e. $(U_{j+1} - U_j)$ must equal $(U_{j+2} - U_{j+1})$ etc.).

The constant term has the same denominator as the effective gain. but the numerator is dependent on the error fuzzy sets and the output values. For the constant term to be zero, the numerator, $E_{j+1}U_j - E_jU_{j+1}$, must be zero. This requires that the ratio between adjacent error sets and adjacent output sets must be equal, i.e.

$$\frac{E_{j+1}}{E_j} = \frac{U_{j+1}}{U_j}. \tag{5.20}$$

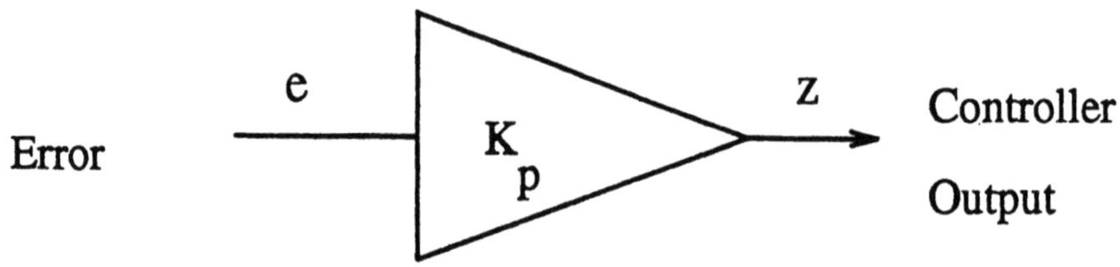

Figure 5 Classical P controller.

The constraint for constant effective gain requires the error sets to have equal differences between peak values for adjacent members and the difference between output values for adjacent rules must also be equal. Therefore, if the effective gain is constant over all ranges of operation,the ratio in equation (20) will be constant and the constant term will be zero.

PFLC Graphical Analysis

The equivalence of the classical P controller and the PFLC can be verified by showing that they have the same output. Graphical analysis of the control action can be applied to the classical controller as well as the FLC. Therefore, the classical P controller as a special case of the PFLC can be visually verified by comparing the output graphs of the two controllers.

Consider the classical P controller shown in Figure 5 and expressed as:

$$z = K_p \cdot e$$

where K_p is the proportional gain. Since K_p is constant, the controller output is a line on a graph. Figure 6 is an example of P controller output where K_p is one and the output is unbounded. However, an actual physical implementation of the P controller has a maximum output. The controller in Figure 7 uses a gain limiter to restrict the maximum output. Since the input error is assumed to be normalized, the actual output is $\min(K_p, K_p e)$. K_p is the gain of the controller as well as the maximum output. Figure 8 shows the controller output with gain of 1 and a maximum output of ± 1.

The control surface for the classical P controller is identical to the PFLC when all peak values of the input and output fuzzy sets are evenly spaced. As shown in Figure 8, the output for PFLC with 7 fuzzy sets for error is linear

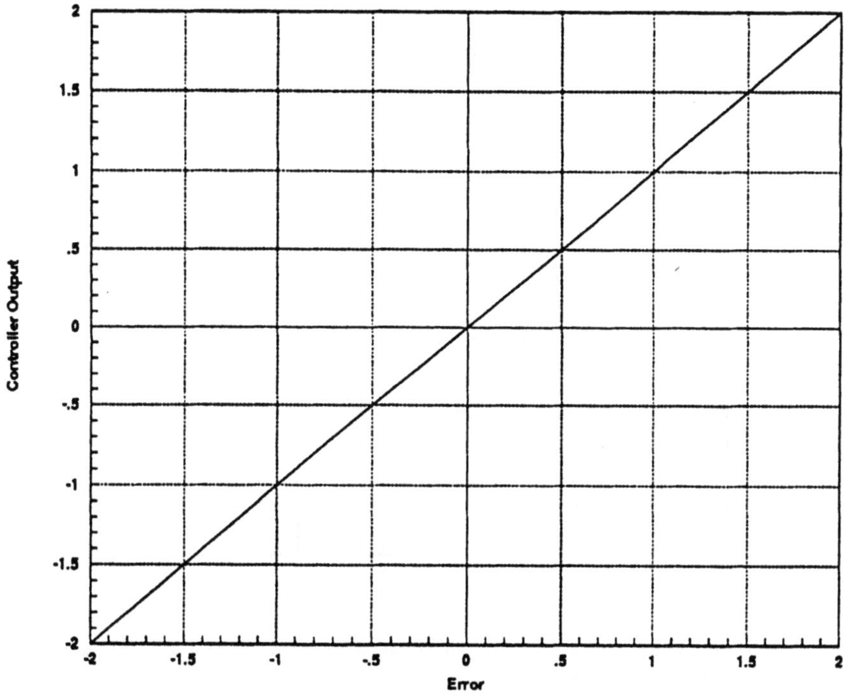

Figure 6 Classical P controller output.

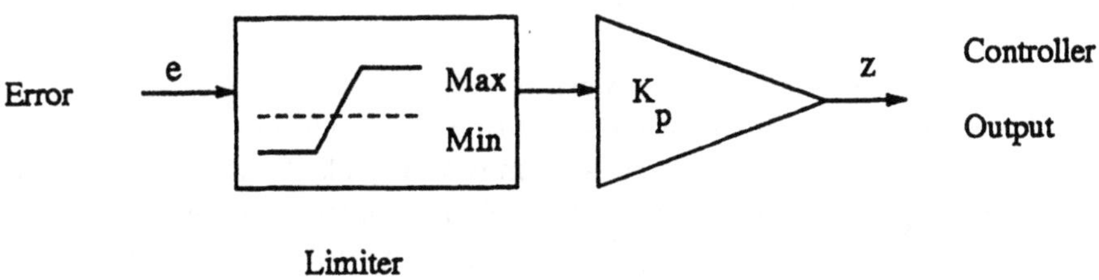

Figure 7 Classical P controller with a gain limiter.

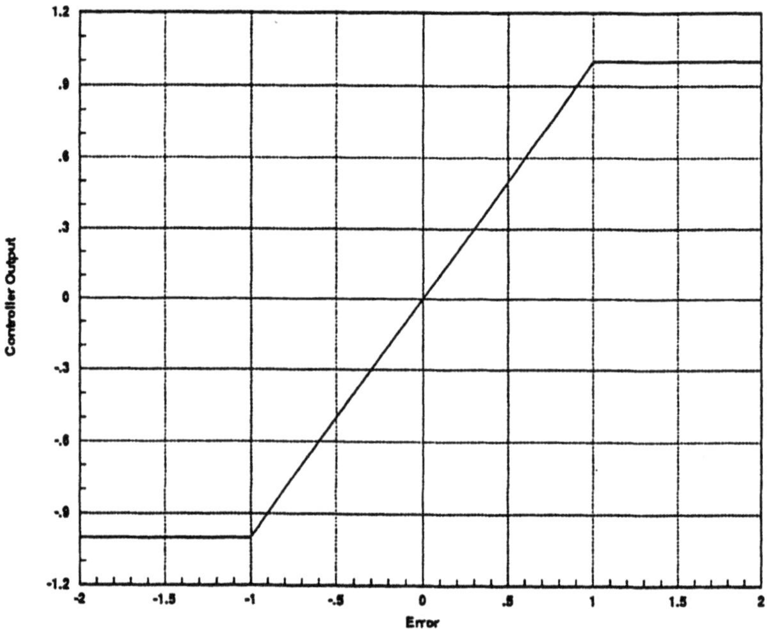

Figure 8 Output of a gain limited classical P controller and a PFLC with evenly spaced fuzzy sets.

with a maximum value of ± 1. The slope of the line for each interval is one and the constant term is zero. This plot is identical to the output plot of the classical P controller shown in Figure 8. Thus,by comparing the output plots, it is verified that the PFLC with equally spaced rules is equivalent to the classical P controller.

PFLC Numerical Example

In this section, the equivalence between the classical P controller and the PFLC is empirically verified with a simulation. Figure 9 shows the block diagram of a unity feedback control system with the plant transfer function as $G_p(s) = \frac{1}{s(s+3.6)}$ and the transfer function of the controller that gives a critically damped step response as $G_c(s) = 3.24$. An FLC was also designed with seven equally spaced error, change in error and output sets. The controller gain were the same as the classical P controller. Figure 10 shows that the response of the system controlled by the PFLC and the gain limited classical P controller are identical.

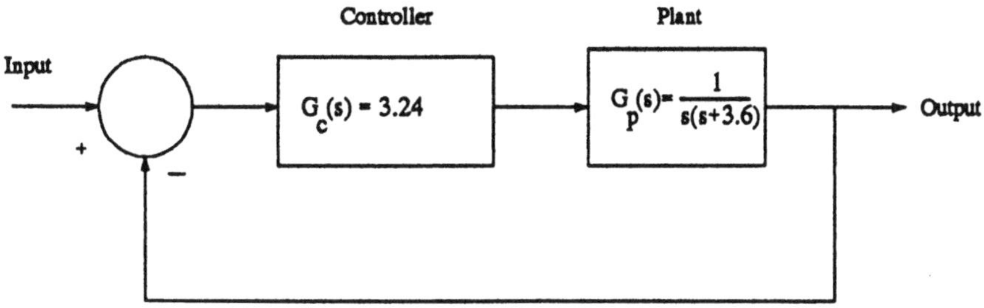

Figure 9 Block diagram of unity feedback control system with P controller.

5.2 Classical PD Controller: A Special Case of PDFLC

The PDFLC as a special case of the classical controller can be verified using similar analysis as the PFLC. First, the output equation is examined to determine if all gains can be made the same and if the constant and nonliner term can be set zero. Graphical analysis is used to show that the input-output relationships are equivalent.

PDFLC Output Equation Analysis

The gain terms for the error and change in error of equation (5.9) are dependent on the distance between input fuzzy set peak values, output fuzzy set peak values and the rule base. The common denominator of all terms in equation (9) in the product of the difference between adjacent peak values of the membership functions. Therefore, the first criteria for obtaining constant gain terms is that all paek values of fuzzy sets must be equally spaced so that the denominator remains constant.

The numerators for the error and change in error gain terms consist of the difference between the adjacent output values multiplied by the set's peak value. The active output values are a 2x2 sub matrix of the rule matrix. If the elements in the sub matrix differ by equal amounts along the rows and along the columns, then e and Δe terms are non-zero but $e^{\star}\Delta e$ and constant term will be zero. If all sub matrices have the same difference between the elements, then the coefficients of e and Δe will be constant for all the applicable rules. This region of the rule matrix is denoted by the plain type font in Table IV. The region denoted by the bold type font in Table IV is where the coefficients

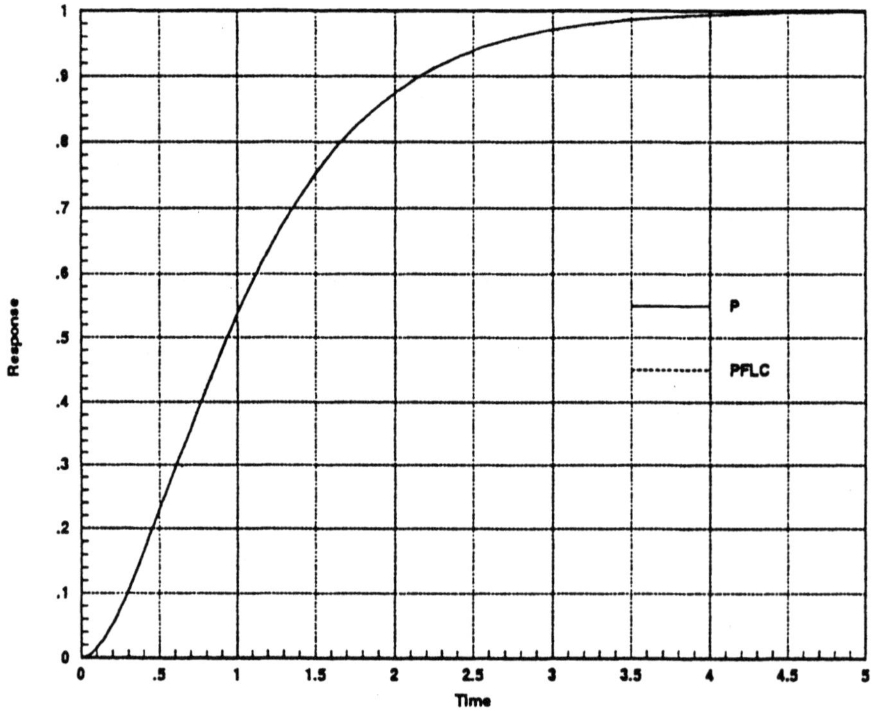

Figure 10 Closed loop system response of the classical P controller and the PFLC.

		Error						
		NB	NM	NS	ZO	PS	PM	PB
	NB	NB	NB	NB	NB	NM	NS	ZO
Change	NM	NB	NB	NB	NM	NS	ZO	PS
in	NS	NB	NB	NM	NS	ZO	PS	PM
Error	ZO	NB	NM	NS	ZO	PS	PM	PB
	PS	NM	NS	ZO	PS	PM	PB	PB
	PM	NS	ZO	PS	PM	PB	PB	PB
	PB	ZO	PS	PM	PB	PB	PB	PB

Table 4 Rule matrix showing linear, nonlinear and constant regions

of e, Δe, and $e^\star \Delta e$ are zero and the constant term is one because the differences between the elements along the rows and along the columns of the 2x2 sub-matrices are zero. The region denoted by the shaded area is a small nonlinear transition region where all four terms of equation (9) have some value in between the two previous regions,. Therefore, except for the small nonlinear region, the PDFLC behaves as a controller with a gain limiter.

PDFLC and Classical Control Surface Analysis

Two controllers are equivalent if they have the same control surface. Surface analysis of the control action can be applied to be the classical controllers as well as the FLC. Therefore, the classical PD controller as a special case of the PDFLC can be visually verified by comparing the two control surfaces .

Consider the classical PD controller shown in Figure 11 and expressed as:

$$z = K_d \cdot \Delta e + K_p \cdot e,$$

where K_d is the derivative gain and K_p is the proportional gain. Since the two control gains are constant, the control surface is a plane as shown in Figure 12. This control surface extends indefinitely in all directions of the plane.

The physical limitation of a controller model is the maximum amplifier output it can provide. The classical controller model in Figure 11 can be modified with a limiter to set a maximum control values as shown in Figure 13. The error input is assumed to be normalized and the change in error input is multiplied by the gain $\frac{K_d}{K_p}$. The output is multiplied by K_p. This

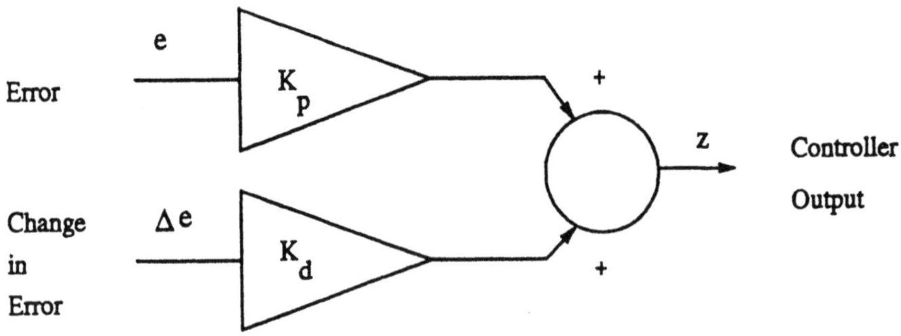

Figure 11 Block diagram of classical PD controller.

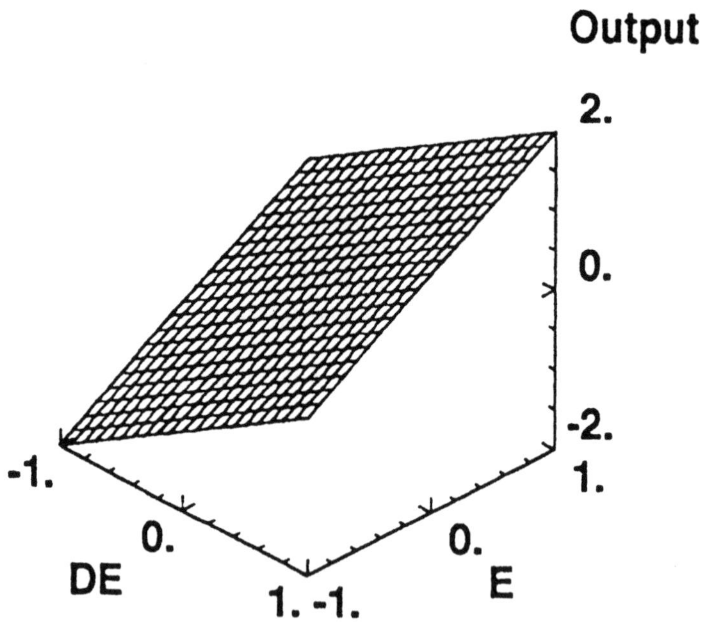

Figure 12 Output of classical PD controller.

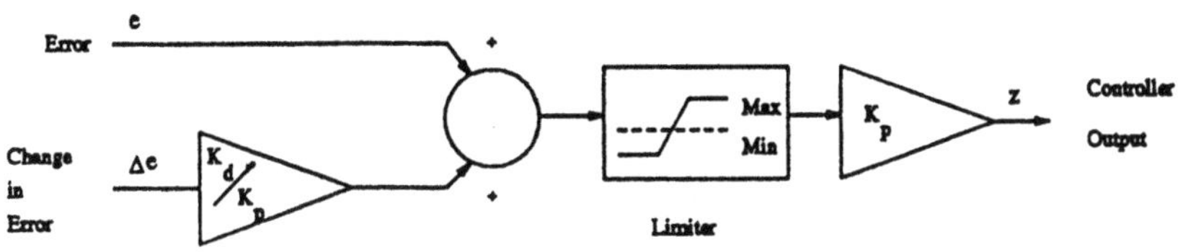

Figure 13 Model of a physical implementation of a PD controller.

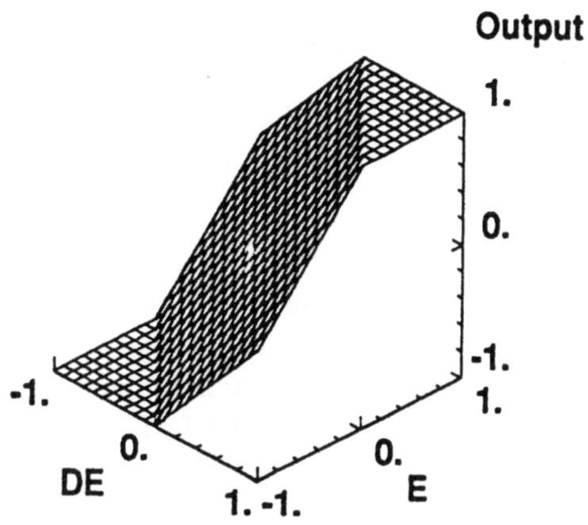

Figure 14 Output of gain limited classical PD controller.

model can be expressed as $\min(K_p, K_p(\frac{K_d}{K_p}\Delta e + e)$. Figure 14 is an example of the controller output for the PD output control surface with both control gains set to one($K_d = K_p = 1$). The limiter values are -1 and 1 which are the same as for the fuzzy controller. As expected, the resultant surface is a plane with cutoff at -1 and 1.

The control surface for the classical controller is nearly identical to the FLC when all peak values of the input and output fuzzy sets are evenly spaced. As an example, Figure 15 shows the control surface of a fuzzy controller with seven sets that have equally spaced peak values for error, change in error and output. The universe of discourse for the sets is between -1 and 1. Therefore, peak values for the seven membership functions of each control variable and the output values are -1, $-\frac{2}{3}$,$-\frac{1}{3}$, 0, $\frac{1}{3}$, $\frac{2}{3}$, and 1, respectively. Note that the

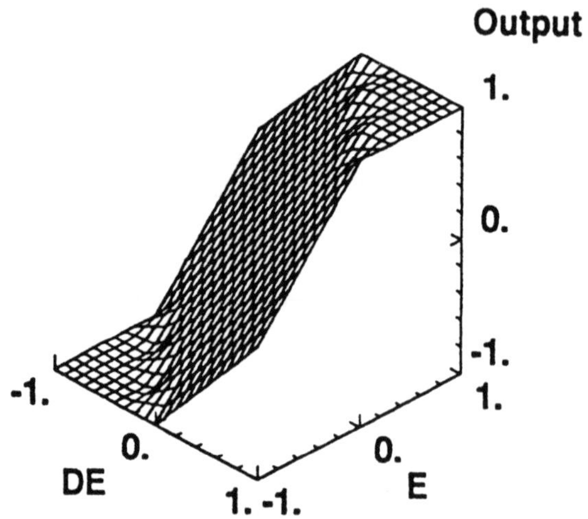

Figure 15 Control surface of PFLC with 7 error, change in error and output fuzzy sets.

control surface is mostly planar with a small nonlinear region at each limit. In the planar region, the control surface is equivalent in gradient and orientation to the classical PD control surface. If the small nonlinear region is ignored, the control action of both controllers are equivalent. As the number of input and output sets and corresponding rules increase, the nonlinear region becomes less significant [1, 2].

All gains for the simulated classical PD controller and PDFLC were set to one. However, if equivalent gains were applied to each controller, the control surface shape and orientation changes accordingly. For the PDFLC, modifying the input variable's gain is the same as expanding its universe of discourse. The widths of all membership functions still remain equivalent and all gains will effectively remain constant.

Numerical Example

In this section, the equivalence between the classical PD controller and the PDFLC is empirically verified with a simulation. Figure 16 shows the block diagram of a unity feedback control system with the plant transfer function given by $G_p(s) = \frac{1}{s(s+3.6)}$ and the transfer function of the controller that gives critically damped step response as $G_c(s) = 16.4s + 100$. An FLC was also designed with seven equally spaced error, change in error and output sets.

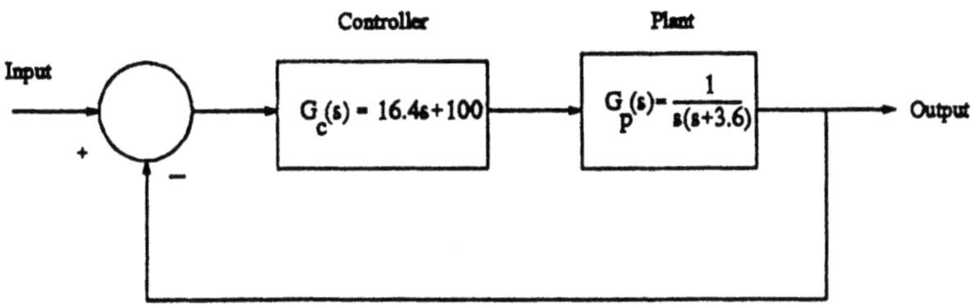

Figure 16 Block diagram of a feedback control system.

The controller gains were the same as the classical PD controller. Figure 17 shows that the response of the system with a PDFLC and the gain limited classical PD controller are identical but the unlimited output classical PD controller has a little faster response. The outputs of each controller as shown in Figure 18 indicates that the unlimited PD controller is able to produce a large control action and gives a faster response.

Now consider a system with the plant transfer function as $G_p(s) = \frac{1}{s(s+25)}$ and the transfer function of the controller that gives a critically damped step response as $G_c(s) = 3.94s + 100$. Figure 19 shows that the closed loop system response for all three controller are identical. The change in error does not reach as large a value as the previous example so the maximum controller output is never reached. Figure 20 shows that the outputs of the three controller are similar.

6 SUMMARY

This study uses mathematical expressions for fuzzification and defuzzification to derive an input-output equation for a specific PFLC and PDFLC. these equations demonstrate that the PDFLC and PDFLC have similar form as their classical counterparts. However, unlike their classical counterparts, the FLC is piecewise linear.

A detail analysis shows that the classical controller is a special case of the FLC. First, the mathematical expression of the PFLC and PDFLC is studied. The PFLC equation consists of an effective gain and a constant term. By

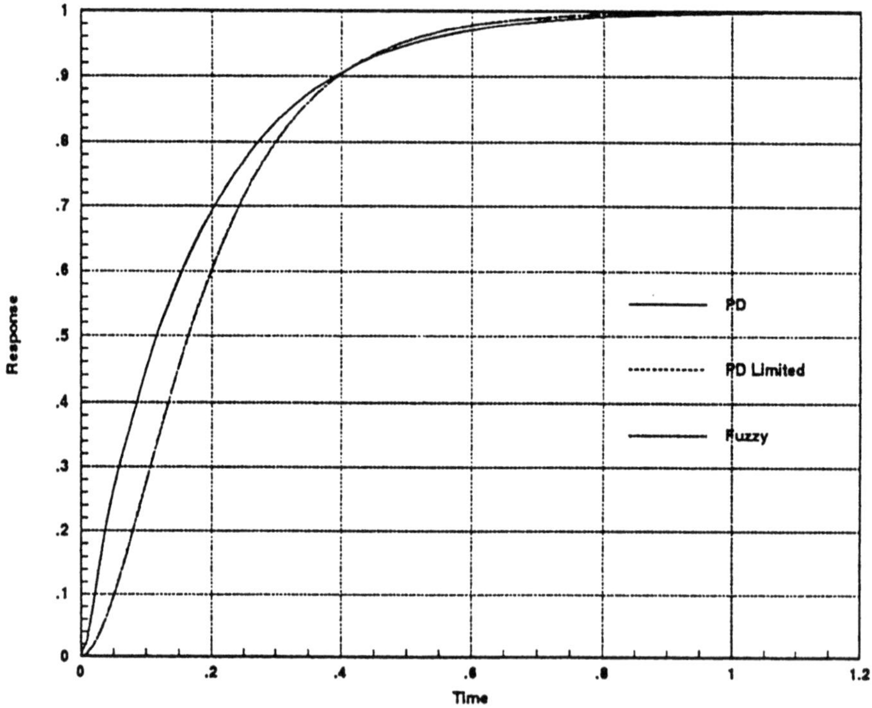

Figure 17 Closed loop response for plant controlled by classical PD controller and PDFLC.

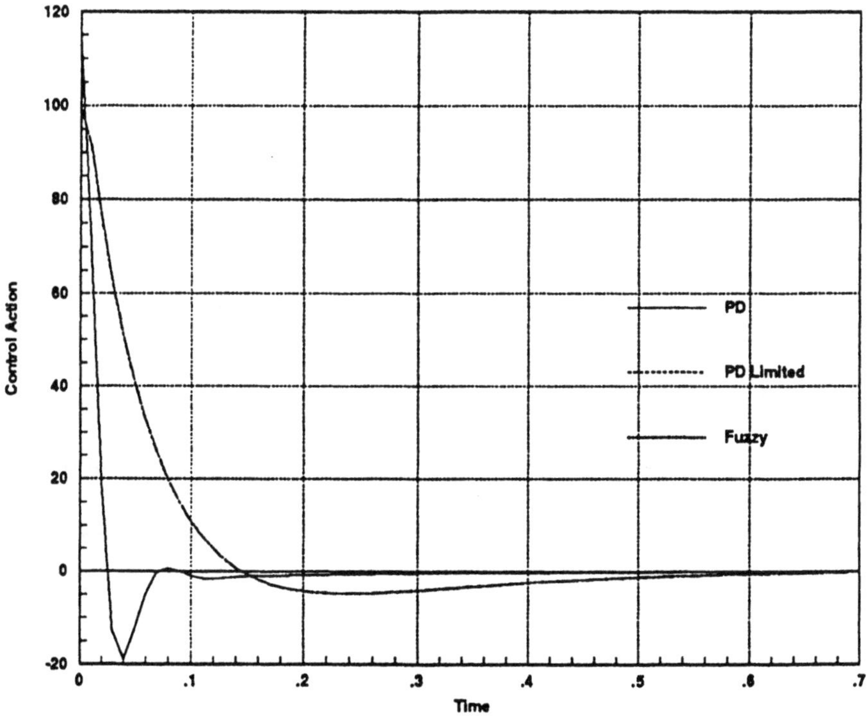

Figure 18 Classical PD controller and PDFLC output for example.

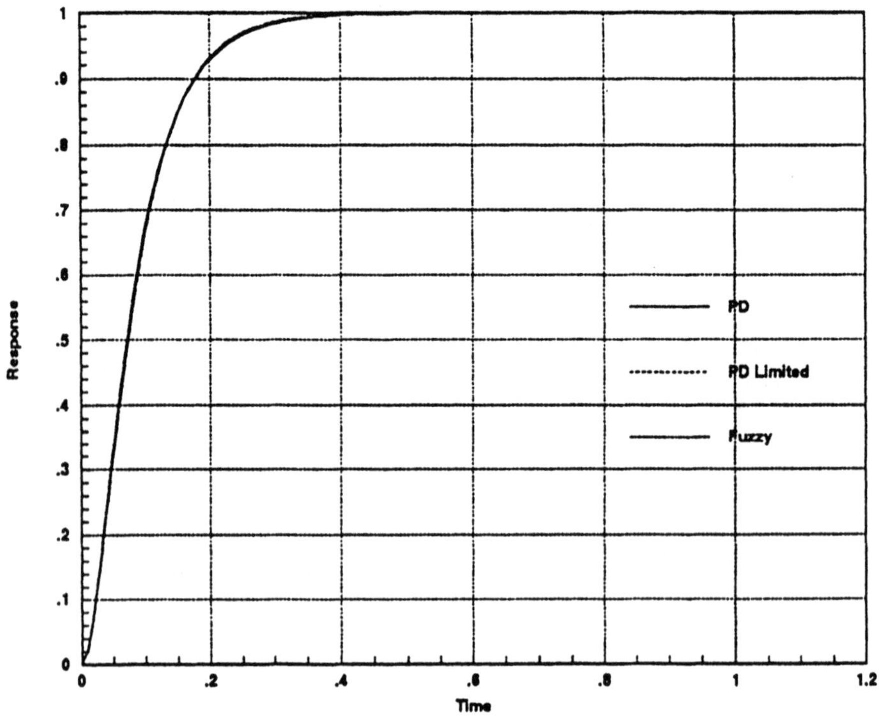

Figure 19 Closed loop response for plant controlled by classical PD controller and PDFLC.

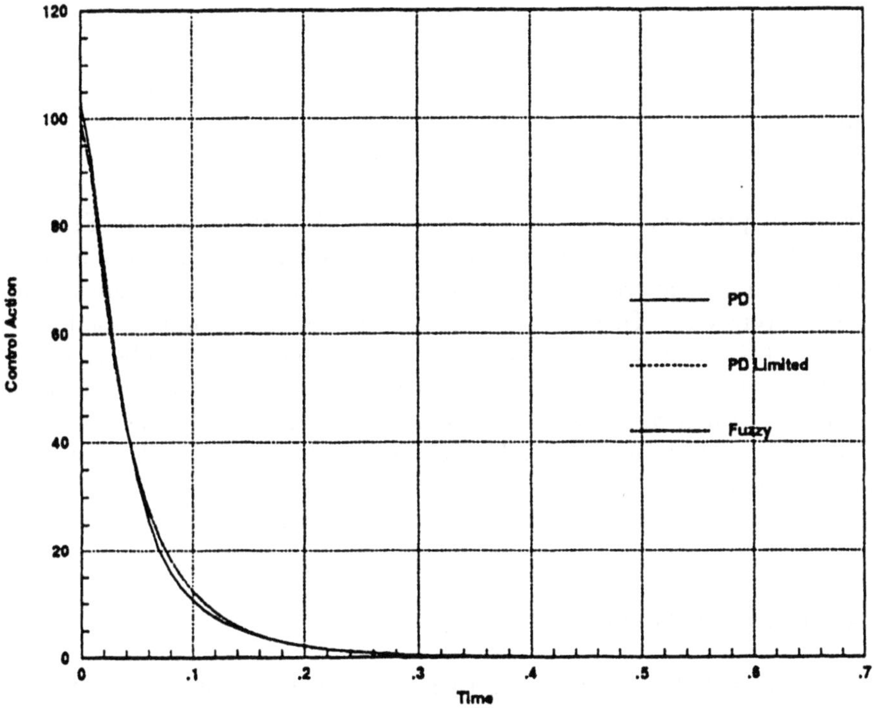

Figure 20 Classical PD controller and PDFLC output for example.

using equally spaced fuzzy sets, the gain can be made constant and the constant term can be made zero over all ranges of operation. The PDFLC term has two effective gains, a constant term and a nonlinear term. Like the PFLC, the gains can be made constant and the constant term can be made zero over all ranges of operation. The nonlinear term only affects the output where the rule base transitions from increasing values for output to maximum output values. The second analysis approach uses graphical illustration to study the control surface. For the PFLC, the controller output plot was exactly the same as the classical P controller. The PDFLC output control surface was very close to that of the classical PDPLC. There was a slight nonlinear region where the plot transitions from the linear region to be maximum output. It was shown that the nonlinear region could be minimized by increasing the number of the fuzzy sets and rules.

By showing the equivalence between the classical controller and the FLC controllers. a baseline of design and performance can be established for the FLC. The performance of the FLC at a minimum has to be the same as the classical controller. Therefore, the improved performance can be established for the FLC. The performance of the FLC at a minimum has to be the same as the classical controller. Therefore, the improved performance as has been demonstrateed in various applications [8, 9, 10] is not strictly due to the use of a FLC, but how the membership functions and rules of the FLC are implemented.

REFERENCES

[1] Thomas Berhm, *"Fuzzy Logic Controller: Analysis and Design," Master Thesis, Wright State University, Winter, (1994).*

[2] J. J. Buckly and H. Ying, *"Fuzzy Controller Thery: Limit Therorems for Linear Fuzzy Control Rules," Automatica, pp 469-472, (1989).*

[3] Lee, Chuen Chien, *"Fuzzy Logic in Control Systems: Fuzzy Logic Controller - Part I," IEEE Trans. Systems, Man Cybernetics, Vol. 20, pp 404-418, (1990).*

[4] Lee, Chuen Chien, *"Fuzzy Logic in Control Systems: Fuzzy Logic Controller - Part I," IEEE Trans. Systems, Man Cybernetics, Vol. 20, pp 419-435, (1990).*

[5] Gholamerza Langari, *"A Framework for Analysis and Synthesis of Fuzzy Linguistic Control Systems," Ph.D thesis, University of California at Berkeley, December, (1990).*

[6] E.H. Mamdani, *"Application of Fuzzy Algorithms for Control of Simple Dynamic Plant," Proc. IEE 121 Vol. 12 pp. 1585-1588, (1974).*

[7] M. Mizumoto, *"Realization of PID Controls by Fuzzy Control Motheds," IEEE Control System Magazine, (1992).*

[8] Kuldip S. Rattan, B. Chiu, V Feliu and H.B. Brown Jr., *"Rule Based Fuzzy Control of A Single-Link Flexible Manipulator in The Presence of Joint Froction and Load Changes," American Control Conference, Pittsburgh, PA, June, (1989).*

[9] D. Sabharwal and K. Rattan, *"Design of A Rule Based Fuzzy Controller for The Pitch Axis of An Unmanned Research Vehicke" NAECON Proceedings, Dayton, (1992).*

[10] M. Sugeno, *"Industrial Applications of Fuzzy Control," North-Holand, (1985).*

[11] L. A. Zadeh, *"Outline of A New Approach To The Analysis of Complex Systems and Decision Processes," IEEE Trans. Systems, Man Cybernetics, Vol. 3, pp 23-44, (1973).*

[12] L. Zheng, *"A Practical Guide to Tune of Proportional and Integral (PI) Like Fuzzy Controllers," IEEE Control System Magazine, (1992).*

6

REAL TIME FUZZY LOGIC CONTROLLER FOR BALANCING A BEAM-AND-BALL SYSTEM

Nowell Godfrey, Hua Li, Yuandong Ji*, William Marcy

Computer Science Department, College of Engineering
Texas Tech University, Lubbock, TX 79409, USA
** Department of Systems Engineering*
Case Western Reserve University, Cleveland, OH 44106, USA

ABSTRACT

Controlling a nonlinear system in real time is a challenging task which usually involves extensive mathematical formulation and intensive computation. In many cases, a linearization of the system model has to be derived first before the design of a controller, which limits the validaty of the system model. In addition, the detailed system parameters have to be known in order to perform such linearization, which can be difficult or even may not be practical in some real world applications. In this study, we demonstrated the design of a real-time fuzzy logic controller for a typical nonlinear system control application, balancing a beam-and-ball system in real time. Unlike the optimal or conventional control technique, the fuzzy logic controller requires no explicit system parameters, such as mass, torque etc., and it is characterized by its simplicity. A hardware prototyping system is built and the experimental results demonstrate the robustness of the controller. The comparative study reveals that the fuzzy logic controller outperforms optimal control based algorithm and in most cases outperforms trained human operators as well.

1 INTRODUCTION

Fuzzy logic control has been gaining popularity in many engineering applications as an alternative to traditional control. Some interesting research results and industrial applications have been reported. Examples include the design of an *adaptive fuzzy system* for backing up a truck-and-trailer [5], accessing the equipments reliability [7], controlling chemical reactors and processes [4], [8], realization of the function of conventional PID controllers [1], optimizing the performance of air-conditioning systems [9], and many others. Some of the common features of these works are (1) controlling highly dynamic, time-varying, non-linear systems, (2) coping with random disturbances and high degree of uncertainty, and (3) handling the lack of formal mathematical model of the process to be controlled.

In this study, we describe our recent work in controlling a nonlinear system, a beam-and-ball system based on fuzzy logic control technique. Unlike most of the conventional and optimal control techniques, our fuzzy logic controller requires no explicit system parameters, and it is characterized by its simplicity and robustness. A comparison between optimal control and fuzzy logic control is made. Simulations, experiments based on in-house built prototype system are given.

2 PROBLEM FORMULATION

A beam-and-ball system is a typical nonlinear system to demonstrate optimal control and Kalman filter concept. The system configuration of the in-house built prototype system is given in Figure 1. Its actual prototype is given in Figure 2.

The control objective is to balance the beam and bring the ping-pong ball to the center of the beam.

The mathematical model of the beam balancing system [2] is,

$$J\frac{d^2\theta}{dt^2} = Mgr sin\phi, \tag{6.1}$$

where J is the moment of inertia of the ball, θ (in radian) is the angle displacement (from the center of the beam) of the ball, ϕ (in radian) is the titling angle of the beam, M is the mass of the ball, and g is the normal gravity acceleration. Let $y = r\theta$, thus y is the displacement of the ball from the center of the beam, and we have

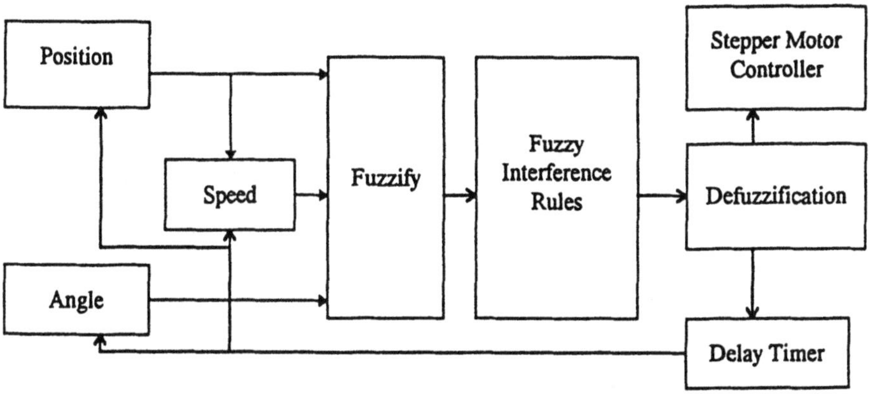

Figure 1 The block diagram of the system configuration.

$$\frac{d^2y}{dt^2} = asin\phi, \tag{6.2}$$

where $a = \frac{Mgr^2}{J}$. Our objective is to design a controller which drives a step motor output angle ϕ to control the system. We would like to drive the ball to any desired position r on the beam, *i.e.*, the system output $y = r$, and keep the ball at that position. Note that both y and θ are measurable by sensors. It is a common practice to linearize the model if optimal control strategy is adopted. After linearization of equation (6.2), we have

$$\frac{d^2y}{dt^2} = au, \tag{6.3}$$

where $u = \phi$ is the system control signal. The state space representation of this system is

$$\dot{x}(t) = Ax(t) + Bu(t) + Fw(t), \tag{6.4}$$

$$y(t) = Cx(t) + Gv(t), \tag{6.5}$$

Figure 2 The in-house built prototype system.

where

$$x(t)=\begin{bmatrix} x_1(t) \\ x_2(t) \end{bmatrix}=\begin{bmatrix} y(t) \\ \dot{y}(t) \end{bmatrix}, A=\begin{bmatrix} 0 & 1 \\ 0 & 0 \end{bmatrix}, B=\begin{bmatrix} 0 \\ a \end{bmatrix}, F=\begin{bmatrix} \beta_1 & 0 \\ 0 & \beta_2 \end{bmatrix},$$

and $C = [1,0]$, $G = \beta_3$. Both $w(t)$ and $v(t)$ are introduced to model the errors from linearization, actuator control, and sensors measurement. They are mutually independent and uncorrelated, and both have the standard Gaussian distribution, $N(0,1)$. Note that the parameters β_1, β_2, and β_3 are not known and will be reserved as tuning parameters of the controller.

Now we have the following optimal control problem: find $u(t)$, subject to quations (6.4) and (6.5), such that the objective function J is minimized,

$$J = E[(x(T) - \begin{bmatrix} r \\ 0 \end{bmatrix})' S(x(T) - \begin{bmatrix} r \\ 0 \end{bmatrix})+$$

$$\int_{t_0}^{T} ((x(t) - \begin{bmatrix} r \\ 0 \end{bmatrix})' Q(x(t) - \begin{bmatrix} r \\ 0 \end{bmatrix}) + u'(t) R u(t)) dt], \qquad (6.6)$$

where $E(.)$ is the expectation, M' is the transpose of matrix M. The weighting matrices $S \geq 0$, $Q \geq 0$, and $R¿0$, will be determined in the design process.

3 THE DESIGN OF FUZZY LOGIC CONTROLLER

From the formulation given by equation (6.6), it is clear that in order to achieve the control goal, optimal control algorithm has to be designed based on the knowledge of the system parameters. In the situations where the system parameters are unknown or changing as functions of time, or high random disturbances exist, the control algorithm design and the computation task can be formidable. To demonstrate fuzzy logic control, we choose the beam-and-ball system as a time-varying system. This can be achieved by clipping a stack of big paper clips on the beam of the system.

3.1 Design Formulation

In order to achieve real-time control of the system, a fuzzy logic controller is designed, which consists of three major stages: selection of sensory inputs and fuzzification, design of fuzzy inference rules, and computation of the control signal.

Input Variable Selection and Fuzzification

We have defined three input parameters and a control signal. The inputs are displacement, $\{d(t_i)\}$, an angular position of the beam, $\{\alpha(i)\}$, and the calculated speed from the displacement of the ball, $\{I(t_i)\}$, where

$$I(t_i) = d(t_i) - d(t_j), \qquad (6.7)$$

for $i, j = 0, 1, ..., n$, and $i \neq j$. The control action, C is a set of control sequence $\{c(t_i)\}$ for $i = 0, 1, ..., n$.

For each of these parameters, we quantize them into six levels and use linguistic variables, positive (or negative) large $\pm L$, positive (or negative) median $\pm M$, and positive (or negative) small $\pm S$, to describe each level. Hence, we have $D=\{D_{\pm L}, D_{\pm M}, D_{\pm S}\}$, $\alpha=\{\alpha_{\pm L}, \alpha_{\pm M}, \alpha_{\pm S}\}$, $I=\{I_{\pm L}, I_{\pm M}, I_{\pm S}\}$, and $C=\{C_{\pm L}, C_{\pm M}, C_{\pm S}\}$.

Then fuzzy membership functions for each linguistic variable are defined on the experimental basis. Given in Figure 3 are membership functions, $\mu_{D_{+L}}$, $\mu_{D_{+M}}$, and $\mu_{D_{+S}}$, for displacement, D_{+L}, D_{+M}, and D_{+S} respectively. Also given in this figure are the membership functions for angular position $\alpha(i)$, and calculated speed $I(t_i)$. Note that The membership functions of calculated speed and the angular position are even functions whose negative parts are the mirror images of the positive part.

Design of Fuzzy Inference Rules

Performing $D \times \alpha \times I$, we cover all the possible combinations of the inputs. Using Zadeh's method [10] and using connective "AND," we have derived 216 inference rules. The effort has been made to fine turn the rules to reflect the requirements of the control objective. The rules after fine-tuning are given in Table 1 through Table 6.

ANGLE = NL		SPEED					
		NL	NM	NS	PS	PM	PL
POSITION	NL	PL	PL	PL	PL	PL	PL
	NM	PL	PL	PL	PL	PM	PM
	NS	PL	PL	PM	PM	PS	NS
	PS	PL	PL	PM	PS	NS	NM
	PM	PL	PL	PS	NS	NM	NL
	PL	PL	PM	PS	NM	NL	NL

Table 1 Fuzzy Inference Rules for NL Angle

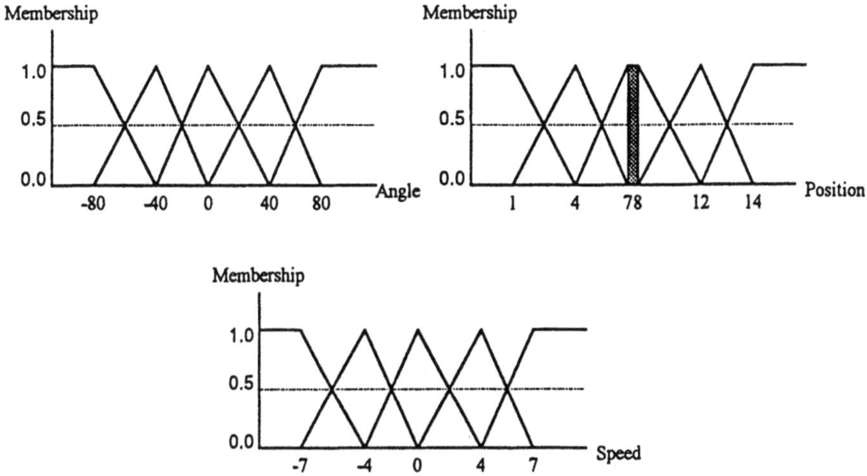

Figure 3 The membership functions for displacement, angular position, and calculated speed.

ANGLE = NM		SPEED					
		NL	NM	NS	PS	PM	PL
POSITION	NL	PL	PL	PL	PM	PM	PM
	NM	PL	PL	PM	PM	PM	PM
	NS	PL	PL	PM	PM	PS	PS
	PS	PL	PM	PS	NS	NM	NL
	PM	PL	PM	PS	NS	NM	NL
	PL	PL	PM	PS	NS	NM	NL

Table 2 Fuzzy Inference Rules for NM Angle

The membership value for each control action for each inference rule is determined by the *min* operation, as $\mu_{C_{+L}}=\min(\mu_{D_{+L}}, \mu_{\alpha_{+L}}, \mu_{I_{+L}})$ for rule 1, etc.

ANGLE = NS		SPEED					
		NL	NM	NS	PS	PM	PL
POSITION	NL	PL	PL	PM	PS	PS	NS
	NM	PL	PL	PM	PS	NS	NS
	NS	PL	PM	PS	NS	NM	NL
	PS	PL	PM	PS	NS	NM	NL
	PM	PM	PS	PS	NS	NM	NL
	PL	PM	PS	PS	NM	NL	NL

Table 3 Fuzzy Inference Rules for NS Angle

ANGLE = PS		SPEED					
		NL	NM	NS	PS	PM	PL
POSITION	NL	PL	PL	PM	NS	NS	NM
	NM	PL	PM	PM	NS	NS	NM
	NS	PL	PM	PS	NS	NM	NL
	PS	PL	PM	PS	NS	NM	NL
	PM	PS	PS	NS	NM	NL	NL
	PL	PS	NS	NS	NM	NL	NL

Table 4 Fuzzy Inference Rules for PS Angle

ANGLE = PM		SPEED					
		NL	NM	NS	PS	PM	PL
POSITION	NL	PL	PM	PS	NS	NM	NL
	NM	PL	PM	PS	NS	NM	NL
	NS	PL	PM	PS	NS	NM	NL
	PS	NS	NS	NM	NM	NL	NL
	PM	NM	NM	NM	NM	NL	NL
	PL	NM	NM	NM	NL	NL	NL

Table 5 Fuzzy Inference Rules for PM Angle

Computation Of Control Signal

The actual control signal is computed as linguistic variables for each set of input signal from $D \times \alpha \times I$. Then defuzzification is performed by the following center of gravity algorithm,

ANGLE = PL		SPEED					
		NL	NM	NS	PS	PM	PL
POSITION	NL	PL	PL	PM	NS	NM	NL
	NM	PL	PM	PS	NS	NM	NL
	NS	PM	PS	NS	NM	NL	NL
	PS	PS	NS	NM	NM	NL	NL
	PM	NM	NM	NL	NL	NL	NL
	PL	NL	NL	NL	NL	NL	NL

Table 6 Fuzzy Inference Rules for PL Angle

$$C(t+1) = \frac{\sum_{i=1}^{216} w_i C(t_i)}{\sum_{i=1}^{216} w_i}, \tag{6.8}$$

where w_i is the corresponding membership value, $C(t_i)$ is the control signal derived from inference rule i at time t which is the current time. $C(t+1)$ is the overall control signal at new time step $t+1$.

3.2 The In-House Built Prototype System

To test the control algorithm, a prototype was in-house built. The prototype system consists of three major subunits: (1) The mechanical apparatus which host a stepper motor, a motor-driven movable beam, and the fixtures for operating environment lighting. The stepper motor used for this apparatus is 20 volt 4-phase DC stepper, with maximum current of 0.22 A. The minimum step size is 1.8 degree. The need for the operating environment lighting is due to the need of the registration of ball position by using photo diodes. These photo diodes, TIL31B, will generate lesser potential when the light intensity decreases due to the ball's occlusion of the lighting source. (2) The I/O interfacing board which host 8255A interface unit and digital sensing circuits for calculating the displacement position of the ball, the angular position of the beam from the analog sensory input of the apparatus. The board is the major linkage between the host microprocessor and the apparatus. And (3) The fuzzy control engine which was written in C and assembly language and implemented on Texas Instruments Business Pro Portable machine with 80286 CPU and 80287 math coprocessor running at 8 Mhz. Please note the use of TI business Pro, an old machine, for this real-time fuzzy logic control project was intentional. It will allow us to better demonstrate a non-trivial, real-time nonlinear system control can be achieved by using fuzzy logic algorithm with relatively modest computational power.

The Apparatus

The apparatus was built with a 36 inch beam. There are 16 photo diodes as photo sensors to register the ball displacement position in real time. These sensors are mounted on the beam with equal spacing of 2 inch on each side of the beam. At the center of the beam, two sensors are placed closer together with less than one inch distance. The apparatus was designed in such way that it can be operated in two different modes: *human control mode* and *computer control mode* based on fuzzy logic controller. The human control mode is used as a reference for the later comparative study against the designed real-time fuzzy logic controller. The human control is achieved through an analog joystick, MACH-I, manufactured by CH Products. The different x-y position of the joystick gives different resistance value which can be used as potential meter to convert the position information to electrical potential value. The resistance vs. the position of the joystick exhibits fairly good linear relationship with 0 Ohm at the top left corner position (as the maximum positive y position and minimum negative x position) and 95K Ohm for both x and y at the bottom right corner position (as the minimum negative y and maximum positive x.) When the joystick is set free at its center position ($x = 0$, $y = 0$), the resistance values for x is 52K Ohm and is 40K Ohm for y. This resistance distribution is used as the reference to design A/D conversion to 8 bit data to cover the entire dynamic range of the joystick.

An optical rotary encoder, Clarostat 601V, was used for the registration of angular position of the beam. The encoder outputs two square waves in quadrature at a rate of 128 pulses per channel per revolution. The outputs are TTL compatible and it operates at 5V DC at 30 milli A maximum. The encoder's output was feed into the digital circuit for the further coding process before it reaches the interface, 8255A.

Interface Design

The interface of the apparatus to the host computer, TI Business Pro, was designed. The function of parallel interfacing unit was achieved by using 8255A Programmable Peripheral Interface. This interface consists of 1 control register and 3 separately addressable ports, denoted as Port A, B, and C. By using the combination of bus signal A0, A1, and $\bar{RD}$, $\bar{WR}$, $\bar{CS}$, one is able to select port for communication and directions (from the bus to the interface or vice versa). In our design Port A and B are used to receive digitized sensory inputs. Position displacement goes to Port A. Angular position and joystick input go to Port B, while Port C is used to send control signals to the stepper motor. Using this parallel interfacing technique, we are able to achieve the data transfer rate at about 3.3 Mbyte/second (at 300 nanosecond per byte speed.) This rate, as we estimated and later confirmed by our experimental result, is about 10 times faster than what is needed to control the beam-and-ball system in real time.

In the prototype design, we have put two redundant systems together on a single board, FLoc1, to be sure that the system will operate and function properly at any time. The first system on FLoC1 board is built by using standard off-the-shelf components, and the second system on the same board is built by using Xilinx FPGA. The functional flow diagram described the interface operation is given in Figure 4.

The hardware prototype board, FLoC1, was designed. The schematics of the design is given in Figure 5. The print circuit board layout of both copper side and solder side of the board are given in Figure 6 (a) and (b) respectively. The photo of the board, which shows the assembled and tested board, FLoC1, is given in Figure 7.

4 COMPARISONS TO OTHER CONTROL STRATEGY

Comparative study was performed to objectively evaluate the performance of the fuzzy logic controller. The study includes the comparison of fuzzy logic control to human control, and the comparison of fuzzy logic control to optimal control.

4.1 Comparisons To Human Control Mode

The prototype system was built to operate in two different modes: human control mode and the fuzzy logic control mode. The human control was achieved through the operation of joystick to balance the beam-and-ball system. Comparison to human control was made as one performance evaluation index. When evaluating the performance, the number of control actions needed to balance the system, the beam swing range for each individual control action, and the number of over controls (the control action results in overdriving the ball from one side of the beam to the other side of the beam) are used to characterize the performance.

The performance of human control exhibits big difference from experienced operator to un-experienced, un-trained operator. In order to make good comparison, we have conducted the comparative study against a set of random selected subjects which include both the experienced operators, including the designers of the controller, and un-experienced operators, including graduate students, department secretary, and engineers. In general, the trained operator performs much better than others. The data given in this section is from trained operators. Figure 8 shows manually controlled angular position of the beam. As can be seen from this data set, the human control generates quite a lot overdriving actions before finally reaching the balanced state, or the stable state where its angle is equal to zero and stays at zero for the same time period of fuzzy logic control. For fuzzy logic control, we use 20 additional sampling intervals, equal to 20x30 milli second, as the minimum required time for balanced result. It should be pointed out that the unit for angle position given in the figure was the counts generated by the optical encoder with one count equal to 0.143 degree.

The fuzzy logic controller was designed to execute control action based on both the input from beam angle and the ball position. Hence, the manual control position measurement from the same experiment is provided in Figure 9.

As can be seen in Figure 9, the control makes the ball rolling most of the time on one side (with sensors from 8 to 12) of the beam. The control process started while ball located at the other side of the beam (with sensor from 0 to 6.) The control objective was to bring the ball to the center and stay there for 20 additional sampling intervals. With about 21 crossings of the center, the beam was balanced.

For fuzzy logic controller once it is fine tuned, it usually takes less number of control actions and produces less number of center crossings to balance the beam. Figure 10 and 11 illustrate two typical experimental results. In Figure 10, the angle of the beam started at zero then there was an intentionally designed sudden random change of the angle to drive the system out of balance for the controller to start its control action. This starting condition is generated by the host computer and it is the same condition for human control. Then after four major control actions corresponding to the angle counts reduction from about positive 20 to less than positive 10, then to the negative range, it then reached the balanced state. Comparing to the manual control, this takes considerably less control actions and the range of the angle is also smaller.

4.2 Comparisons To Other Control Strategy

In order to achieve the same control objective by optimal control let us first consider the case where $r = 0$, or to balance the beam and bring the ball to the center of the beam. This is a regulator problem and is the well known Linear Quadratic Gaussian (LQG) problem [3]. The solution to such problem can partially compensate the error induced by linearization.

Using the separation theorem, the optimal control is given by

$$u(t) = -R^{-1}B\prime K(t)\hat{x}(t), \tag{6.9}$$

where $K(t)$ is the symmetric solution of the Riccatti equation,

$$\dot{K}(t) = -K(t)A - A\prime K(t) + K(t)BR^{-1}B\prime K(t) - Q, \tag{6.10}$$

with a terminal condition $K(T) = S$. Note that the integration of equation (6.10) must be backwards. The state estimate $\hat{x}(t)$ in (6.9) is given by the Kalman filter

$$\frac{d\hat{x}(t)}{dt} = A\hat{x}(t) + Bu(t) + P(t)C\prime G^{-1}(y(t) - C\hat{x}(t)), \tag{6.11}$$

where $P(t)$ is the solution of the following differential equation,

$$\dot{P}(t) = AP(t) + P(t)A\prime - P(t)C\prime G^{-1}CP(t) + (FF\prime)^{\frac{1}{2}} \tag{6.12}$$

with initial condition $P(0) = P_0$.

Now consider $T \to \infty$, in order to make the problem tractable, we have to assume the system is time invariant. In fact, in our fuzzy logic control mode, the system is not time invariant. We have successfully controlled the system after placing a stack of heavy paper clips on the beam. We have both $(A, Q^{\frac{1}{2}})$ and (A, C) observable, and both (A, B) and $(A, (FF\prime)^{\frac{1}{2}})$ controllable. The steady-state solution exists and is given by (6.9), with $K > 0$ is the solution of

$$KA + A\prime K - KBR^{-1}B\prime K + Q = 0. \tag{6.13}$$

The state estimate $\hat{x}(t)$ in (6.10) is given by the constant gain Kalman filter ($P(t) = P$ in equation (6.11)). Where $P > 0$ is the symmetric solution of the algebraic filtering Riccati equation.

$$AP + PA\prime - PC\prime G^{-1}CP + (FF\prime)^{1/over2} = 0.$$

Now let's consider the Kalman filter design. Since the position, $y(t) = x_1(t)$, is to be controlled, we define $Q = \begin{bmatrix} 1 & 0 \\ 0 & 0 \end{bmatrix}$, $R = 0.1$ for moderate penalty of using control energy. The remaining unknown parameters, β_1, β_2, and β_3 (with the restriction $\beta_3 > 0$), determine the behavior of the filter, which can be used to fine tune the compensator on line. From equation (6.9)-(6.12), we have

$$K = \frac{1}{a\sqrt{10}} \begin{bmatrix} \sqrt{a}\sqrt[4]{40} & 1 \\ 1 & \frac{1}{\sqrt{a}\sqrt[4]{2.5}} \end{bmatrix}$$

$$P = \begin{bmatrix} \sqrt{\beta_3(\beta_1 + 2\sqrt{\beta_2\beta_3})} & \sqrt{\beta_2\beta_3} \\ \sqrt{\beta_2\beta_3} & \sqrt{\beta_2(\beta_1 + 2\sqrt{\beta_2\beta_3})} \end{bmatrix}, \tag{6.14}$$

and the steady state control signal is

$$u(t) = -[\sqrt{10}, \sqrt{5a}\sqrt[4]{10}]\hat{x}(t), \tag{6.15}$$

and the filter is

$$\frac{d\hat{x}(t)}{dt} = \begin{bmatrix} -\sqrt{\frac{\beta_1+2\sqrt{\beta_2\beta_3}}{\beta_3}} & 1 \\ -a\sqrt{10} - \sqrt{\frac{\beta_2}{\beta_3}} & -a\sqrt{5a}\sqrt[4]{10} \end{bmatrix} \hat{x}(t) + \begin{bmatrix} \sqrt{\frac{\beta_1+2\sqrt{\beta_2\beta_3}}{\beta_3}} \\ \sqrt{\frac{\beta_2}{\beta_3}} \end{bmatrix} y(t) \quad (6.16)$$

with $\hat{x}(0)=E[x(0)]$. Note the above two equations are also a dynamic system in state space representation, with $y(t)$ as its input and $u(t)$ as its output. This observation will simplify the simulation of the control system as described later.

Having discussed the case for $r = 0$, we now generalize the control system to include $r \neq 0$, i.e., a nonzero set point. For this, consider the coordinate transformation

$$z = \begin{bmatrix} z_1 \\ z_2 \end{bmatrix} = \begin{bmatrix} x_1 - r \\ x_2 \end{bmatrix} = x - \begin{bmatrix} r \\ 0 \end{bmatrix}. \quad (6.17)$$

The LQG problem becomes

$$\min_{u(t)} E[z\prime(T)Sx(T) + \int_{t_0}^{T} (x\prime(t)Qx(t) + u\prime(t)Ru(t))dt]. \quad (6.18)$$

Hence, the system is

$$\dot{z}(t) = Az(t) + Bu(t) + Fw(t),$$

$$\tilde{y}(t) = y(t) - r = Cx(t) + Gv(t). \quad (6.19)$$

Following the same approach of solving the regulator LQG problem, a steady state compensator can be obtained as

$$\frac{d\hat{x}(t)}{dt} = \begin{bmatrix} -\sqrt{\frac{\beta_1+2\sqrt{\beta_2\beta_3}}{\beta_3}} & 1 \\ -a\sqrt{10} - \sqrt{\frac{\beta_2}{\beta_3}} & -a\sqrt{5a}\sqrt[4]{10} \end{bmatrix} \hat{z}(t)+$$

$$\begin{bmatrix} \sqrt{\frac{\beta_1+2\sqrt{\beta_2\beta_3}}{\beta_3}} \\ \sqrt{\frac{\beta_2}{\beta_3}} \end{bmatrix} (y(t) - r(t)), \tag{6.20}$$

with $\hat{z}(t) = E[z(0)]$ and $u(t)$=-$[\sqrt{10}, \sqrt{5a^4}\sqrt{10}]\hat{z}(t)$.

The experiments have been conducted to verify the design of the optimal controller. As pointed out in this section, optimal control strategy is based on the known parameters of the system. When the actual prototyping system, a beam-and-ball system, is built, no attempt has been made to measure these system parameters, such as the torque, the mass M, etc. due to the fact that in many real control applications many of these parameters may not be known. But in order to perform the simulation of the optimal controller, we have to make some assumptions about the system parameters. This is certainly a big disadvantage comparing to fuzzy logic controller. The system parameter identification can be very difficult. The simulation is conducted by using SIMULINK for Microsoft Windows. The program can simulate nonlinear systems and it retains all of the MATLAB's general purpose functionality. Without lose of generality, assume $a = 5$, from equation (6.20) we have

$$\frac{d\hat{x}(t)}{dt} = \begin{bmatrix} -\sqrt{\frac{\beta_1+2\sqrt{\beta_2\beta_3}}{\beta_3}} & 1 \\ -16.811 & -44.457 \end{bmatrix} \hat{z}(t) + \begin{bmatrix} \sqrt{\frac{\beta_1+2\sqrt{\beta_2\beta_3}}{\beta_3}} \\ \sqrt{\frac{\beta_2}{\beta_3}} \end{bmatrix} (y(t) - r(t)) \tag{6.21}$$

with

$$\hat{z}(0) = E[z(0)], u(t) = -[3.162, 8.891]\hat{z}(t). \tag{6.22}$$

Starting the simulation with $\beta_1 = \beta_2 = \beta_3 = 1$, and $\hat{z}(0) = 0$, we have the compensator as

$$\frac{d\hat{x}(t)}{dt} = \begin{bmatrix} -1.732 & 1 \\ -15.81 & -44.457 \end{bmatrix} \hat{z}(t) + \begin{bmatrix} 1.732 \\ 1 \end{bmatrix} (y(t) - r(t)), \hat{z}(0) = \begin{bmatrix} 0 \\ 0 \end{bmatrix} \tag{6.23}$$

$$u(t) = -[3.162, 8.891]\hat{z}(t). \tag{6.24}$$

The block diagram of the control system is given in Figure 12.

Let the time interval be $[0, 200]$ for $t = 0$ and $T = 200$, and initial disturbance $u(1) = -2$ which means the displacement angle of the ball is 114.6˙, we choose $r(t) as$

$$r(t) = \begin{bmatrix} 0 & t < 100 \\ 5 & t \geq 100 \end{bmatrix}, \tag{6.25}$$

thus the behavior of the control system will give us the taste of the controller performance under the initial disturbance or the ability of tracking changed set point.

The entire system was simulated by using SIMULINK. The output (the ball position) is plotted in Figure 13. Note that the negative peak is due to the strong initial impulse disturbance (rotating the ball by 114˙, the positive peak is the response to the new set point $r = 5$. As we change the set point between 4 and -4, the controller can track the change as well. The reader has to be advised for the interpretation of these *simulation* result of the control technique. It has several major limitations:

1. Since the control algorithm is derived from the close form mathematical equation, the system parameters have to be known. Otherwise the simulation can not be performed. In reality, the identification of these parameters can be difficult. While in the fuzzy logic controller design, there is no need to make such unrealistic assumption.
2. These assumed values are all constant, while for the fuzzy logic controller they can be time varying parameters. In fact, in our fuzzy logic control experiments we have intentionally changed the system configuration by altering the beam mass as much as to its 20 percentage.

Therefore, the conventional control technique can hardly compare to the performance of the fuzzy logic controller.

5 SUMMARY

In this study, we described the design of a real time fuzzy logic controller for nonlineary system, a beam-and-ball system control. Controlling a nonlinear system in real time is a challenging task which usually involves extensive mathematical formulation and intensive computation. Using fuzzy logic technique, we demonstrated that there is no need to perform linearization, and there is no need to require the explicit system parameters, such as torque, mass, etc., to design a real time controller. Simplicity and robustness demonstrated by the experiments are the major attractive features of this design. Comparing to conventional optimal control technique, in order to derive meaningful mathematical model, system parameters have to be given which can be very difficult especially for time varying systems, or the system under heavy random disturbances. In addition, intensive computation is usually involved for real time solution. A hardware prototype system is built and the experimental results demonstrated the robustness of the controller. This demonstration prototype shows that in most cases the fuzzy logic controller outperforms trained human operators.

6 ACKNOWLEDGEMENT

The first two authors would like to express their appreciation to Dr. Larry Masten of FSI International Inc. for the support in part to this project and to the College of Engineering funds through the Center of Robotics and Automation of Texas Tech University. This beam-and-ball system demonstrated in World Congress Neural Network Conference in San Diego in June 1994 and it was given an *Industrial Neural Network Award.* A video tape of the demonstration can be obtained from Prof. Hua Li of Texas Tech University.

REFERENCES

[1] G.M. Abdelnour, C.H. Chang, F.H. Huang, and J.Y. Cheung, "Design of a Fuzzy Controller Using Input and Output Mapping Factors," IEEE Trans. on Systems, Man, and Cybernetics, Vol. 21, No. 5, pp. 952-960, Sept. 1991.

[2] K.J. Astrom and B. Wittenmark, *Computer Controlled Systems, Theory and Design*, Prentice-Hall, Englewood Cliffs, NJ, 1984.

[3] A.E. Bryson and Y.C. Ho, Applied Optimal Control, Hemisphere, 1975.

[4] M. Galluzzo, V. Cappellani, and U. Garofalo, "Fuzzy Control of pH Using NAL," International J. of Approximate Reasoning, No. 5, pp. 505-519, 1991.

[5] S.G. Kong, B. Kosko, "Adaptive Fuzzy Systems for Backing up a Truck-and-Trailer," IEEE Trans. on Neural Networks, Vol. 3, No. 2, pp. 211-223, March 1992.

[6] C.C. Lee, "Fuzzy Logic in Control Systems: Fuzzy Logic Controller, Part II," IEEE Trans. on Systems, Man, and Cybernetics, Vol. 20, No. 2, pp. 419-435, March 1990.

[7] T. Onisawa, "Fuzzy Reliability Assessment Considering the Influence of Many Factors on Reliability," International J. of Approximate Reasoning, No. 5, pp. 265-280, 1991.

[8] M. Parekh, M. Desai, H. Li, and R. Rhinehart, "In-Line Control of Nonlinear pH Neutralization Based on Fuzzy Logic," IEEE Trans. on Components, Packaging, and Manufacturing Technology, Part A, Vol. 17, No. 2, June 1994, pp. 192-201.

[9] T. Tobi, T. Hanafusa, "A Practical Application of Fuzzy Control for an Air-Conditioning System," International J. of Approximate Reasoning, No. 5, pp. 331-348, 1991.

[10] L.A. Zadeh, "Fuzzy Sets," Inform. Control, Vol. 8, pp. 338-353, Academic Press Inc., 1965.

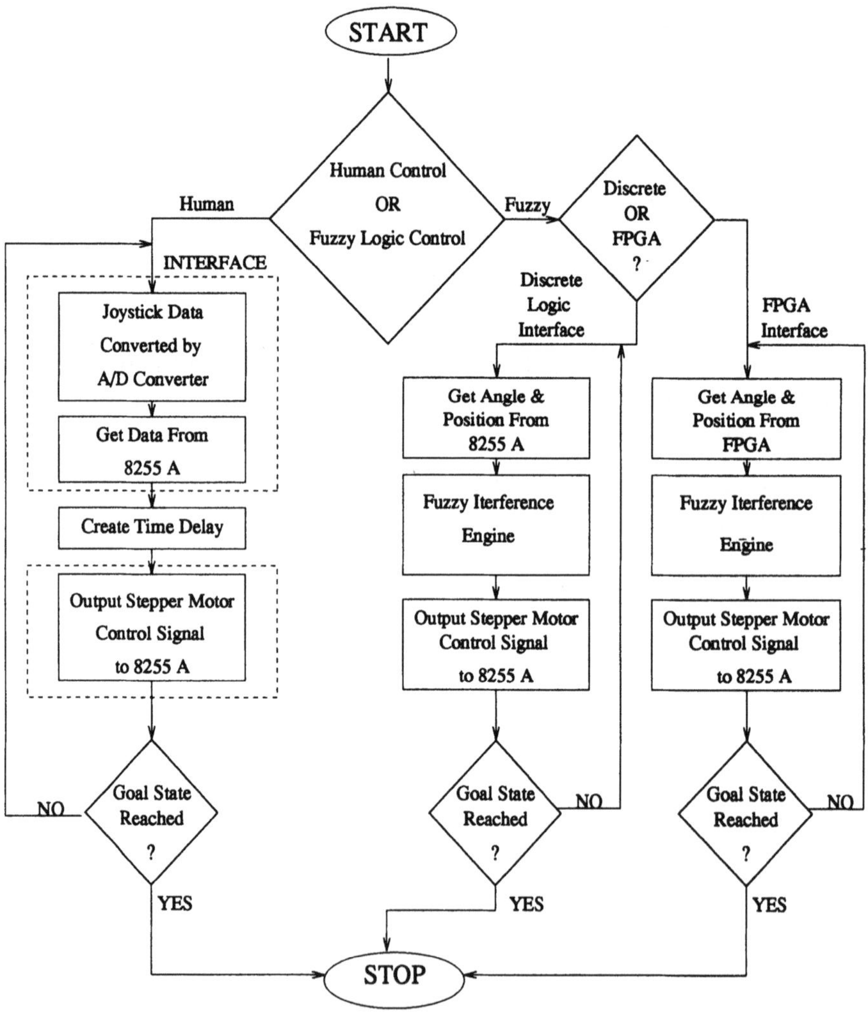

Figure 4 The flow diagram of the functionality of the system

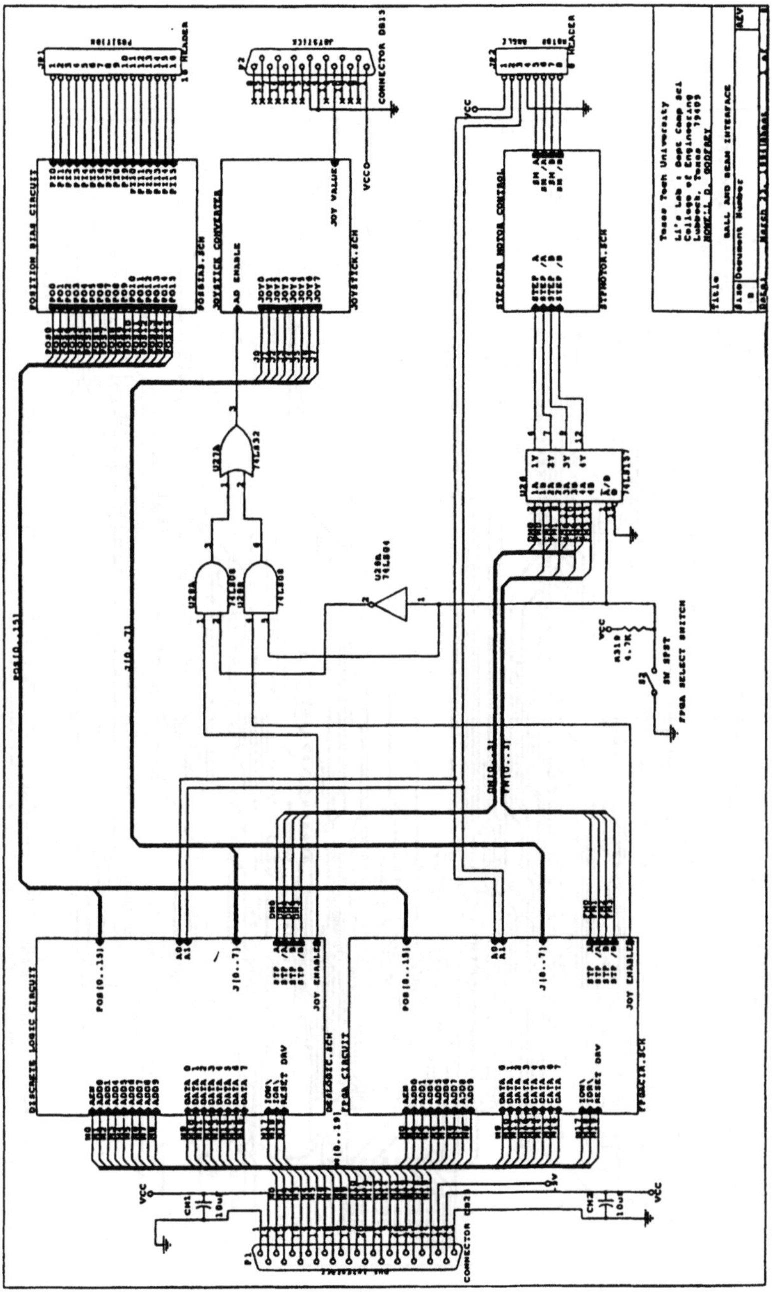

Figure 5 The schematics of the FLoC1 board.

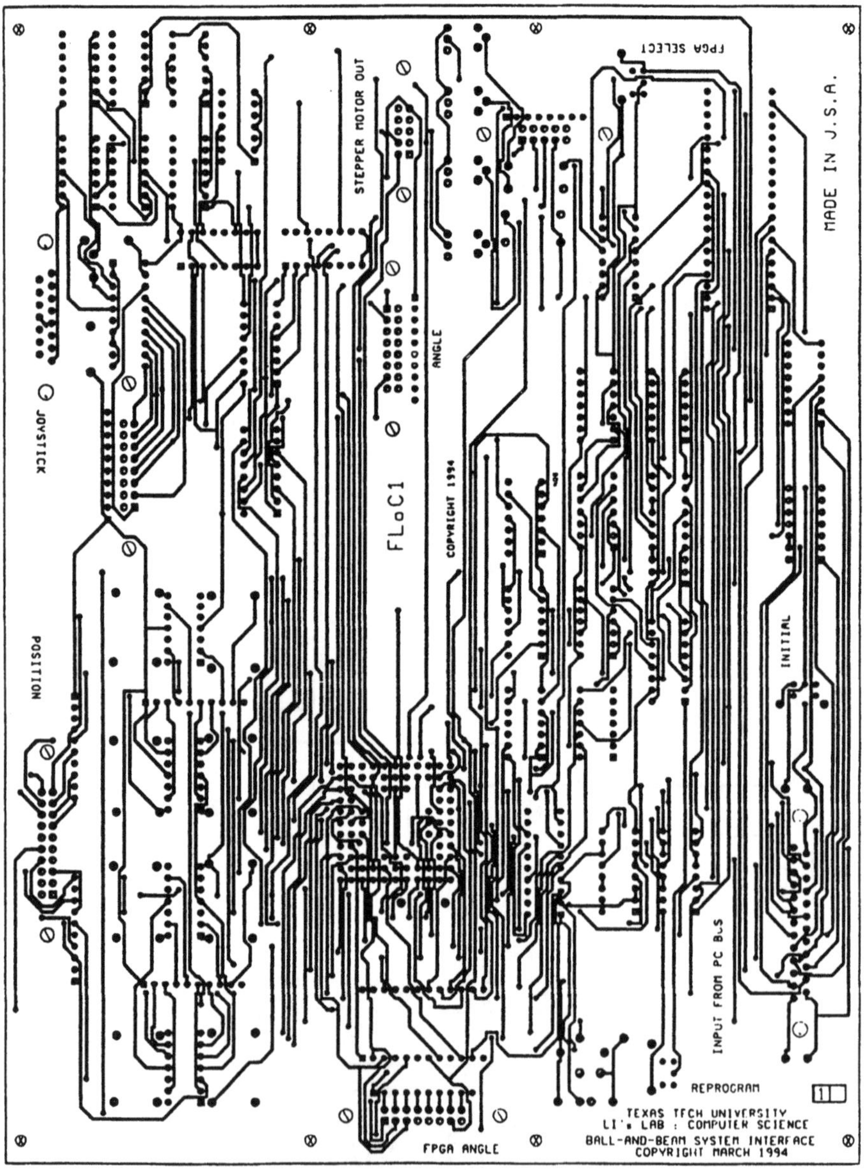

Figure 6(a) The cooper side of the FLoC1 board layout.

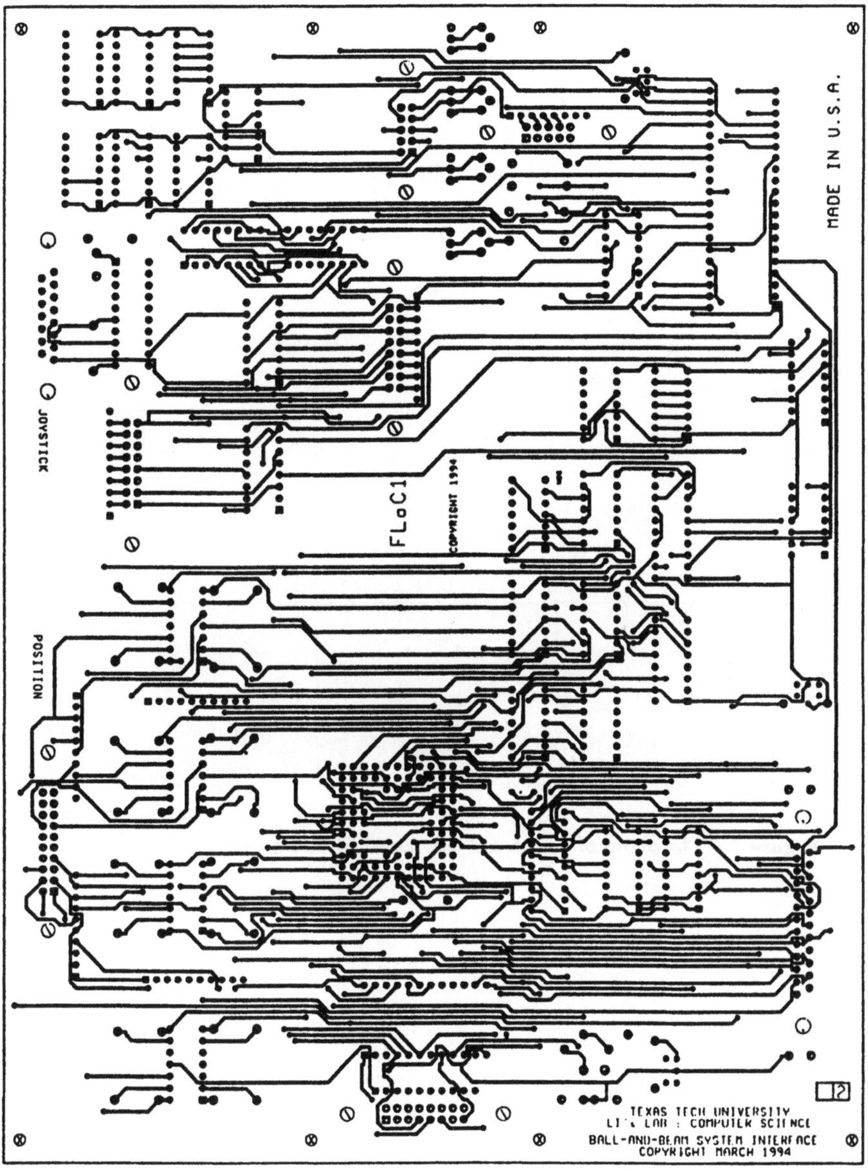

Figure 6(b) The solder side of the FLoC1 board layout.

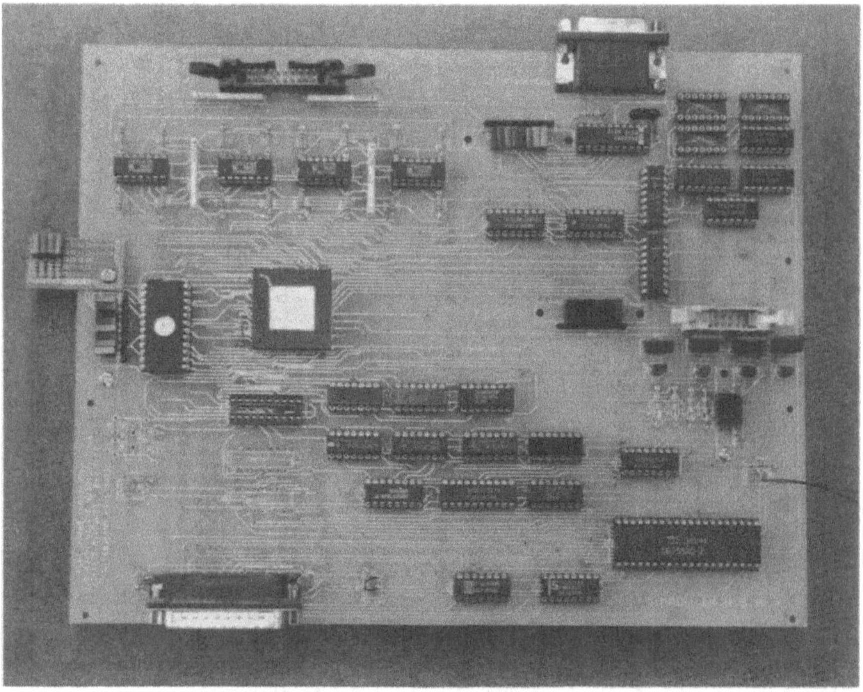

Figure 7 The photo of the assembled, tested FLoC1 board.

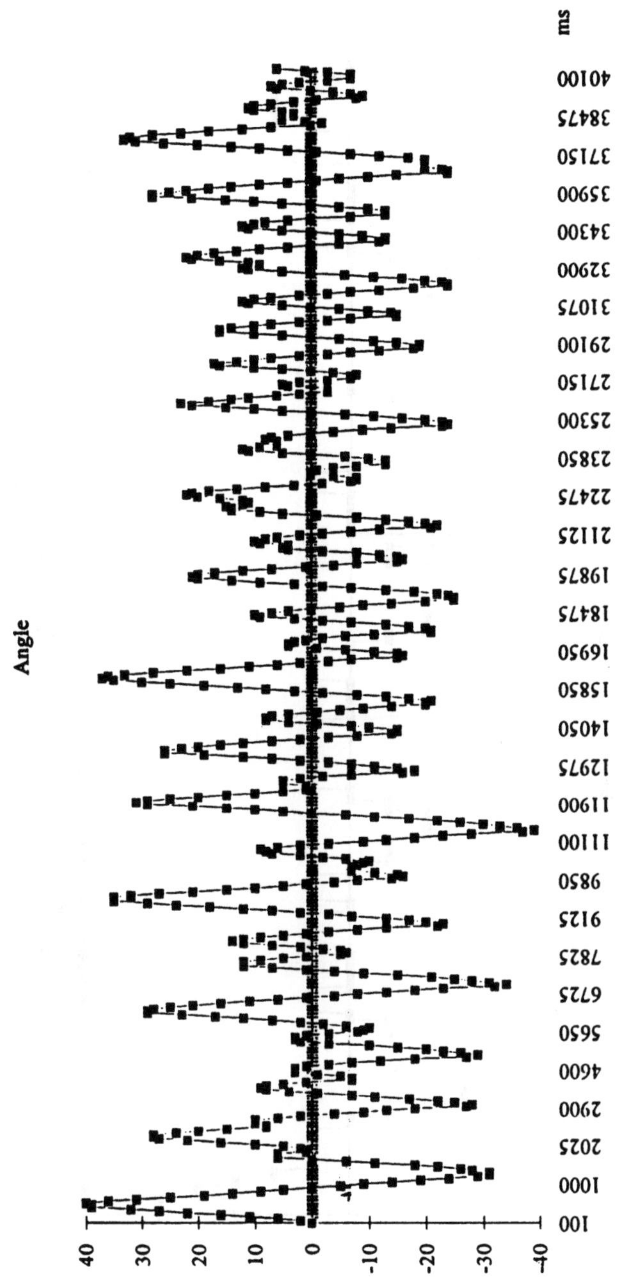

Figure 8 The manually controlled angle movement from a trained operator.

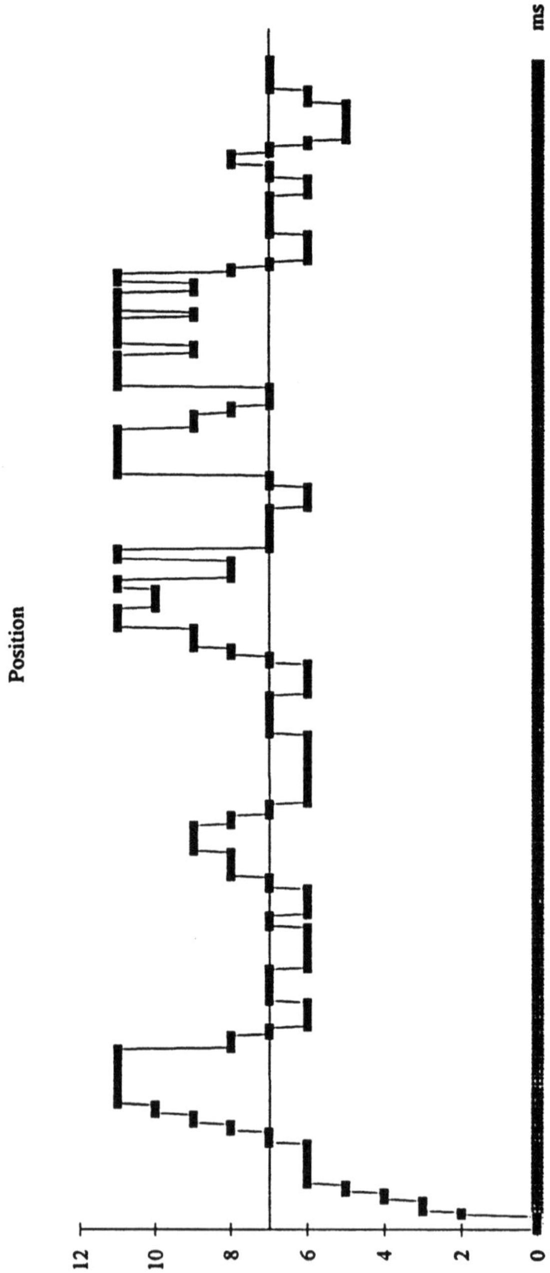

Figure 9 The manually controlled position measurement from a trained operator.

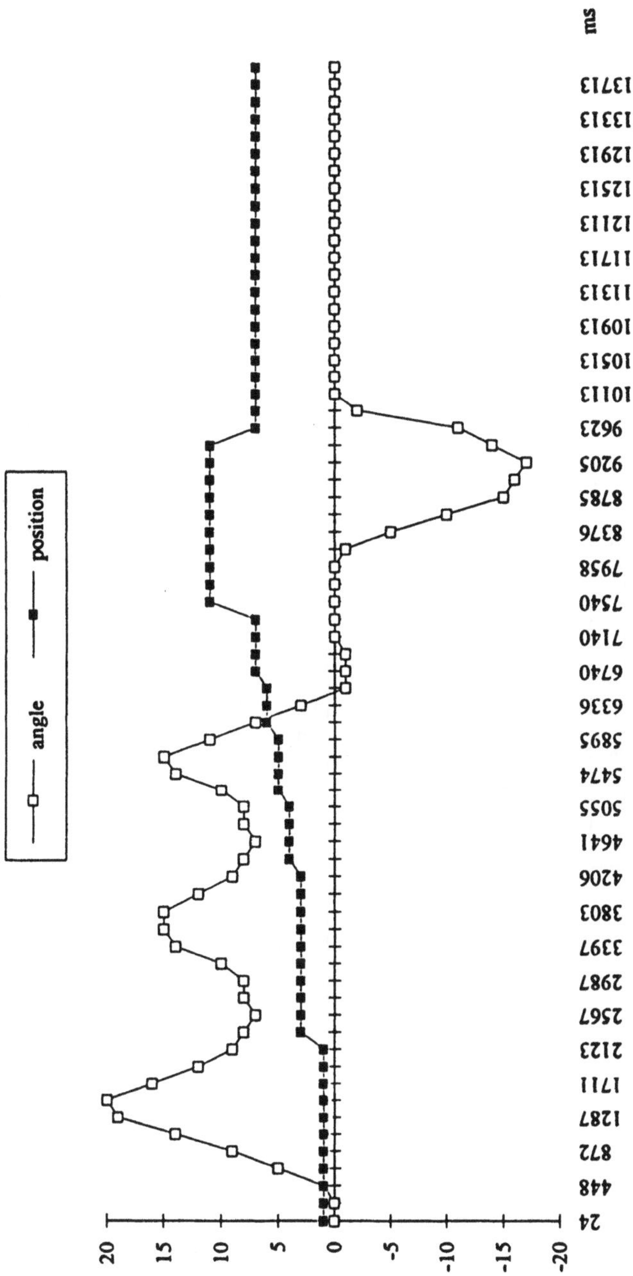

Figure 10 A typical fuzzy logic control result with ball centered.

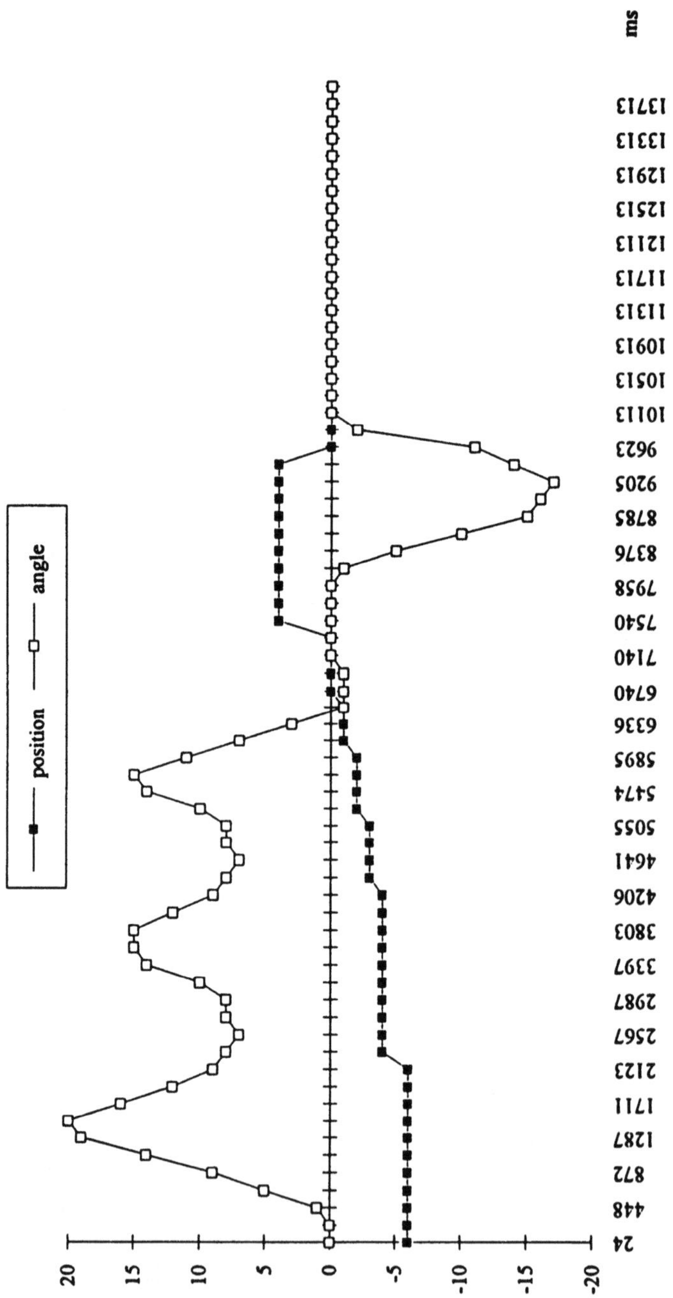

Figure 11 Another fuzzy logic control result where the beam is balanced while the ball is stopped at one end of the beam.

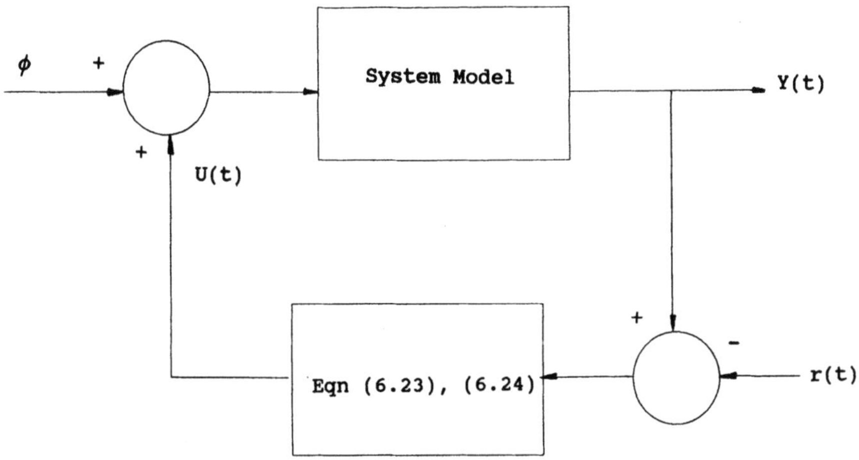

Figure 12 The block diagram of the system with designed optimal controller

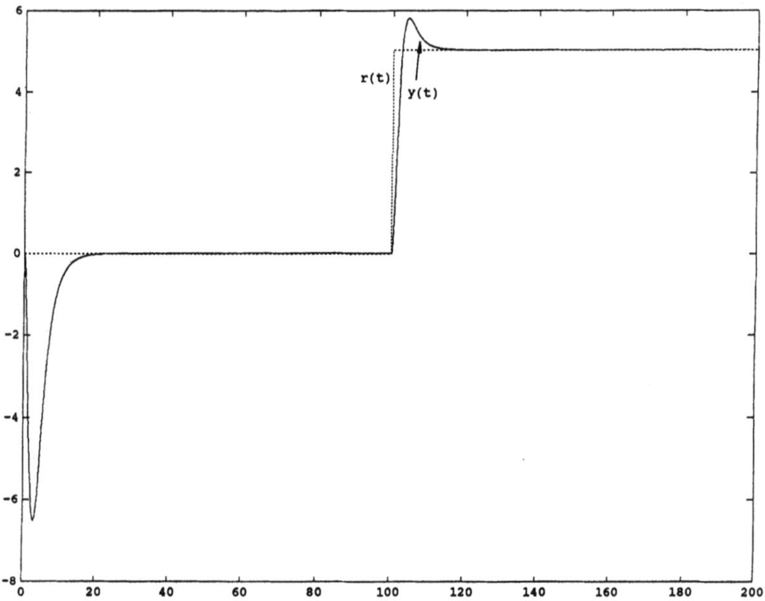

Figure 13 The ball position by the optimal controller (time scale in millisecond).

7

DESIGN OF FUZZY CONTROLLERS BASED ON FREQUENCY AND TRANSIENT CHARACTERISTICS

Kazuo Tanaka

Department of Mechanical Systems Engineering
Kanazawa University
2-40-20, Kodatsuno, Kanazawa 920, Japan

ABSTRACT

This chapter proposes two design procedures for fuzzy phase-lead compensator design based on frequency and transient characteristics. The main feature of fuzzy phase-lead compensator proposed in this chapter is its effective phase compensating characteristics. We derive two important theorems by introducing concept of frequency and transient characteristics. The first theorem is for judging whether a fuzzy phase-lead compensator should be used or not. The second is for realizing a phase-lead compensation. A design procedure (Design Procedure 1) of fuzzy phase-lead compensators for linear system is constructed using the theorem. Furthermore, it is extended to a design procedure (Design Procedure 2) for unknown or non-linear system. We apply Design Procedure 2 to tank level control problem which is a non-linear system with dead time. Simulation results show validity of these design procedures.

1 INTRODUCTION

Fuzzy control was first introduced in early 1970's by Mamdani [1]. However, at present we lack theoretical method for controller design although fuzzy control has been applied to many real industrial processes. Fuzzy control can be widely applied to more complicated, sensitive and dangerous processes such as nuclear plants if we have a systematic analysis and design tool. This chapter presents theoretical compensation methods based on frequency and transient characteristics in fuzzy control system.

The main purpose of controller design is to realize control objective such that transient characteristics, speed of response and damping characteristics are satisfied. The best way of realizing such a design is to introduce concepts of frequency characteristics, or gain crossover frequency and phase margin. Because transient characteristics are strongly related to frequency characteristics. It is known that gain crossover frequency is related to speed of response, and phase margin is related to damping characteristics.

Generally speaking, it is necessary to compensate phase characteristics of control systems in order to improve damping characteristics. But generally phase compensation is not easy. It has been said in the field of fuzzy control that a phase-lead compensation can be achieved if we use a coordinate transformation of e-$\dot{e}$ phase plane, exactly speaking, rotation of e-$\dot{e}$ phase plane [2]-[4], [6]. However, this chapter will show that the coordinate transformation does not aways realize an effective phase-lead compensation. We propose a new coordinate transformation for effective realization.

We derive two important theorems by introducing concepts of frequency and transient characteristics. The first is a theorem for judging whether a fuzzy phase-lead compensator should be used or not. The other is a theorem for realizing a phase-lead compensation. A design procedure (Design Procedure 1) of fuzzy phase-lead compensators for linear controlled objects is constructed using these theorems. Furthermore, it is extended to a design procedure (Design Procedure 2) for unknown or non-linear systems. We apply Design Procedure 2 to tank level control which is a non-linear system with dead time. Let us define symbols which will be used in this chapter.

$G(s)$: A transfer function of a controlled object.
$G_c(s)$: A transfer function of a linear PI controller $\frac{a+bs}{s}$.
$G_c{}^*(s)$: A transfer function of a linear PI controller $\frac{a+bs}{s}$.
ω_{CG}: A gain crossover frequency of open loop transfer function $G_c(s)G(s)$.
θ_m: A phase margin of open loop transfer function $G_c(s)G(s)$.
ω_{0CG}: A desired gain crossover frequency of open loop transfer function.
θ_{0m}: A desired phase margin of open loop transfer function.
ε_p: A overshoot of transient response of control system.
T_p: A time to peak of transient response of control system.
ε_{0p}: A overshoot of desired transient response of control system.
T_{0p}: A time to peak of desired transient response of control system.
$g(\omega)$: A gain in frequency ω of $G_c(s)G(s)$.

$\Psi(\omega)$: A phase in the frequency ω of $G_c(s)G(s)$.
$g^*(\omega)$: A gain in frequency ω of $G^*{}_c(s)G(s)$.
$\Psi^*(\omega)$: A phase in the frequency ω of $G^*{}_c(s)G(s)$.

2 FREQUENCY AND TRANSIENT CHARACTERISTICS

Figure 1 shows an example of Bode disgram. Generally speaking, ω_{CG} and θ_m are related to speed of response and damping characteristics, respectively .

Let us consider the following second order lag system.

$$G(s) = \frac{{\omega_n}^2}{s_2 + 2\zeta\omega_n s + {\omega_n}^2},$$

where ζ is damping ratio and ω_n is undamping natural frequency. It is known that ω_{CG} and θ_m can be represented by ζ and ω_n.

$$\theta_m = 90 - \tan^{-1}\sqrt{0.25\sqrt{4 + \frac{1}{\zeta^4}} - 0.5}, \qquad (7.1)$$

$$\omega_{CG} = \sqrt{\sqrt{4\zeta^4 + 1} - 2\zeta^2}\omega_n. \qquad (7.2)$$

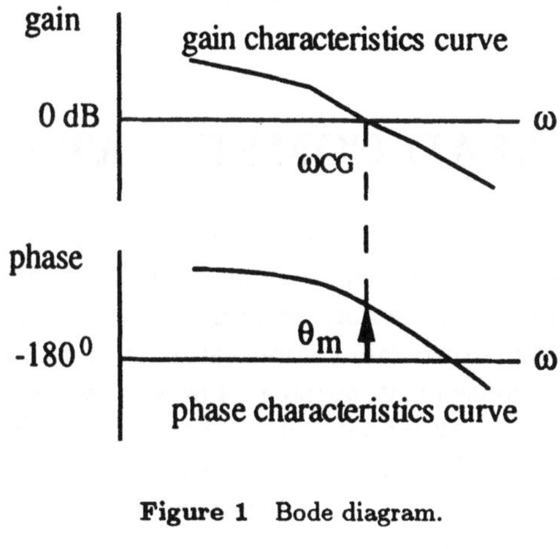

Figure 1 Bode diagram.

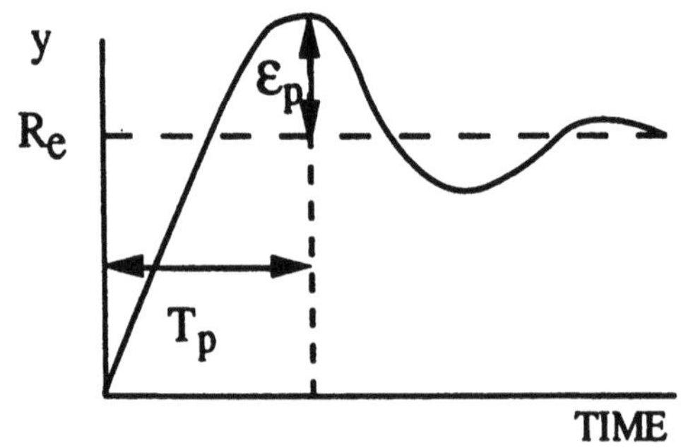

Figure 2 Transient response.

Figure 2 shows a transient response of a second order lag system. ζ and ω_n can be represented by ε_p and T_p as folows.

$$\zeta = \frac{-\frac{1}{\pi}\ln\frac{\varepsilon_p}{100}}{\sqrt{1+(-\frac{1}{\pi}\ln\frac{\varepsilon_p}{100})^2}} \tag{7.3}$$

$$\omega_n = \frac{\pi}{T_p\sqrt{1-\zeta^2}} \tag{7.4}$$

The transient characteristics of a high order lag system with a overshoot ε_p and a time to peak T_p can be approximated by a second order lag system with ζ and ω_n calculated by substituting ε_p and T_p into Eqs. (7.3) and (7.4).

3 PHASE-LEAD COMPENSATION

The easiest way of improving speed of response is to increase gain, that is, to perform gain compensation. When applying the ordinary fuzzy controller whose parameters are scaling factors for premise variables and a consequent variable, we can easily realize gain compensation by increasing the value of scaling factor for the consequent variable. However, overshoot is caused by increasing gain. To avoid overshoot, that is, to improving damping characteristics without losing speed of response, phase compensation is necessary. There is no clear relation between scaling factors of the ordinary fuzzy controller and phase characteristics. It is, therefore, difficult to effectively realize phase compensation in the ordinary fuzzy controller with the scaling factors .

Fujii [6] has shown a transformation matrix of realizing phase-lead compensation. We will show that the matrix does not always realize an effective phase-lead compensation. We derive an important theorem by introducing phase margin and gain crossover frequency in the frequency domain. This theorem gives a new transformation matrix which can realize an effective phase-lead compensation. Furthermore, we construct a fuzzy phase-lead compensator and propose two design procedures of the fuzzy phase-lead compensator [7]. The feature of the compensator is to have parameters for effectively improving phase characteristics.

3.1 Outline Of Fuzzy Phase-Lead Compensation

We derive two important theorems by introducing concepts of frequency and transient characteristics.

Theorem 1 Assume that a controlled object is represented by

$$G(s) = \frac{k}{s^2 + p_1 s + p_2}.$$

Furthermore, assume that ζ_0 and ω_{0n} are calculated by substituting ε_{0p} and T_{0p} into Eqs. (7.3) and (7.4) and that we use a linear PI controller

$$G_c(s) = \frac{a + bs}{s},$$

where

$$a = \frac{{\omega_{0n}}^2(p_1 - 2\zeta_0\omega_{0n})}{k}, \tag{7.5}$$

$$b = \frac{({\omega_{0n}}^2 + 2p_1\zeta_0\omega_{0n})}{k}. \tag{7.6}$$

if

$$\zeta_0\omega_{0n} < \frac{p_1}{2+h}, \tag{7.7}$$

where h is a safe coefficient and h > 1, then the control system (the closed loop transfer function $\frac{G_c(s)G(s)}{1+G_c(s)G(s)}$) can be approximated by the second order lag system

$$\frac{{\omega_n}^2}{s^2 + 2\zeta\omega_n s + {\omega_n}^2}.$$

The proof is given Appendix A. Here h must be theoretically a large number. It is, however, practically sufficient to have $h = 2 \sim 3$. Theorem 1 shows that it is sufficient to use a linear PI controller $G_c(s) = \frac{a+bs}{s}$, where

$$\begin{aligned} a &= \frac{{\omega_{0n}}^2(p_1 - 2\zeta_0\omega_{0n})}{k}, \\ b &= \frac{{\omega_{0n}}^2 + 2p_1\zeta_0\omega_{0n} - 4{\zeta_0}^2{\omega_{0n}}^2 - p_2}{k}, \end{aligned}$$

if Eq. (7.7) is satisfied. In other words, we do not need to design a fuzzy phase-lead compensator if Eq. (7.7) is satisfied. We should attempt to design a fuzzy phase-lead compensator only when Eq. (7.7) is not satisfied. However, we must notice that a desired transient response with ζ_{0p} and T_{0p} may be realized by a linear controller ever if Eq. (7.7) is not satisfied. Because, Eq. (7.7) of Theorem 1 gives a sufficient condition. It is, however, generally difficult to realize a desired transient response by a PI controller in this case.

Theorem 2 If we use

$$G^*{}_c(s) = \frac{a^* + b^*s}{s},$$

instead of

$$G_c(s) = \frac{a + bs}{s},$$

where

$$[a^* \quad b^*] = [a \quad b]\mathbf{T}(\theta_c, \omega_{CG}), \tag{7.8}$$

$$\mathbf{T}(\theta_c, \omega_{CG}) = \begin{bmatrix} \cos(-\theta_c) & -\frac{1}{\omega_{CG}}\sin(-\theta_c) \\ \omega_{CG}\sin(-\theta_c) & \cos(-\theta_c) \end{bmatrix}, \tag{7.9}$$

then the gain crossover and the phase margin of open loop transfer function of $G^*{}_c(s)G(s)$ become ω_{CG} and $\theta_m + \theta_c$, respectively.

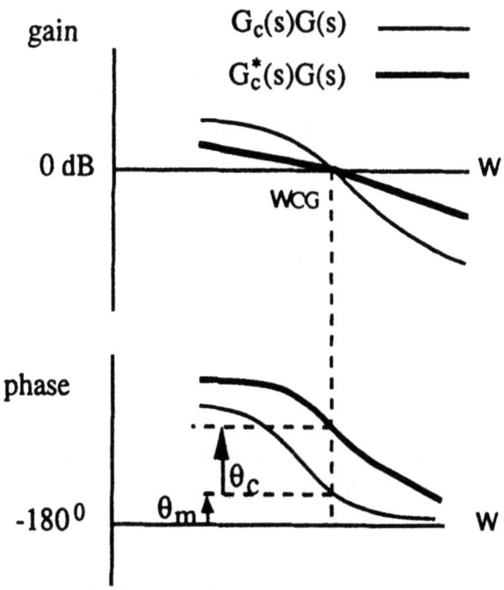

Figure 3 Phase-lead compensation.

The proof of this theorem is given in Appendix B. Figure 3 shows an example of Bode diagrams of $G_c(s)$ and $G^*{}_c(s)G(s)$.

It is found from Figure 3 that a phase-lead compensation with the lead angle θ_c can be realized without changing the gain crossover frequency ω_{CG}, that is,

$$\Psi^*(\omega_{CG}) = \Psi(\omega_{CG}) + \theta_c = \theta_m + \theta_c,$$

if we use

$$G^*{}_c(s) = \frac{a^* + b^* s}{s}.$$

instead of

$$G_c(s) = \frac{a + bs}{s}.$$

However, we should notice that the gain characteristics change in the frequency range except the gain crossover frequency ω_{CG}. Exactly speaking,

$$\begin{aligned} g^*(\omega) &< g(\omega), \quad \omega < \omega_{CG} \\ g^*(\omega) &= g(\omega), \quad \omega = \omega_{CG} \\ g^*(\omega) &> g(\omega). \quad \omega > \omega_{CG} \end{aligned}$$

As mentioned above, it is known that ω_{CG} and θ_m are related to speed of response and damping characteristics, respectively. The main purpose of phase-lead compensation is to design a control system such that damping characteristics are satisfied without losing speed of response, that is, to increase phase margin without changing ω_{CG}. By using the transformation matrix of Eq. (7.9), we can increase phase margin without changing ω_{CG}.

Fujii [6] has reported that a phase-lead compensation in fuzzy control systems can be realized if we use the following transformation matrix,

$$\begin{bmatrix} \cos(-\theta_c) & -\sin(-\theta_c) \\ \sin(-\theta_c) & \cos(-\theta_c) \end{bmatrix}. \tag{7.10}$$

A coordinate transformation achieved by the transformation matrix in Eq. (7.10) means rotation of $e - \dot{e}$ coordinate system. Eq. (7.10) is a special case of Eq. (7.9). In other words, Eq. (7.10) is equivalent to Eq. (7.9) when $\omega_{CG} = 1$, that is,

$$\mathbf{T}(\theta_c, 1) = \begin{bmatrix} \cos(-\theta_c) & -\sin(-\theta_c) \\ \sin(-\theta_c) & \cos(-\theta_c) \end{bmatrix}$$

The phase-lead compensation by Eq. (7.10) is useful only if ω_{CG} =1. However, required design performance with respect to speed of response is not always $\omega_{CG} = 1$. We should use the transformation matrix Eq. (7.9) instead of Eq. (7.10) in order to effectively realize phase-lead compensation.

We construct a fuzzy phase-lead compensator by introducting the transformation matrix of Eq. (7.9) shown in Therom 2.

Rule 1: IF Φ is about "$-\pi$, or 0, or π", THEN

$$\dot{u}_1 = [a \quad b] \begin{bmatrix} e \\ \dot{e} \end{bmatrix} \tag{7.11}$$

Rule 2: IF Φ is about "$-\frac{\pi}{2}$, or 0, or $\frac{\pi}{2}$", THEN

$$\dot{u}_2 = [a \quad b]\mathbf{T}(\theta_c, \omega_{CG}) \begin{bmatrix} e \\ \dot{e} \end{bmatrix} = [a^* \quad b^*] \begin{bmatrix} e \\ \dot{e} \end{bmatrix},$$

where $\Phi = \tan^{-1} \frac{e}{\dot{e}}$. The final output of this controller is calculated as

$$\dot{u} = \frac{w_1 \dot{u}_1 + w_2 \dot{u}_2}{w_1 + w_2}, \tag{7.12}$$

where w_1 is a membership value of the fuzzy set, about "$-\pi$ or 0 or π", of Rule 1 and w_2 is a membership value of the fuzzy set, about "$-\frac{\pi}{2}$ or $\frac{\pi}{2}$", of Rule 2. Figure 4 shows these fuzzy sets, where $\Delta\Phi$ is a premise parameter of these fuzzy sets and $-\frac{\pi}{2} < \Delta\Phi < \frac{\pi}{2}$.

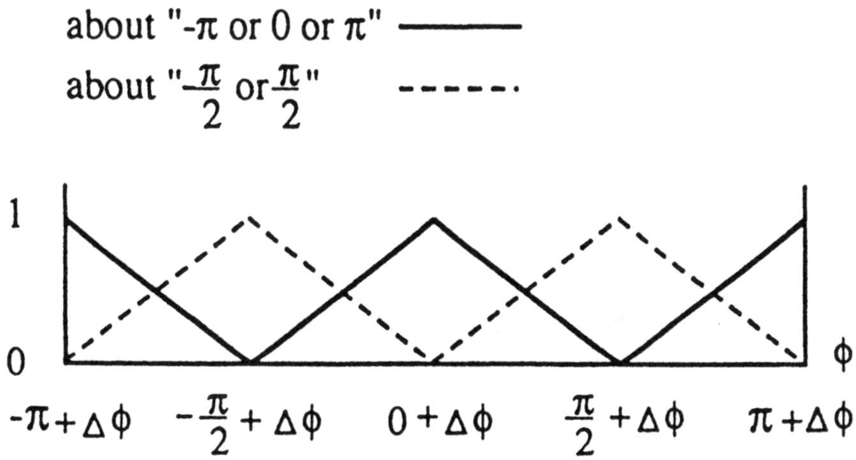

Figure 4 Fuzzy sets.

The parameters of the fuzzy compensator are a, b, ω_{CG} and θ_c in Eq. (7.11) and $\Delta\Phi$ in Figure 4. The controller design is to determine these parameters. Since Rule1 and Rule 2 are related to speed of response and damping characteristics respectively, we should determine a and b such that speed of response is satisfied, that is, $T_p \le T_{0p}$ and ω_{CG} and θ_c such that the damping characteristics are satisfied, where $\varepsilon_p \le \varepsilon_{0p}$. If $T_{0p} \le T_p$ or $\varepsilon_{0p} \le \varepsilon_p$ after determining a, b, ω_{CG} and θ_c, the value of $\Delta\Phi$ can be used as a fine adjustment parameter. If the value of $\Delta\Phi$ is decreased, speed of response is improved. Conversely, if the value of $\Delta\Phi$ is increased, damping characteristics is improved. Therefore, we should decrease it if $T_{0p} \le T_p$. We should increase it if $\varepsilon_{0p} \le \varepsilon_p$. Figure 5 shows fuzzy partitions for each value of the premise parameter $\Delta\Phi$. Each region such that membership value of each rule is more than 0.5 is shown. As shown in Figure 5, the regions by changing the premise parameter $\Delta\Phi$ can be adjusted.

The above points will be considered in design procedures. The design procedures will be given in Section 3.2 and 3.3 for linear, or non-linear systems.

3.2 Design Procedure 1 (The Case Of Linear Plants)

Assume the plant $G(s)$ is linear. The procedure consists of five steps .

[Step 1] Select a desired transient response of control system with a overshoot ε_{0p} and a time to peak T_{0p}. From the value of ε_{0p} and T_{0p}, calculate the values of ζ_0, ω_{0n}, θ_{0m} and ω_{0CG} by Eqs. (7.1) $\sim$ (7.4). From Theorem 1, judge whether a fuzzy phase-lead compensator should be used or not. If a fuzzy compensator should be used, then go to [Step 2].

[Step 2] Assume that $\theta_c = 0$. Then, since $a = a^*$ and $b = b^*$, the fuzzy controller is reduced to a linear controller $G_c(s) = \frac{a+bs}{s}$. The purpose of this step is to determine a and b in Rule 1. The speed of response is not satisfied if we use a and b calculated from Eqs. (7.5) and (7.6), because the value of a real root is close to those of real parts of complex roots. This means that speed of response is affected by the real root. So, calculate a and b by substituting the following $\zeta_{0'}$ instead of ζ_0 into Eq. (7.5) and (7.6).

$$\zeta_{0'} = \frac{p_1}{(3+h)\omega_{0n}}$$

The above equation can be derived by solving

$$\zeta_{0'}\omega_{0n} = \frac{p_1}{2+(1+h)}$$

for $\zeta_{0'}$.

[Step 3] If the linear controller satisfies damping characteristics, that is, $\varepsilon_p \leq \varepsilon_{0p}$, then end, else calcalate the phase margin θ_m and the gain crossover frequency ω_{CG} of $G_c(s)G(s)$. Next, derive a^* and b^* from the transformation matrix $\mathbf{T}(\theta_c, \omega_{CG})$ of Eq. (7.9), where

$$\theta_c = \theta_{0m} - \theta_m$$

[Step 4] Assume that $\Delta\Phi = 0$. Investigate whether $T_p \le T_{0p}$ and $\zeta_p \le \zeta_{0p}$ or not. If both of them are satisfied, then end, else go to [Step 5].

[Step 5] Adjust the premise parameter $\Delta\Phi$. Decreases the value of $\Delta\Phi$ if $T_p > T_{0p}$ and $\varepsilon_p \le \varepsilon_{0p}$. Conversely, increases it if $T_p \le T_{0p}$ and $\varepsilon_p > \varepsilon_{0p}$. If the desired transient response can not be realized by adjusting the premise parameter $\Delta\Phi$, then h = h+1 and go back to [Step 2].

3.3 Design Procedure 2 (The Case Of Non-Linear Plants)

Design Procedure 1 can not be applied to unknown or non-linear systems since frequency characteristics can not be used in this case. However, without loss of generality, Design Procedure 1 can be extended to a design procedure for unknown non-linear controlled objects, because there are clear relations between frequency characteristics and transient characteristics .

[Step 1] Selet a desired transient response of control system with a overshoot ε_{0p} and a time peak T_{op}.

[Step 2] Assume that $\theta_c = 0$. Then, since $a = a^*$ and $b = b^*$, the fuzzy controller is reduced to a linear controller $G_c(s) = \frac{a+bs}{s}$. Determine a and b in Rule 1 such that $T_p \le T_{0p}$.

[Step 3] If the linear controller satisfies damping characteristics, that is, $\varepsilon_p \le \varepsilon_{0p}$, then end, else go to [Step 4].

[Step 4] Assume that $\omega_{0CG} = 1$ and $\Delta\Phi = 0$. Increase gradually the value of θ_c of Rule 2 until $\varepsilon_p \le \varepsilon_{0p}$. This means that a^* and b^* of Rule 2 are adjusted. Investigate whether $T_p \le T_{0p}$ and $\zeta_p \le \zeta_{0p}$ or not. If both of them are satisfied, then end, else go to [Step 5].

[Step 5] Adjust the premise parameter $\Delta\Phi$. The value of $\Delta\Phi$ decreases if $T_p > T_{0p}$ and $\varepsilon_p \le \varepsilon_{0p}$. Conversely, it increases if $T_p \le T_{0p}$ and $\varepsilon_p > \varepsilon_{0p}$. If the desired transient response can not be realized by adjusting the premise parameter $\Delta\Phi$, select other value of a and b such that the value of T_p is smaller, where $\theta_c = 0$, that is, $a^* = a$ and $b^* = b$, and go back to [Step 3].

4 DESIGN EXAMPLES

We illustrate some examples of the above design methods.

4.1 Example 1

Let us consider the following plant

$$\ddot{y} = -10\dot{y} - 16y + 2.45u. \tag{7.13}$$

We design a fuzzy compensator by Design Procedure 1. Fujii [6] reported that a desired transient response is realized by a fuzzy controller. However, we show that the desired transient response is easily realized by a simple linear controller. In other words, it is sufficient to use a linear PI controller in this case.

In [6], ε_{0p} =6. [%] and T_{0p} = 1.2[sec.]. From Eqs. (7.1)~(7.4), we obtain

$$\zeta = 0.68, \qquad \omega_{0n} = 3.51,$$
$$\theta_{0m} = 63.34[deg],$$
$$\omega_{0CG} = 2.35[rad/sec].$$

Then,

$$\zeta_0\omega_{0n} = 2.39 < \frac{p_1}{2+h} = 2.5,$$

where h=2. Eq. (7.7) of Theorem 1 is satisfied. Therefore, we can realize the desired transient response by using a linear controller

$$G_c(s) = \frac{a+bs}{s},$$

where a = 25.62 and b= 8.36. Figure 7 shows the simulation result. It is found from Figure 7 that the *PI* controller realizes the desired transient characteristics.

4.2 Example 2

Let us consider the following plant.

$$\ddot{y} = -4\dot{y} - 4y + 3u. \tag{7.14}$$

We design a fuzzy compensator by Design Procedure 1.

[Step 1] We select a desired transient response as follows.

$$\varepsilon_{0p} = 5.0[\%] \text{ and } T_{0p} = 1.5[sec.].$$

From Eqs.(7.1) ~(7.4), we obtain that

$$\zeta_0 = 0.69, \qquad \omega_{0n} = 2.89,$$
$$\theta_{0m} = 64.63[deg],$$
$$\omega_{0CG} = 1.89[rad/sec].$$

Then,

$$\zeta_0\omega_{0n} = 1.85 > \frac{p_1}{2+h} = 1.0.$$

where h = 2. Eq. (7.7) of Theorem 1 is not satisfied. It is, therefore, difficult to realize the desired transient response by a linear PI controller.

[Step 2] We obtain $\zeta_0' = 0.277$. Therefore, a= 6.70 and b= 2.74.

[Step 3] $\omega_{0CG} = 1.89$,
$\theta_m = 39.94$ [deg].
$\theta_c = \theta_{0m}$ - θ_m = 64.63 - 39.94 = 24.69.
Therefore, $a^* = 3.92$ and $b^* = 3.97$.

[Step 4] ~ [Step 5] $T_p = 6.9$[sec.] and $\varepsilon_p = 0.02[\%]$ when $\Delta\Phi = 0$ [rad]. So, we adjust the premise parameter $\Delta\Phi$. $T_p = 1.5$[sec.] and $\varepsilon 0p = 4[\%]$ when $\Delta\Phi = \frac{-60\pi}{180}$[rad]. Figure 6 shows the simulation result.

4.3 Example 3

Now let us consider the following non-linear plant,

$$\ddot{y} = -4\dot{y} - 4y^2 + 3u(1 - \sin(0.1\pi\dot{u})). \tag{7.15}$$

We design a fuzzy compensator by Design Procedure 2.

[Setp 1] We select a desired transient response as, $\varepsilon_{0p} = 10[\%]$ and $T_{0p} = 2$[sec.]

[Step 2] $a := 2$ and $b = 0.5$.

[Step 3]~[Step 4] $\theta_c = 40$ [deg], therefore, $a^* = 1.21$ and $b^* = 1.67$.

[Step 5] $T_p = 2.0$ [sec.] and $\varepsilon_p = 10[\%]$ when $\Delta\Phi = 0$[rad]. Figure 9 shows the simulation result.

5 APPLICATION TO TANK LEVEL CONTROL

We apply Design Procedure 2 to tank level control which is a non-linear system with dead time. Figure 8 shows the simulation model of tank system, where h1 and h2 denote the level of tank 1 and tank 2, respectively. D1 and D2 denote the cross sectional areas of tank 1 and tank 2, respectively. A1 and A2 denote the cross section areas of each pipe, respectively. L denotes the dead time. It is assumed in the simulation that the plant model is unknown. The tank system can be described as follows.

$$\dot{h}_1(t) = \frac{q_1(t-L) - q_2(t)}{D_1}, \tag{7.16}$$

$$q_2(t) = A_1\sqrt{2gh_1(t)}, \tag{7.17}$$

$$\dot{h}_2(t) = \frac{q_2(t) - q_3(t)}{D_2}, \tag{7.18}$$

$$q_3(t) = A_2\sqrt{2gh_2(t)}. \tag{7.19}$$

Figure 9 shows fuzzy control system, where r is the setpoint,

$$K_1 = w_1 a + w_2 a^*, \tag{7.20}$$

$$K_2 = w_1 b + w_2 b^*. \tag{7.21}$$

Eqs. (7.20) and (7.21) can be easily derived from Eqs. (7.11) and (7.12). The values of K_1 and K_2 change corresponding to the values of e and $\dot{e}$ because w_1 and w_2 depend on e and $\dot{e}$.

The control purpose is to keep h_2 at a constant value of setpoint. Figure 10 shows the control result of the fuzzy phase-lead compensator designed by Design Procedure 2 with L = 0.

The controller parameters are obtained as follows,

$$a = 0.4, b = 0.2, \theta_c = 60[deg], \Delta\Phi = 60[deg].$$

As motioned above, θ_c is a parameter for improving the phase characteristics. Figure 11 shows the effect of the parameter θ_c, where $a = 0.4$, $b = 0.2$, $\Delta\Phi = 0$. It can be found that we can improve the damping characteristics by increasing the valye of θ_c. This means that the phase margin can be increased by increasing the value of θ_c.

Figure 12 shows the effect of the parameter $\Delta\Phi$, where a= 0.4, $b = 0.2$, and $\theta_c = 20$. Figure 13 shows the effect of the parameter $\Delta\Phi$, where $a = 0.4$, $b = 0.2$, and $\theta_c = 60$. It can be seen from Figures 12 and 13 that the parameter $\Delta\Phi$ has the same effect as the parameter θ_c when the value of θ_c is not small. We should notice that the fuzzy phase-lead compensator is reduced to a linear controller when $\theta_c = 0$. Therefore, there is no effect of the parameter $\Delta\Phi$ only when $\theta_c = 0$.

Next, let us consider the case of L > 0. It is necessary to compensate the dead time in the case of L > 0. We show that the dead time can be compensated by the parameter $\Delta\Phi$. Figure 14 shows the effect of the parameter $\Delta\Phi$, where $a = 0.4$, $b = 0.2$, and $\theta_c = 60$. We can say that the parameter $\Delta\Phi$ has the effect of compensating the dead time. Figure 15 shows the control result when $a = 0.4$, $b = 0.2$, $\theta_c = 60$ [deg], $\Delta\Phi = 90$ [deg.].

6 CONCLUSION

The compensation methods of fuzzy control systems based on frequency and transient characteristics have been discussed. We have derived two important theorems. One is a theorem for judging whether a fuzzy phase-lead compensator should be used or not. The other is a theorem for realizing a phase-lead compensation in fuzzy control systems. A design procedure of fuzzy phase-lead compensators for linear controlled objects has been proposed using these theorems. Moreover, it has been extended to a design procedure for unknown or non-linear controlled objects. We applied one of the design procedures to tank level control which is a non-linear system with dead time. Simulation results have shown validity of these design procedures.

REFERENCES

[1] E.H. Mamdani, "Applications of fuzzy algorithm for control of simple dynamic plant," Proc. IEE, Vol. 121, No. 12, pp. 1585 - 1588 (1974).

[2] M. Sugeno, "Fuzzy control," Nikkankogypu Publ. Co., (1988).

[3] K. Tanaka and M. Sano, "A new tuning method of fuzzy control," Proceeding of 4th IFSA World Congress, Vol. 1, pp. 207-210, (1991).

[4] A. Ishigame et. al., "Design of electric power system stabilizer based on fuzzy control theory," IEEE International Conference on Fuzzy Systems, pp. 973-980, (1992).

[5] K.Tanaka, M. Sano and T. Takema, "A phase-lead compensation of fuzzy control systems," Journal of Japan Society for Fuzzy Theory and Systems, Vol. 4, No. 3, pp. 163-170, (in Japanese).

[6] A. Fujii, T. Ueyama and N. Yoshitani, "Design of fuzzy controller using frequency response," Proceeding of 5th Fuzzy System Symposium, pp. 115-120, (1989)(in Japanese).

[7] K. Tanaka and M. Sano, "Design of fuzzy controller based on frequency and transient characteristics," Proceeding of 2th IEEE International Conference on Fuzzy Systems. Vol. 1, pp. 111-116, (1993).

[8] K. Tanaka and M. Sano, "Analysis and design of fuzzy controller in frequency domain," Proceeding of IEEE International Conference on IECON'93, (1993)(To appear).

Appendix A: The Proof Of Theorem 1

From $G(s)$ and $G_c(s)$, we obtain

$$1 + G_c(s)G(s) = s^3 + p_1 s^2 + (p_2 + kb)s + ka. \tag{7.22}$$

Assume that the characteristics equation of control system is represented as

$$(s + m)(s^2 + 2\zeta_0 \omega_{0n} s + {\omega_{0n}}^2). \tag{7.23}$$

The characteristics roots of Eq. (A.2) are

$$\begin{aligned} s_1 &= -m \\ s_2 &= -\zeta_0\omega_{0n} + \omega_{0n}\sqrt{1 - {\zeta_0}^2}, \\ s_3 &= -\zeta_0\omega_{0n} - \omega_{0n}\sqrt{1 - {\zeta_0}^2}. \end{aligned}$$

By substituting Eqs. (7.5) and (7.6) into Eq. (A.1) and by comparing Eq. (A.1) with Eq. (A.2), we can derive

$$s_1 = -m = 2\zeta_0\omega_{0n} - p_1.$$

In order that the control system can be approximately by second order lag system

$$\frac{{\omega_n}^2}{s^2 + 2\zeta\omega_n s + {\omega_n}^2}$$

that is, that the influence of the root of s_1 can be ignored, the real value of s_1 must be enough smaller than those of s_2 and s_3. Therefore,

$$2\zeta_0\omega_{0n} - p_1 < -h\zeta_0\omega_{0n},$$

where h is a big position number (h > 1). By solving the above inequality for $\zeta_0\omega_{0n}$, we can derive

$$\zeta_0\omega_{0n} < \frac{p_1}{2 + h}.$$

Appendix B: The Proof Of Theorem 2

Assume that $g_c(\omega)$ and $g^*{}_c(\omega)$ are gains of $G_c(s)$ and $G^*{}_c(s)$ in the frequency ω, respectively. Moreover, assume that $\Psi_c(\omega)$ and $\Psi^*{}_c(\omega)$ are phase of $G_c(s)$ and $G^*{}_c(s)$ in the frequency ω, respectively. Then, we obtain

$$\begin{aligned} g_c(\omega) &= 20\log\sqrt{a^2+b^2\omega^2}, \\ g^*{}_c(\omega) &= 20\log\sqrt{a^{*2}+b^{*2}\omega^2}, \\ \Psi_c\omega &= \tan^{-1}\frac{b\omega}{a}, \\ \Psi_c{}^*\omega &= \tan^{-1}\frac{b^*\omega}{a^*}. \end{aligned}$$

From $g_c(\omega_{CG}) = g^*{}_c(\omega_{CG})$,

$$a^{*2} + b^{*2}\omega_{CG} = a^2 + b^2\omega_{CG}. \tag{7.24}$$

From $\Psi_c{}^*(\omega_{CG}) = \Psi_c{}^*(\omega_{CG}) + \omega_c$,

$$b^* = \frac{a^*}{\omega_{CG}}\tan\Psi_c{}^*(\omega). \tag{7.25}$$

By substituting Eq. (B.2) into Eq. (B.1), we obtain

$$a^* = a\cos(-\theta_c) + \omega_{CG} b\sin(-\theta_c).$$

Moreover, we obtain

$$b^* = b\cos(-\theta_c) - \frac{a}{\omega_{CG}}\sin(-\theta_c).$$

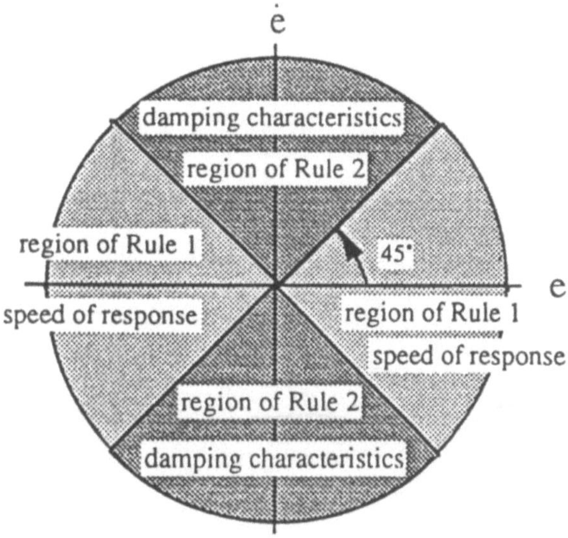

$\Delta\phi = 0$ [rad]

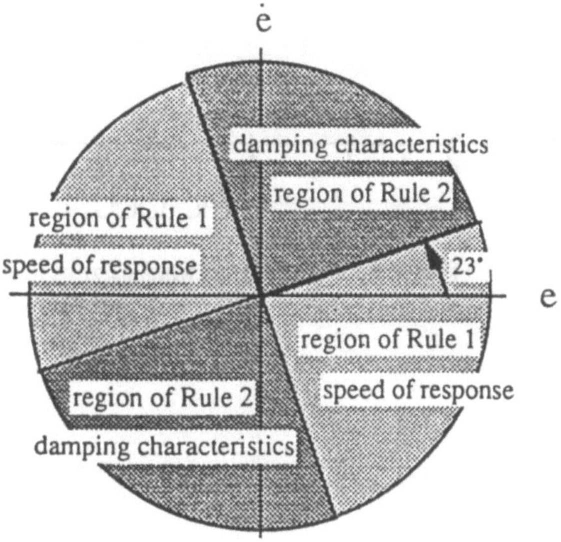

$\Delta\phi = -22\pi/180$ [rad]

Figure 5 Fuzzy partition.

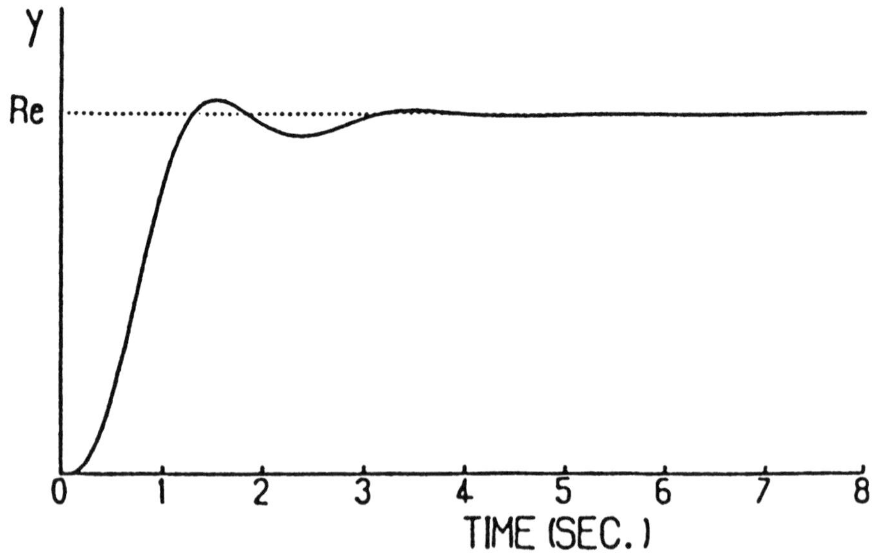

Figure 6 Control result (Example 2).

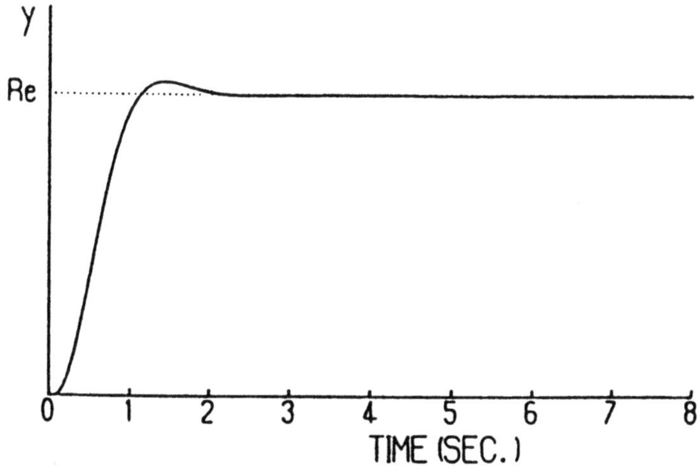

Control result. (Example 1)

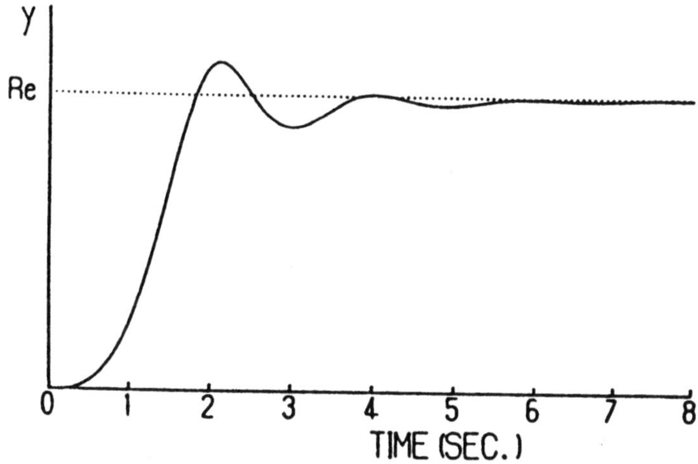

Control result. (Example 3)

Figure 7 Control results (Example 1 and 3)

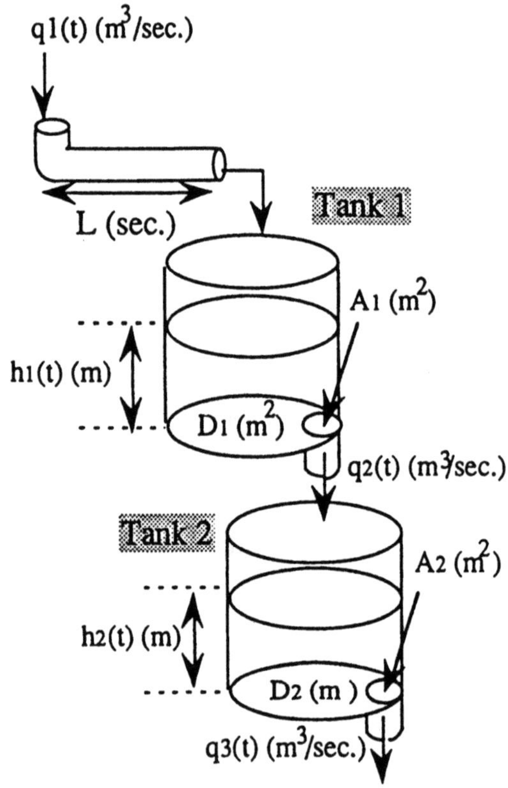

Figure 8 Two tank system.

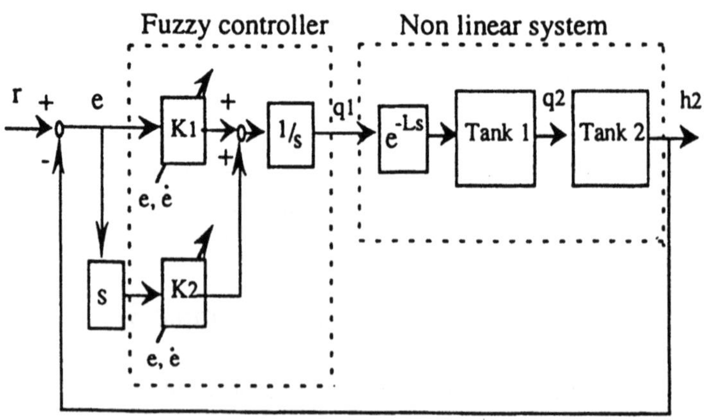

Figure 9 Fuzzy control system.

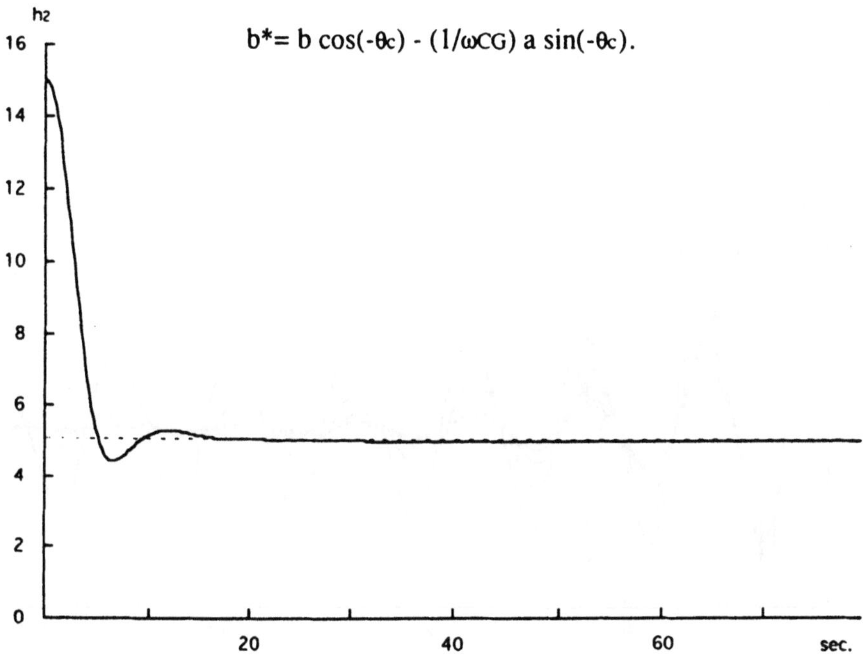

Figure 10 Control result (L = 0).

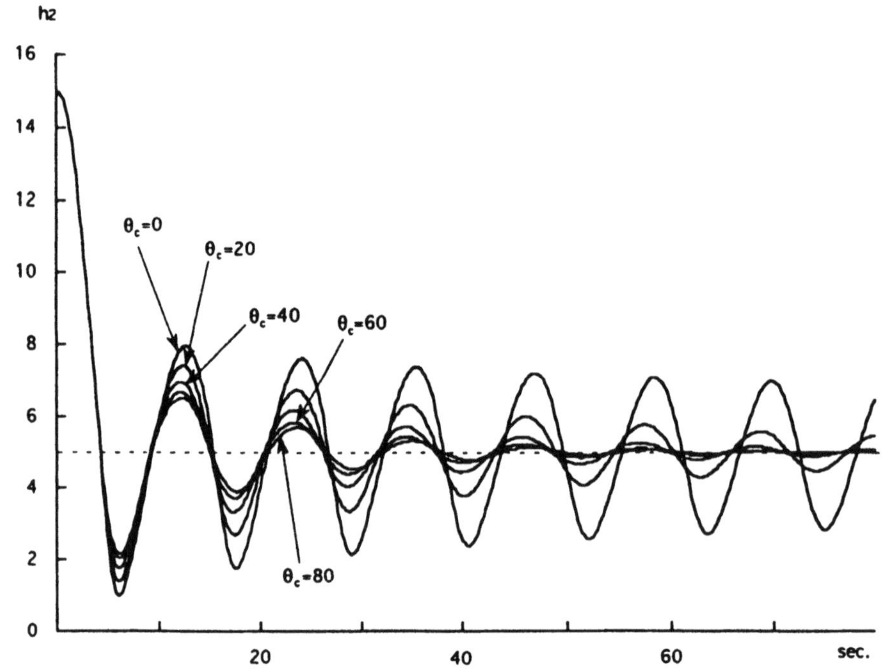

Figure 11 Effect of the parameter ω_c.

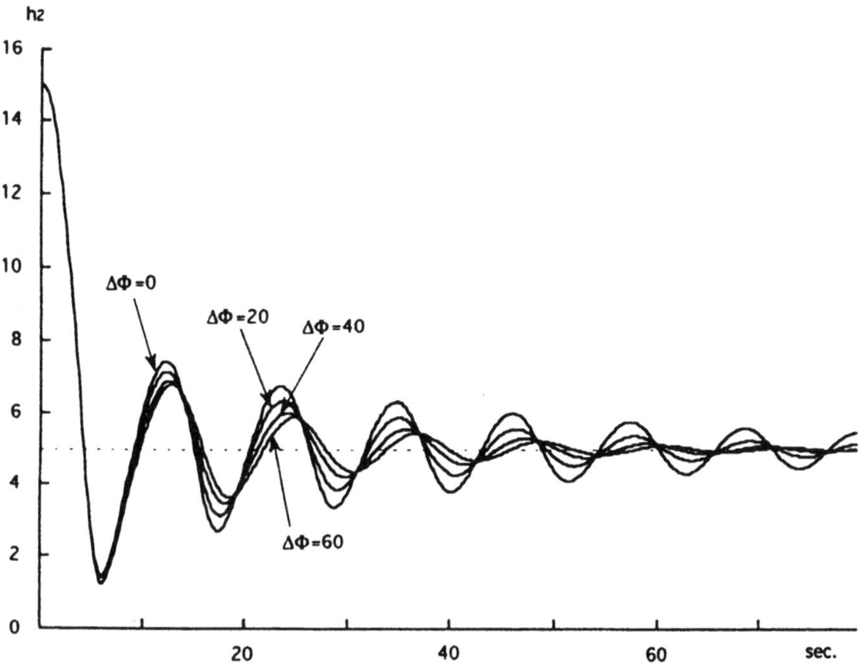

Figure 12 Effect of the parameter $\Delta\Phi$ ($\theta_c = 20$).

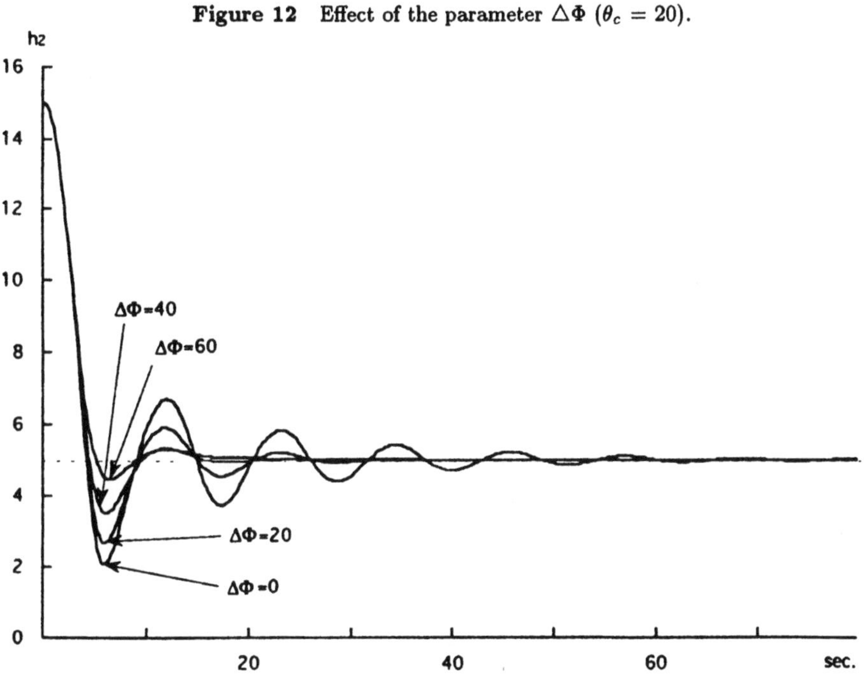

Figure 13 Effect of the parameter $\Delta\Phi$ ($\theta_c = 60$).

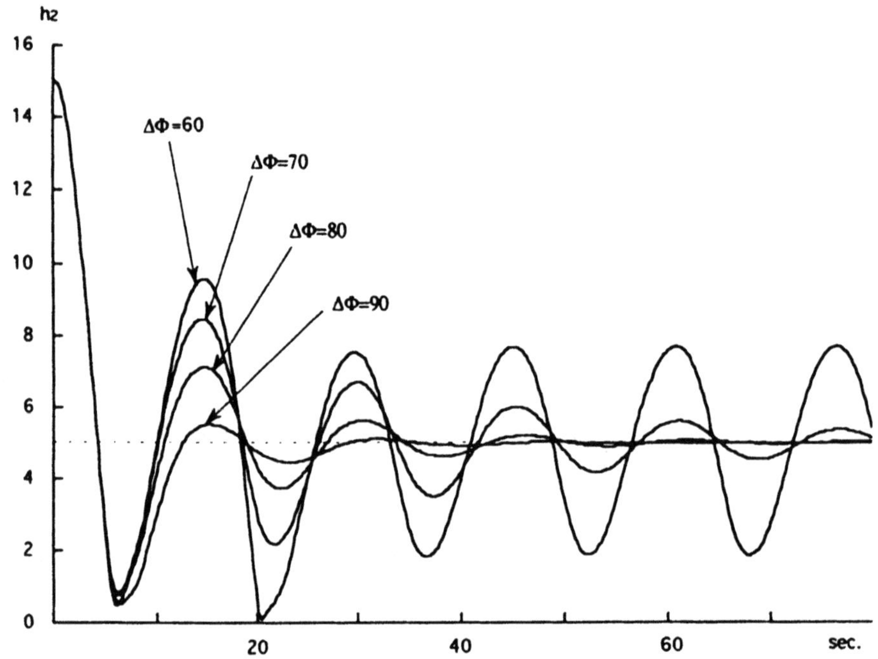

Figure 14 Effect of the parameter $\triangle\Phi$ for dead time.

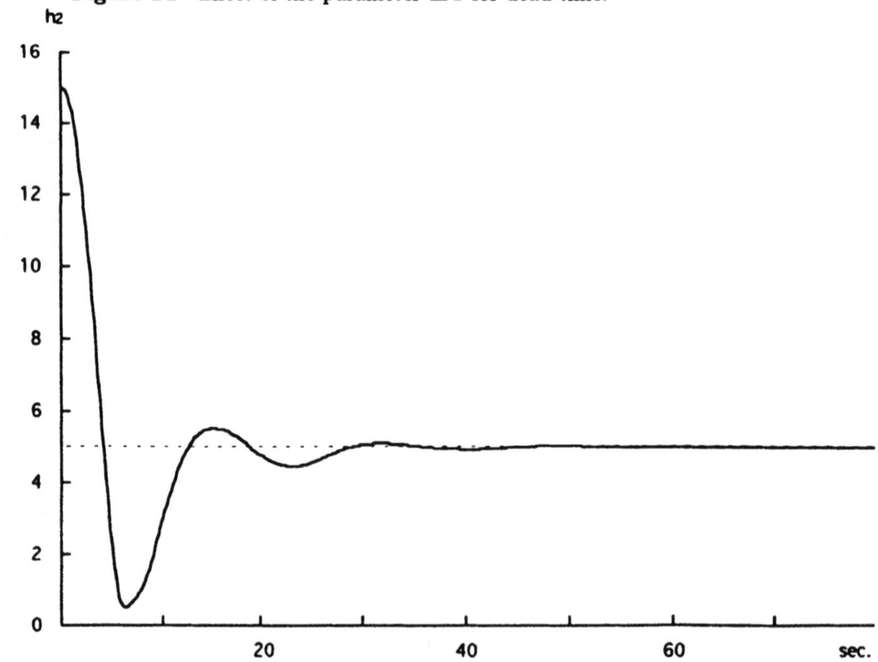

Figure 15 Control result (L = 3).

8

FUZZY INFERENCE INTEGRATING 3D MEASURING SYSTEM WITH ADAPTIVE SENSING STRATEGY

Koji Shimojima, Toshio Fukuda, Fumihito Arai, and Hideo Matsuura

Department of Mechano-Informatics and Systems, Nagoya University
Furo-cho, Chikusa-ku, Hagoya 464-01, Japan

ABSTRACT

This chapter deals with a 3-D measurement system applied to a curved metal surface carving system, and a sensor integration method based on fuzzy inference and adaptive sensing strategy. The measurement system consists of two different sensors. One is a LED displacement sensor, while the other is a vision system. The LED displacement sensor's spot-light is used as a part of the vision system based on the active stereo sensing method. In addition, the LED displacement sensor's outputs are used for calibrating camera parameters. Therefore, we can calibrate the camera parameters easily. Then, we use neural networks to compensate the output of the image processing for some errors, such as camera parameter's error and lens distortion. By utilizing the neural networks, we can use a vision system accurately. We use a sensor integration method based on the fuzzy set theory. Fuzzy inference's input consists of information on the change in the sensor output and the position change of the sensor system, together with the environmental data of measurement. Vision system can be used under various environmental condition, but it takes long time. The LED displacement sensor can obtain the information quickly, but it can only be used in limited environmental condition. In order to measure the object quickly and accurately, the sensor integration system has an adaptive sensing strategy. The sensing strategy depends on a relation between the state of the object, e.g. color, temperature, etc., and the sensor specification. The proposed system is shown to be effective through extensive experiments.

1 INTRODUCTION

Most of the present automated manufacturing systems need the CAD data or force-torque sensory data to carve work pieces [1], [2]. Since the CAD data do not express the profile of the used work piece, therefore the systems using the CAD data cannot carve it well, while the systems using the force-torque sensory data lacks the capability of recognizing the desirable shape from the sensory input. In addition, it is very difficult to analyze force or torque accurately, which can not make carving precisely.

To solve these problems, we proposed an automated carving system with an integrated measurement system [3]. The system measured each target's surface shapes and made the carving path from the measurement data. We also proposed a sensor integration system for the measuring system [4]. In this measuring system, we used some eddy current sensors. This system had two problems. One was that this system could not measure the rough surface because the sensors had to be very close to the target surface due to the sensor's characteristics. The other was that the system could not recognize the position on the target surface where the system was measuring. Recognition of the measuring position is very important for precise carving because the carving path is made from the measuring path; therefore, the measuring path and the carving path are closely linked together.

For these problems, we proposed a measurement system and its integration method [5]. The system consists of two LED displacement sensors (this sensor can measure displacement at a distance of 40 mm) and two CCD cameras (for the measuring point's recognition on the target surface). The LED displacement sensor is not only used as a displacement sensor but also as a marker of the measuring point for the vision system. Therefore, the measurement system can measure the surface shape of the target at a distance and recognize the measuring point.

The vision system consists of a CCD camera and a LED displacement sensor, with which the active stereo method is applied. This vision system has two characteristics. One is that the calibration of the camera parameters is very easy because the system calibrates the camera parameters by the measurement data of the LED displacement sensor. The other is that a neural network (NN) is utilized as a compensator of the vision system's output errors, such as camera parameters' error and lens distortion. By utilizing the neural network, we can use the vision system as accurately as possible.

Improving the accuracy of detecting, measuring, and inspecting the product, the sensory system would have multi various sensors, as well as realizing the measuring the complicated product that cannot be measured by the single sensor.

In order to manage multi various sensory data, the sensor integration/fusion system [6], [31], [37] has been studied. Multiple sensory systems have a major problem: how to integrate/fuse sensory data to produce the more reliable and accurate information. Various sensor integration/fusion methods have been reported so far: hypothesis testing by sensor models and Bayesian approach [7], multiple hypothesis approach [24], confidence distance matrix [Luo 1988], estimation by performance and cost criteria [9], estimation of cost by Bayesian method [10], probabilistic approach [Sabater 1991], Bayes-maximum entropy method, energy-minimize curve method [22], using multi-input hidden Markov model [18], using Kalman filter [11], [25]-[28], [32], [34], [36], [43], Kalman filter with dummy parameters [35], intentional observation with information criterion [47], multi-agent approach with logical sensor [22], [30], using recursive term utility function [19], using maximum likelihood method [50], weighting scheme based on the information theory [20], using neural networks [38]-[40], [48], using mappings between sensor [41], using fault tolerance network and the Subsumption Architecture [29], using knowledge database [45], using hierarchical structure [44], using the physical network [49], a multicriteria approach [21], the state-based sensor [42], using the tree data-structure [33], and a fuzzy logic [12], [17].

However these approach have some problems. Mathematical approach, e.g. Bayesian, Kalman filter, e.t.c are difficult to use in unstable and unknown environments, neural network approach are difficult to change sensor units.

In the previous system, we used a SIS using the fuzzy inference to integrate the measurement data. These SISs evaluated the suitability for the environment of each sensor, e.g., the intensity of a spot-light's reflection for the LED displacement sensor, and then used the appropriate sensor for sensors' environments. This SIS, however, could not evaluate internal conditions of sensors, e.g., the sensitivity for the intensity of the spot-light's reflection for the LED displacement sensor and camera parameters for the vision system.

The measurement system proposed here is mounted at the tip of the manipulator. Then changes in outputs of a sensor are caused by the position changes of the manipulator. Therefore, SIS's inputs consist of information on both the change in sensor outputs and the changes of the manipulator's position, together with the environmental data of a sensor. Then the SIS can evaluate both the suitability for sensors' environments and the internal condition of a sensor in order to integrate the measurement data.

The vision system can obtain various information and be affected a little by the change of sensor's environments, but the system takes much time to measure. On the contrary, the LED displacement sensor takes a short time to measure a distance, but the sensor can only obtain the distance information and is affected by the changing of the environments. In order to use the sensory system effectively (speedy and accurately), the sensory system has the adaptive sensing strategy with the SIS. The sensing strategy is depends on the output of the SIS for each sensor as a suitability for sensing an object.

For this sensor integration system with the adaptive sensing strategy, the measurement system can measure the object as accurately as possible under any sensors' environments and internal conditions. The effect of this SIS is shown through extensive experiments.

2 3-D MEASUREMENT SYSTEM

In order to measure the rough surface in safety, measuring must carry out at a distance from the object. The automated manufacturing system based on the the measuring system like our proposed system, recognition of the height of the measurement system along the z axis (see Figure 1) is very important for precise carving, because the height of the measurement system along the z axis in measuring is that of the air cutter along the z axis in carving.

For these problems, we propose a measurement system, which can measure displacement from a distance and recognize the point where the system measures. The fundamental measurement system consists of a set of a CCD camera and a LED displacement sensor. In this chapter, we use a pair of the measurement system in order to measure the angle around the y axis. Figure 1 shows the relation among sensors and the object in the operational space. The specifications of the measurement system are as follows:

LED displacement sensor:	measurable region	35mm to 45mm
	measurement accuracy	20 μm
CCD camera 1:	the focal distance	7.5mm
CCD camera 2:	the focal distance	15mm
Image processing board:	512 pixel x 512 pixel	256 gray scale

For the combination of these sensors, the measurement system can measure the rough surface shape and recognize the height of the measurement system along the z axis accurately.

Accuracy of the LED displacement sensor is greatly effected by a reflection of an object surface. Accuracy of the vision system does not depend on the reflection of object surface but the shape of the LED displacement sensor's spot-light. In order to integrate these sensor outputs in consideration of each sensors' specification, we use a sensor integration system based on the fuzzy inference. Figure 2 shows the outline of the sensor integration system.

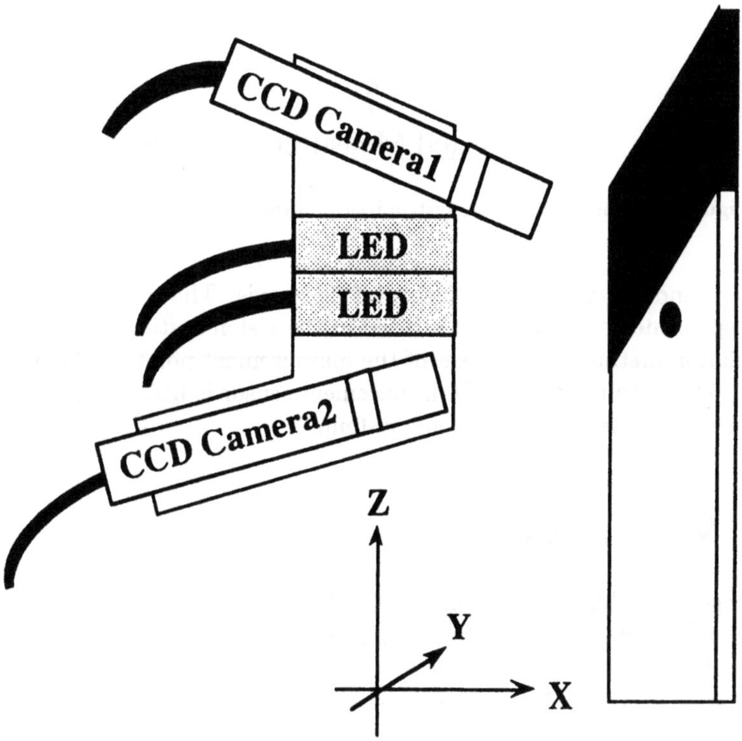

Figure 1 Distribution of sensors and coordinate system.

2.1 LED Displacement Sensor

The measurement system has 2 LED displacement sensors. The sensors measure distance to the object surface along the x axis and the angle between

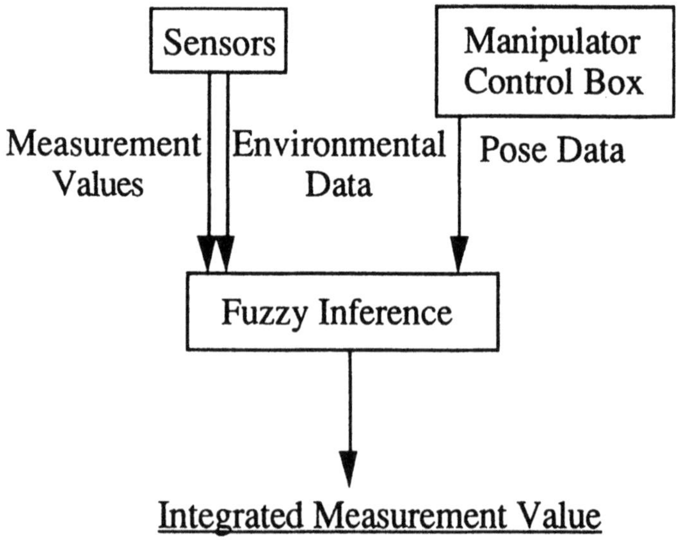

Figure 2 Sensor integration system based on the fuzzy inference.

the sensor and the object surface around the y axis. The spot-light of the LED displacement sensor is also utilized in the vision system based on the triangulation method as a marker of the measurement point on the object surface. The output of the LED displacement sensor is utilized for calibrating the camera parameters of the vision system.

2.2 Vision System

The image data obtained through the CCD camera is utilized for measuring distance to the object surface both along the x axis and the z axis. A large number of studies have been made on the vision system based on the triangulation method. These vision systems are classified into two types, such as a passive stereo vision and an active stereo vision. The passive stereo method [13], e.g., the binocular vision and the photometric stereo, has the stereo matching problem. Accuracy of the measurement is related with that of the correspondences among different images. Furthermore, it takes much time to compute the distance. The active stereo method, e.g., using spot-light, slit-light and pattern-light [14]-[15], can compute the distance from a single image, but this algorithm needs an artificial source of light such as a laser.

In this chapter, considering the image processing speed and accuracy in measurement, we apply the active stereo method to the 3D measurement system where the source of light is the LED displacement sensor. The position of the sensor system along the z axis is determined by recognition of both spot-lights' positions and the border line between the carving part and the no-carving part on the object. Figure 3 shows the spot-light's detection flow chart and Figure 4 shows the border line detection flow chart.

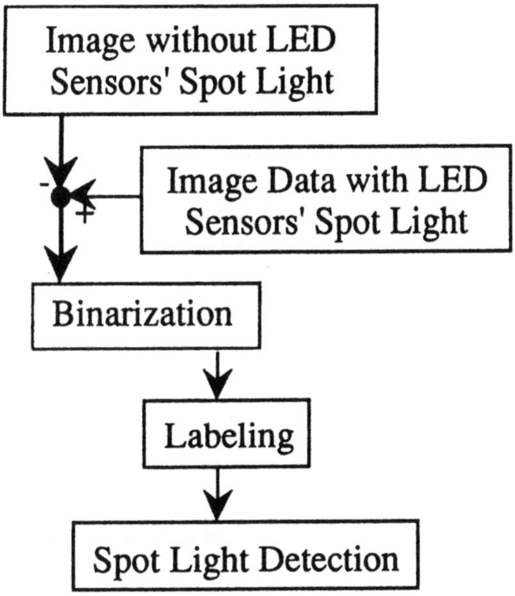

Figure 3 Flow chart of the spot-light detection.

Active Stereo Method

Distance between the sensor and the object is measured from the spot-light's position of the LED displacement sensor based on the triangulation method (see Figures 5 and 6). Distances along the x axis (L_i) and the z axis (Z_i) between the sensor and the object are calculated as follows:

$$L_i = Z_0 \tan(q_0 + q_i), \tag{8.1}$$

$$Z_i = \frac{L_i}{\tan(q_0 + q_i)}, \tag{8.2}$$

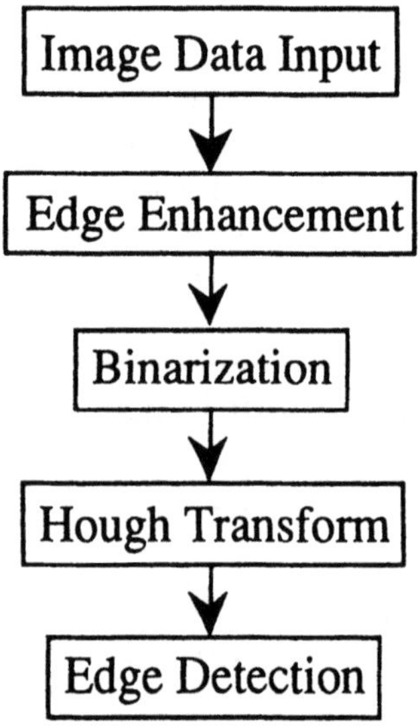

Figure 4 Flow chart of the border line detection.

$$\tan(q_i) = \frac{dF_x F_{x_i}}{f_i}. \tag{8.3}$$

where F_{x_i}, f_i, dF_x, q_0, and Z_0 means the gravity center position of the spot-light, the focal distance of the CCD camera, a size of a pixel, the angle and the distance between the CCD camera and the LED displacement sensor respectively.

Camera Calibration

In order to use the vision system accurately, camera parameters have to be calibrated accurately. In this vision system, the source of light is the LED displacement sensor, and outputs of the LED displacement sensor are utilized for the camera calibration data. Therefore, camera calibration is very easy and accurate. In this system, we calibrate q_0, Z_0, and dF_x.

The sensor system is mounted at the tip of the manipulator. Data for the camera calibration is obtained by moving the manipulator. Data consist of both outputs of the LED displacement sensor and the gravity center position of the spot-light at each measurement point. We defined the camera parameters, which make the total error minimum between LED displacement

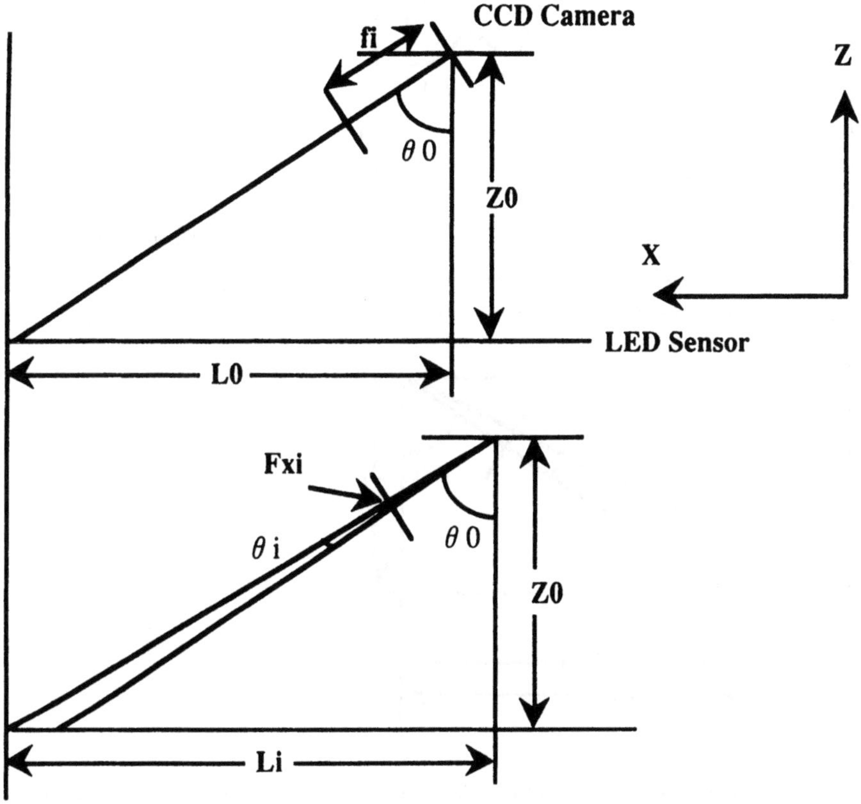

Figure 5 Measurement of the distance along the x axis based on the triangulation method.

sensor's outputs and vision system's outputs which are calculated by eqs.(8.1) to (8.3).

Spot-light Recognition And Compensation By Neural Networks

In this chapter, we use two neural networks for different purposes: (1) recognition of the spot-light, (2) compensation of the vision system. The measurement system consists of two LED displacement sensors. Therefore,

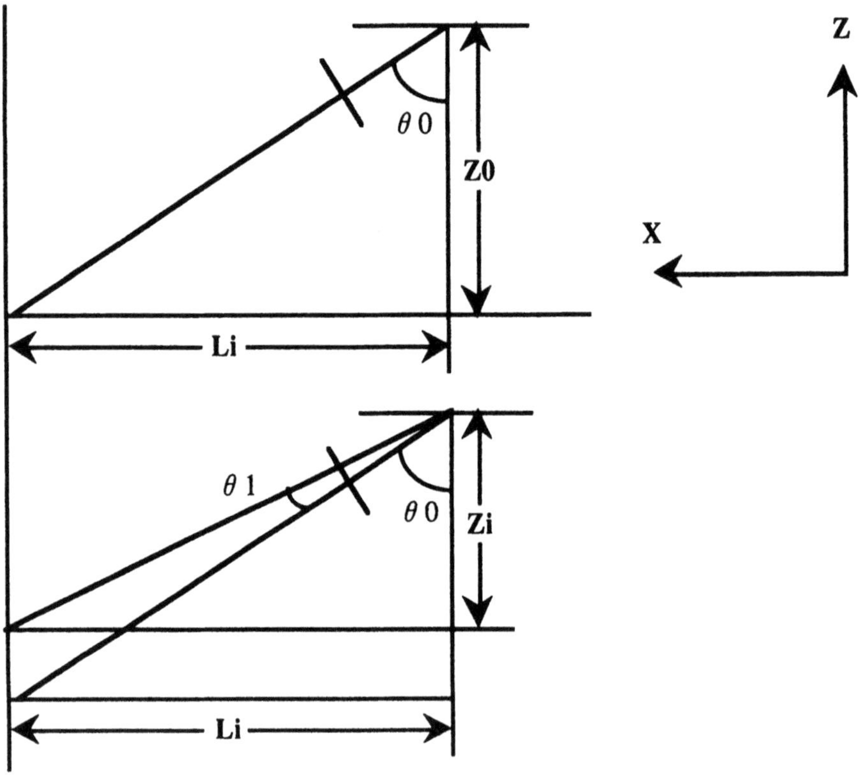

Figure 6 Measurement of the distance along the z axis based on the triangulation method.

two large areas on each image are extracted as LED displacement sensors' spot-light. If there are two extracted areas, it is very easy to recognize the correspondence between extracted areas and LED displacement sensors. In the case of only one extraction from an image, it is difficult to recognize the correspondence without many rules about the correspondence because the gravity center position is moved by the distance between the CCD camera and the object. We use a neural network (NN) for recognition of the correspondence between extracted areas and LED displacement sensors because NNs have a function of making a correlative map between inputs and outputs by learning. The structure of the NN is 3 layers (input layer 4 units, hidden layer 10 units and output layer 4 units).

The lens distortion is one of the problems in the measurement system. In addition, there is some error in camera calibration. In this system, we utilize a NN for compensation of the lens distortion and camera calibration's error since NNs have a function of interpolation of the learning data. The structure of the NN is 3 layers (input layer 1 unit, hidden layer 30 units, and output layer 1 unit) and we use NN for each spot-light of the LED displacement sensor.

An input of the NN is the gravity center position of the LED displacement sensor and an output of the NN is compensation value of the gravity center position. We use the back propagation algorithm [16] for learning. Equations concerning the input/output of NN are as follows:

$$I_n = \frac{G_n}{512}, \tag{8.4}$$

$$T_n = \frac{(R_n - G_n + 40)}{80}. \tag{8.5}$$

where *In*, *Gn*, *Tn*, and *Rn* means the input of the NN, the gravity center position of the LED displacement sensor, the training data, and the gravity center position calculated from the distance between sensors and the object respectively. In this chapter, we determine the maximum of compensation as 40 pixels. Figure 7 shows the construction of the vision system.

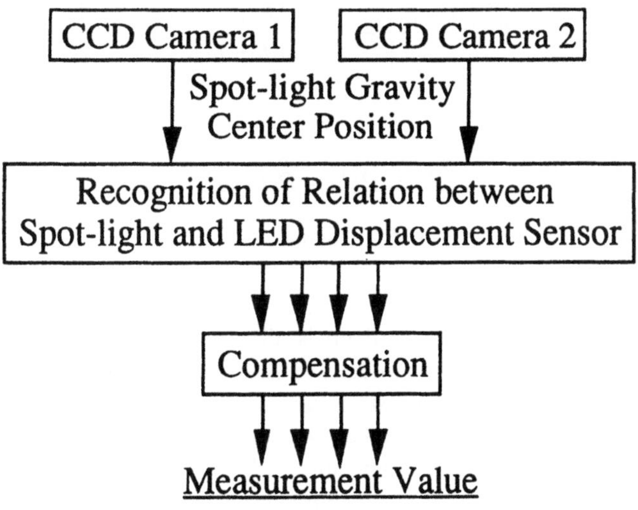

Figure 7 Construction of the fuzzy inference.

2.3 Integration Of Measurement Values By The Fuzzy Inference

In this measurement system, we can obtain 3 measurement values at the same measurement point. One is a LED displacement sensor, while the others are two CCD cameras.

The measurement system is mounted at the tip of the industrial 5 axis manipulator. Therefore, if the manipulator can move accurately, we can select or integrate the measurement values by comparing the variation of each sensor output with the variation of manipulator's positions. In ordinary industrial manipulator systems, because of characteristics of a actuator, accuracy of linkage setting, friction, and backlash, there is a difference between desired trajectory and actual trajectory, and we cannot measure a magnitude of the difference.

In order to integrate the measurement values, we propose the SIS based on the fuzzy inference. We consider the rate of the variation of manipulator's positions to the variation of sensors' outputs as one of the inputs of the fuzzy inference. The rate is expressed in eq. (8.6) as X_1.

If $P(t) - P(t-1) \neq 0$:

$$X_1 = \frac{\|S(t) - S_i(t-1) - (P(t) - P(t-1))\|}{\|P(t) - P(t-1)\|}, \tag{8.6}$$

If $P(t) - P(t-1) = 0$:

$$X_1 = P(t) - P(t-1). \tag{8.7}$$

where $P(t)$ means the position of the sensor system at time t, $Si(t)$ means the output of the SIS.

The LED displacement sensor's output is affected by the intensity of a spot-light's reflection, while the output of the vision system is affected by the aspect value of the spot-light on the image and accuracy of camera calibration not the intensity. Another fuzzy inference's input X_2 is the affected factor of a sensor, such as the intensity of the spot-light of the LED displacement sensor and the aspect value of the vision sensor. The normalized value of X_2 means the environmental data of measurement and if X_2 is nearly equal to 1, the condition is suitable for a sensor, and in the case of X_2 nearly equal to 0,

		X_1				
		S	MS	M	MB	B
X_2	S	M	MS	MS	S	S
	MS	M	M	MS	MS	S
	M	MB	M	M	MS	MS
	MB	MB	MB	M	M	MS
	B	B	MB	MB	M	M

Table 1 Fuzzy rules

the condition is almost prohibitive for a sensor. The fuzzy inference's output is used as the suitability of each sensor for measurement and the SIS's output is determined by comparing the suitability of all sensors.

Table 1 expresses the fuzzy rules. In the table, the suitability of a sensor for measuring is higher in order of S, MS, M, MB, and B. For example, if the X_1 is S (X_1 is around zero) and X_2 is B (the environment is suitable for the sensor) then fuzzy inference's output is B (the estimated suitability is very high). We determined the membership functions and rules by operator's experience.

3 ADAPTIVE SENSING STRATEGY

A sensor has a character individually. In our measurement system, the vision system can measure a distance in various sensor environment and obtain the various information such as the information of the object surface, however it takes a long time to obtain the information. The LED displacement sensor can measure quickly, but the sensor's output is affected by the change of its environment and the suitable environment for the sensor is limited .

When the sensory system uses both the vision system and the LED displacement sensor, the sensory system can measure the object accurately, however it takes a long time. If the object is suitable for the LED displacement sensor and the sensor is active, the vision system is not necessary. On the contrary, the object is not suitable for the LED displacement sensor, the system must use the vision system to keep the accuracy.

In order to measure the object quickly and accurately, the sensory system must change the sensing strategy. The determination of the sensing strategy is based on the suitability of sensors that are calculated by SIS. When the suitability of the LED displacement sensor is high, i.e. the object is suitable for the sensor, the sensory system uses only the LED displacement sensors. In other hand, the suitability of the LED displacement sensor is low, i.e. the object is not suitable for the sensor, the system uses only use the vision system. In other case, both suitability are almost same, the system uses both sensors and integrated the sensory data based on the suitability of sensors.

Adaptive sensing strategy can make the sensory system to measure accurately and quickly by using the vision system when the measuring system recognizes the object is not suitable for the LED displacement sensor. Figure 8 shows the adaptive sensing strategy.

4 EXPERIMENTS AND RESULTS

Experiments were carried out by using the industrial 5 axis manipulator and the sensor system (see Fig. 9) set at the tip of the manipulator. A measurement object was a sheet 1.2 mm thick. We attached papers with the sheet. One is white, others are gradation color from black to white (see Fig.10).

Three types of experiments were performed: 1) camera calibration, 2) compensation for the vision system utilizing NN, and 3) sensor integration system with adaptive sensing strategy.

4.1 Calibration And Compensation

Parameters of the vision system were calibrated by two ways. One was that the manipulator was moved to several different points in order to move the sensor system and then the camera parameters calibrated by the outputs of the LED displacement sensor of each point. The other was that the X-Z table was moved to several different points in order to move the object, and the parameters calibrated by the output of the X-Z table with 10 μm accuracy.

Two NNs were trained to compensate for the measurement error of the vision system. One was used for the vision system calibrated by the LED

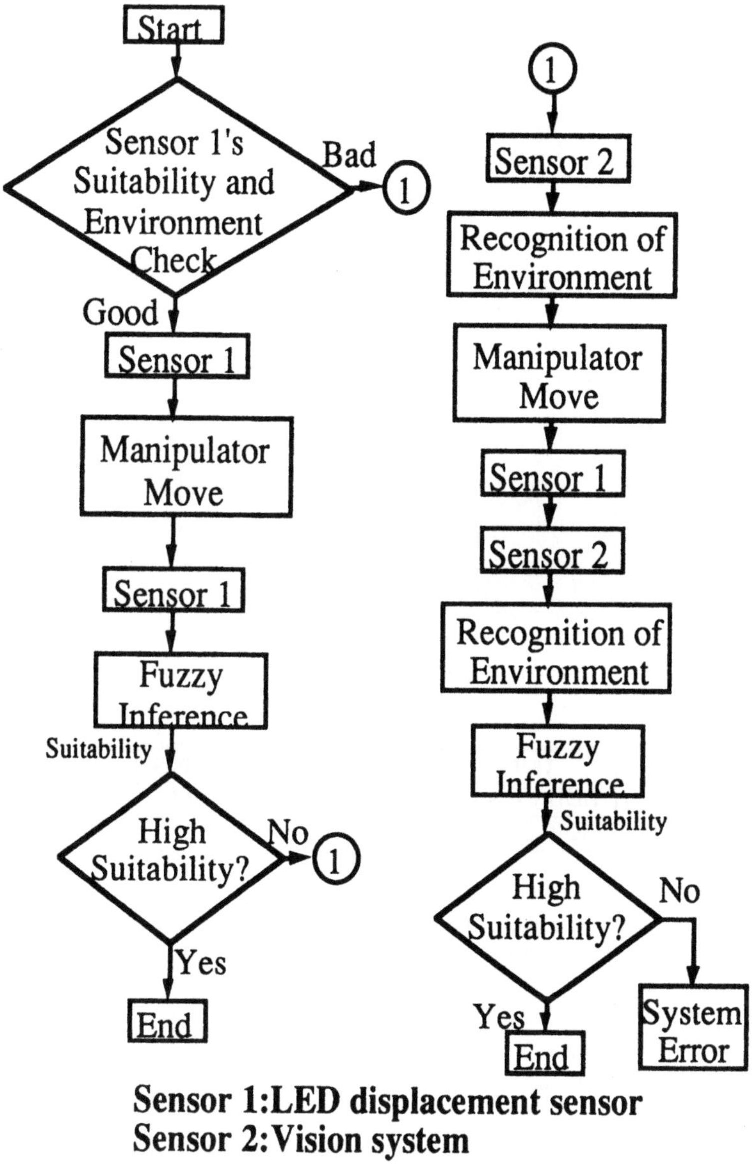

Figure 8 Adaptive sensing strategy.

displacement sensor. The NN was trained by the output of the LED displacement sensor when the object was in the area of the LED displacement sensor's measurable region and by the manipulator's pose data when the object was out of the measurable range. The other was used for the vision

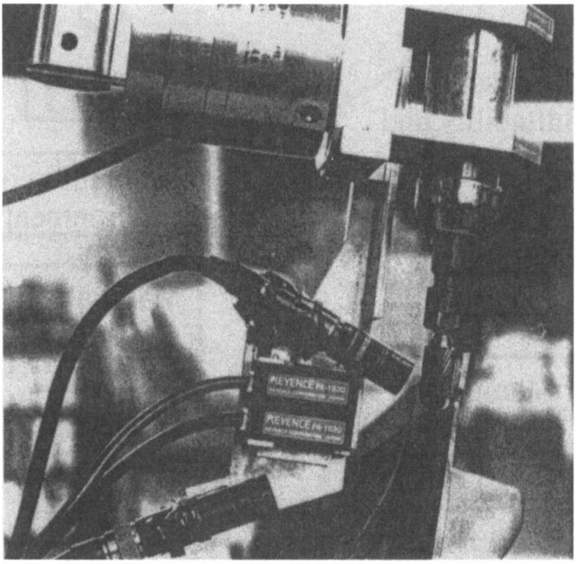

Figure 9 Experimental system.

test	With NN	Without NN
LED Sensor	0.20/0.08	1.04/0.13
X-Z Table	0.22/0.02	0.50/0.17

Table 2 Average measurement error (mm) Measurement range (30mm to 100mm)/the calibration area

system calibrated by the X-Z table and trained by the outputs of the X-Z table. The number of each training data set was 56.

Experiments were carried out by moving the X-Z table to move the object. Figures 11 and 12 show results of measurement of the vision system calibrated by the LED displacement sensor and the X-Z table with/without NNs' compensation. Figures 13 and 14 show the error between the reference data obtained from the X-Z table and the measurement values of Figs. 10 and 11. Table 2 shows the average error between the measurement value and the reference.

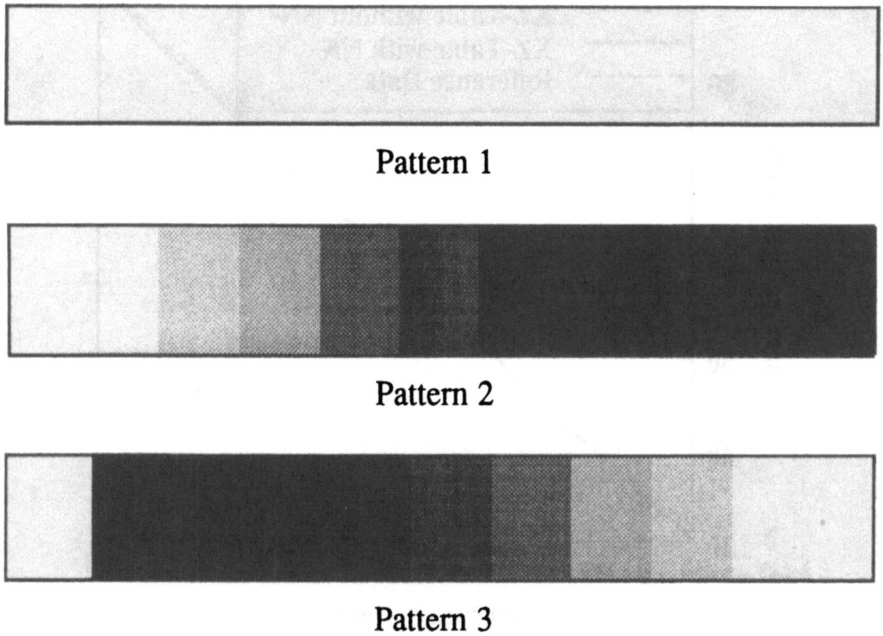

Figure 10 The Object of the experiments.

Experimental results showed that accuracy of calibration by the X-Z table was about 2 times as high as that by the LED displacement sensor. Utilizing NN's compensator, accuracy of the vision system calibrated by the LED displacement sensor was about 5 times as high as that without NN. In the case of the X-Z table, accuracy with NN was about 2.5 times as high as that without NN. The reason of improvement of accuracy was that the vision system can measure distance between the sensors and the object accurately in all measurable area by NN's compensation for the lens distortion and camera calibration's error.

The measurement errors were increased above 70mm because the vision system was based on the triangulation method and accuracy fell with the distance. Furthermore, the intensity of the spot-light was low and the system could not detect the gravity center position of the spot-light accurately.

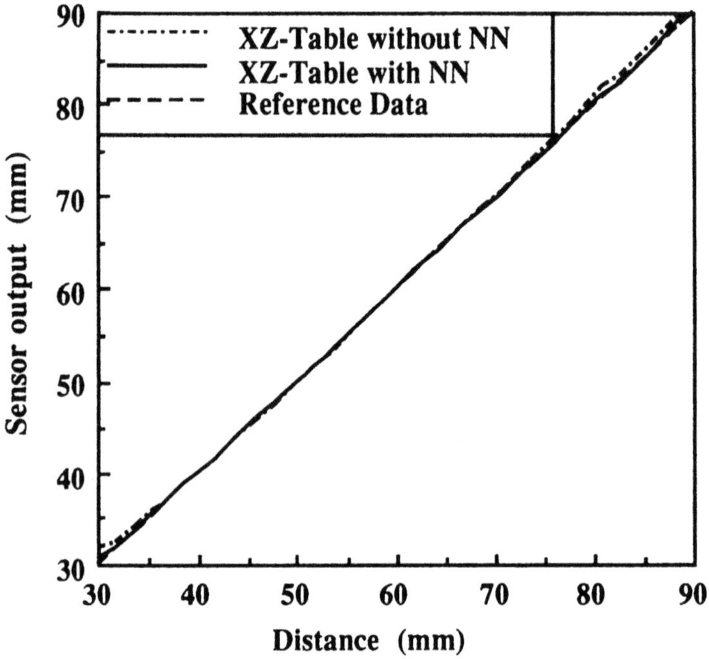

Figure 11 Results of measuring (X-Z table).

4.2 Experiments Of Sensor Integration System With Adaptive Sensing Strategy

We experimented with the SIS with the adaptive sensing strategy. Figures 15 to 22 show experimental results. Figures 15, 17, 19 and 21 show the output of SIS and measurement error. Figure 16, 18, 20, and 22 show SIS's selected sensor and the suitability of the LED displacement sensor estimated by the SIS.

Figures 15 and 16 show the experimental results with the pattern 1 (white color object). The object's color is white therefore the suitability of the LED displacement sensor is very high. Figure 16 shows that the suitability of the LED displacement sensor estimated by the SIS is high, and the SIS only used the LED displacement sensor through measuring the object.

Figures 17 and 18 show the experimental results with the pattern 2. The suitability decreases, because the object's color is change from white to black. Figure 18 shows that the estimated suitability decreases with changing of the

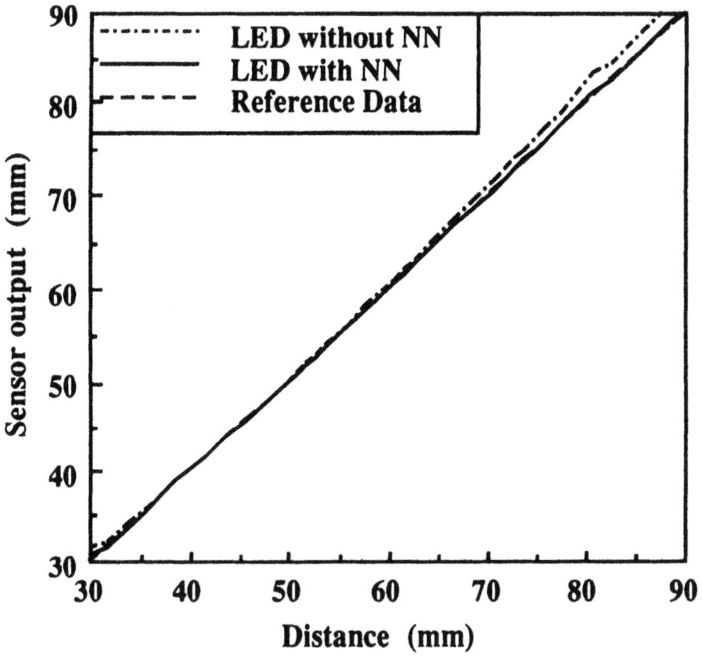

Figure 12 Results of measuring (LED displacement sensor).

object's color. When the suitability was low, the SIS only used the vision system.

Figure 19 and 20 show the results with the pattern 3. The middle of the object is black or dark gray therefore the suitability of the LED displacement sensor decreases and then increases. Figure 20 shows the estimated suitability decreased and then increased with the changing of the color. When the estimated suitability is high, the SIS only used the LED displacement sensor. Then the estimated suitability decreased, the SIS used only the vision system. Around 90 mm, the estimated suitability was increasing, then the SIS used both the LED displacement sensor and the vision system. Then the estimated suitability increased, the SIS only used the LED displacement sensor again.

Figures 21 and 22 show the results in the case of the object's color is white (pattern 1) and the LED displacement sensor is breakdown. The suitability of the LED displacement sensor is as high as in the case of Figs. 15 and 16, however the estimated suitability of the LED displacement sensor is low and the SIS only used the vision sensor. Because the SIS recognized the breakdown of the LED displacement sensor. Experimental results show that

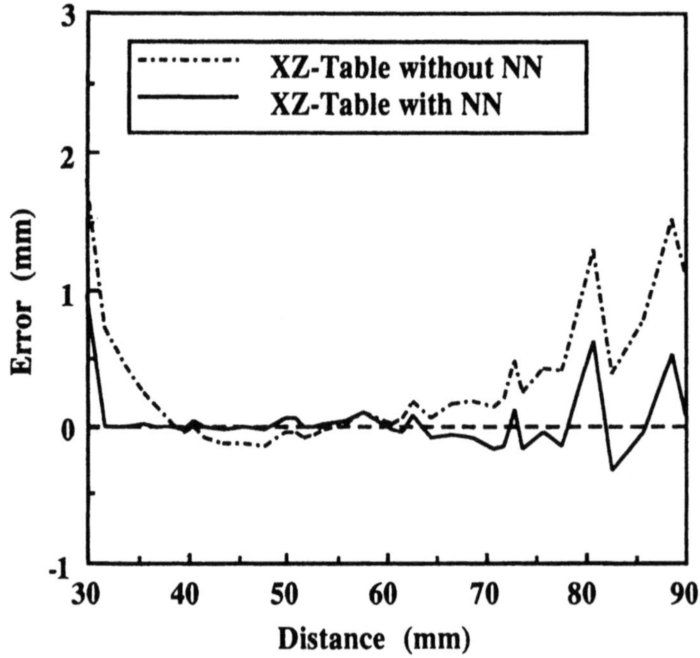

Figure 13 Measurement error against the reference data (X-Z table).

the SIS use the vision system when the SIS estimated the LED displacement sensor was not suitable for measuring the object or the LED displacement sensor was breakdown. The SIS keeps the average of measurement error under 4.0×10^{-5}m.

5 CONCLUDING REMARKS

In this chapter, we presented a 3D measurement system and its integration method with adaptive sensing strategy, which is one of the major parts of the automated carving system. The proposed method was:

- Camera calibration method,
- NN's compensation for a vision system,

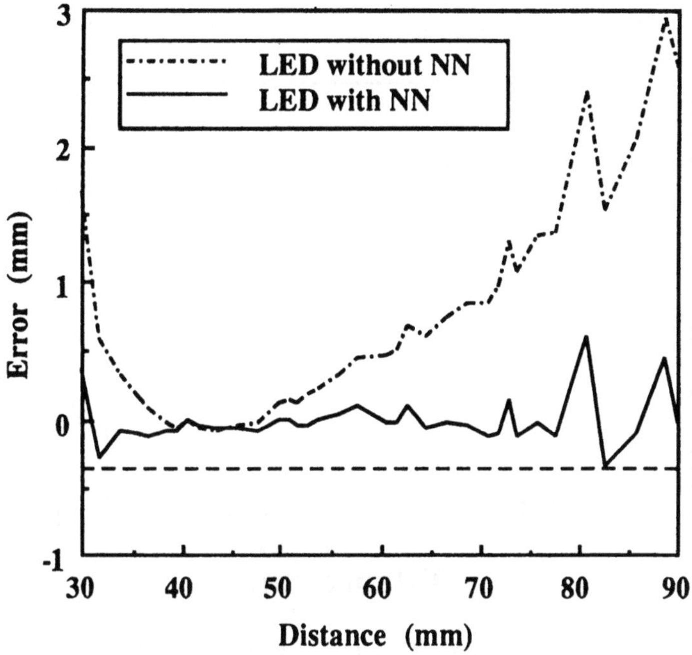

Figure 14 Measurement error against the reference data (LED displacement sensor).

- Sensor integration system based on the fuzzy inference with information of sensors' environments and the manipulator's pose.

- Adaptive sensing strategy.

We also showed effectiveness of the proposed SIS through some experiments as follows:

- Utilizing outputs of the LED displacement sensor, camera parameters can be calibrated easily.

- Using NN's compensator for the lens distortion and camera calibration's errors, the vision system can measure accurately.

- Utilizing the fuzzy inference with moving data of the sensor system and environmental data of measurement, the measurement system can measure

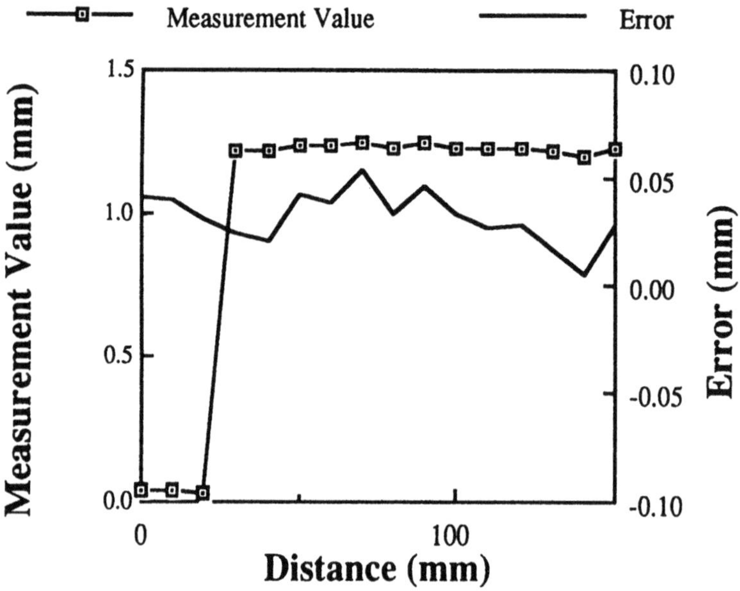

Figure 15 Experimental result (pattern 1) measurement value and error.

accurately and is stable against variation of environments and internal parameters of sensors.

- Utilizing the adaptive sensing strategy, the SIS uses the vision system which takes long time to measure when the SIS estimates the LED displacement sensor cannot measure the object, and then the system achieves fast and accurate measuring.

Future works are: 1) improvement of the image processing time, 2) parameter calibration of the manipulator for accurate sensor data integration, and 3) automatic extraction of sensor specification and making the fuzzy rules and membership function.

REFERENCES

[1] F.M.Proctor, R.J.Norcross, K.N.Murphy, "Automating Robot Programming in the Cleaning and Deburring Workstation of the AMRF," SME, Technical Paper, pp.1/11, (1989).

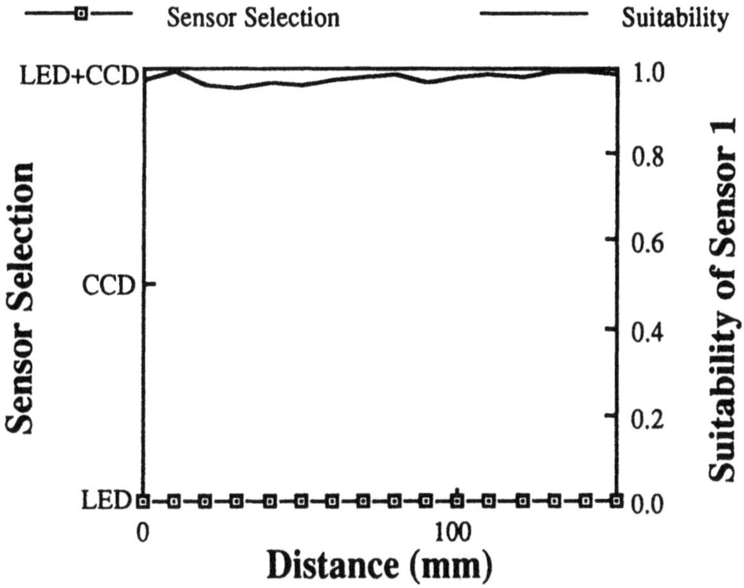

Figure 16 Experimental result (pattern 1) sensor selection and estimated suitability.

[2] K. Kashiwagi, K. Ono, E. Izumi, T. Kurenuma, K. Yamada, "Force Controlled Robot for Grinding," IEEE Int'l Workshop on Intelligent Robots and Systems(IROS '90), pp.1001/1006, (1990).

[3] T.Fukuda, K.Shimojima, F.Arai, H.Matsuura, "Multisensor Integration System based on Fuzzy Inference and Neural Network," J. of Information Sciences, Vol. 71, No.1 and 2, pp. 27/41, (1993).

[4] T.Fukuda, K.Shimojima, F.Arai, H.Matsuura, "A Multi-Sensor Integration System with Fuzzy Inference and Neural Network," Pacific Rim Int'l Conf. on Artificial Intelligence 90(PRICAI '90), pp. 859/864, (1990).

[5] K.Shimojima, T.Fukuda, F.Arai, H.Matsuura, "Fuzzy Inference Integrated 3-D Measuring System with LED Displacement Sensor and Vision System," J. of Intelligent and Fuzzy System, Vol.1, No.1, pp. 63/72, (1993).

[6] R.C. Luo, M.G. Kay, "Multi-sensor integration and fusion in intelligent

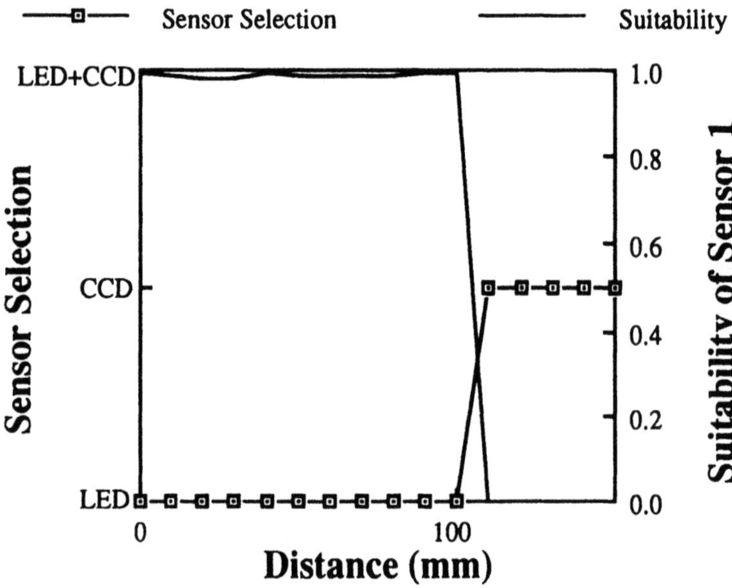

Figure 17 Experimental result (pattern 2) sensor selection and estimated suitability.

systems," IEEE Trans. on System, Man, and Cybernetics, Vol. 19, No. 5, pp. 901/931, (1989).

[7] H. F. Durrant-Whyte, "Sensor models and multi-sensor integration," Int'l J. Robot. Res., Vol. 7, No.6, Dec., pp. 97/113,(1988).

[8] R. C. Luo, M. Lin, "Dynamic multi-sensor data fusion system for intelligent robots," IEEE J. Robot. Automat., Vol. 4, No. 4, pp. 386/396, (1988).

[9] Y. F. Zheng, "Integration of multiple sensors into a robotic system and its performance evaluation," IEEE Trans. Robotics and Automat., Vol. 5, No. 5, pp. 658/669 (1989).

[10] J. M. Richardson, K. A. Marsh, "Fusion of multi-sensor data, Int'l. J. Robot. Res., Vol. 7, No. 6, pp 78/96,(1988).

[11] M. Abdlghafour, T. Chandra, M. A. Abidi, "Data fusion through fuzzy

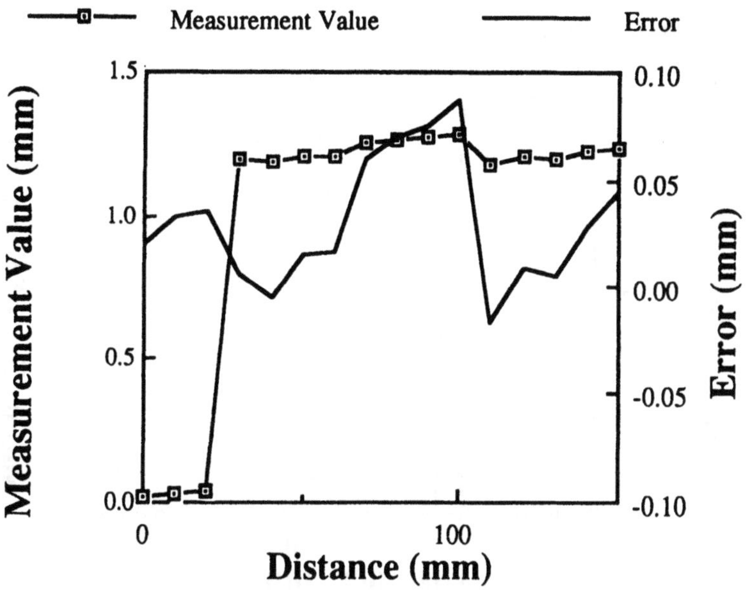

Figure 18 Experimental result (pattern 3) measurement value and error.

logic applied to feature extraction from multi-sensor images," IEEE Int'l Conf. on Robotics and Automation, Vol.2, pp. 359/366, (1993).

[12] T. Kanade, M. Okutomi, "A Stereo Matching Algorithm with an Adaptive Window: Theory and Experiment," IEEE Int'l Conf. on Robotics and Automation, pp.1088/1095, (1991).

[13] K. Sato, S. Inokuchi, "Three-dimensional surface measurement by space encoding range imaging," J. of Robotic System, vol.2, no.1, pp.27/39, (1985).

[14] R. Gutsche, T. Stahs, F.M. Wahl, "Path Generation with a Universal 3d Sensor," IEEE Int'l Conf. on Robotics and Automation, pp. 838/843, (1991).

[15] D.E. Rumelhart, J.L. McClelland, and The PDP Research Group, "Parallel Distribute Processing, The MIT Press, USA, (1986)

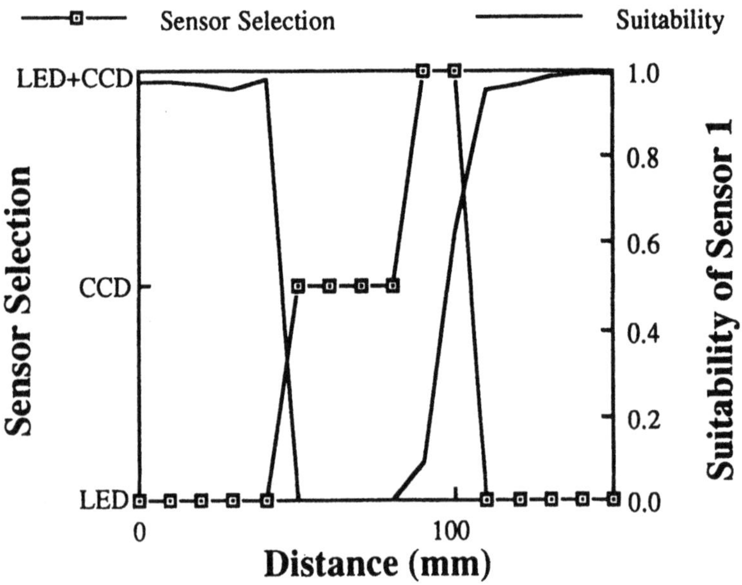

Figure 19 Experimental result (pattern 3) sensor selection and estimated suitability.

[16] M.A. Abidi, "A regularized multi-dimensional data fusion technique," Proc. of IEEE Int'l Conf. on Robotics and Automation, pp. 2738/2744, (1991).

[17] T. Aono, M. Ishikawa, "Auditory-visual fusion using multi-input hidden markov model," Proc. of the IMACS/SICE Int'l Symposium on Robotics, Mechatronics and Manufacturing System'92, pp.1085/1090, (1992)

[18] O. Basir, H.C. Shen, "Sensory Data Integration: A Team Consensus Approach," Proc. of IEEE Int'l Conf. on Robotics and Automation, pp. 1683/1688, (1992).

[19] O. Basir, H.C. Shen, "Aggregating interdependent sensory data in multi-sensor systems," Proc. of 1993 IEEE/RSJ Int'l Conf. on Intelligent Robots and System, pp.377/383, (1993 a).

[20] O. Basir, H.C. Shen, "Goal-driven task assignment and sensor control in multi-sensor systems: a multicriteria approach," Proc. of IEEE Int'l

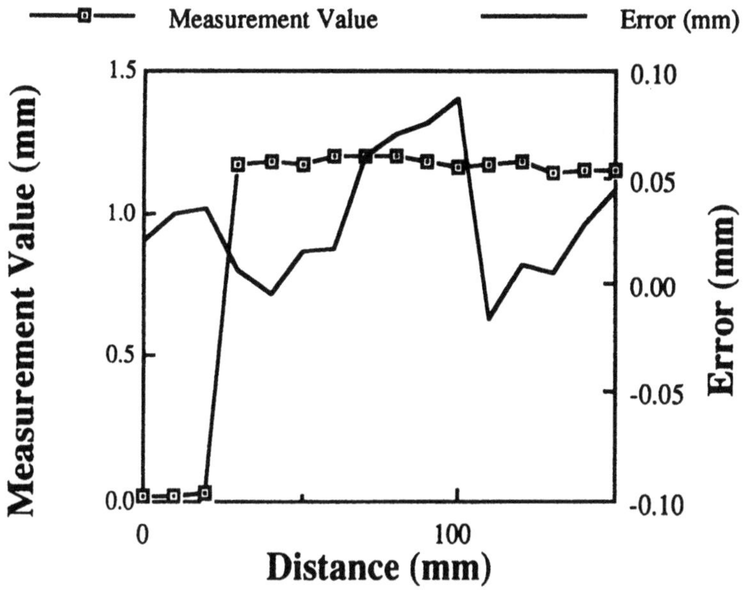

Figure 20 Experimental result (pattern 1, LED sensor is breakdown) measurement value and error.

Conf. on Robotics and Automation, Vol. 2, pp. 559/566, (1993 b).

[21] M. Beckerman, "A bayes-maximum entropy method for multi-sensor data fusion," Proc. of IEEE Int'l Conf. on Robotics and Automation, pp. 1668/1674, (1992).

[22] V. Berge-Cherfaoui, B. Vachon, "A multi-agent approach of the multi-sensor fusion," Proc. of Fifth Int'l Conf. on Advanced Robotics, pp.1264/1274, (1991)

[23] I.J. Cox, J.J. Leonard, "Probabilistic data association for dynamic worl modeling: a multiple hypothesis approach," Proc. of Fifth Int'l Conf. on Advanced Robotics, pp.1287/1294, (1991)

[24] F. Dessen, "Sensor integration using an event driven state estimator," Proc. of Intelligent autonomous system, pp. 897/906, (1989)

[25] C. Duriev, H. Clergeot, "A statistical approach to geometric robot

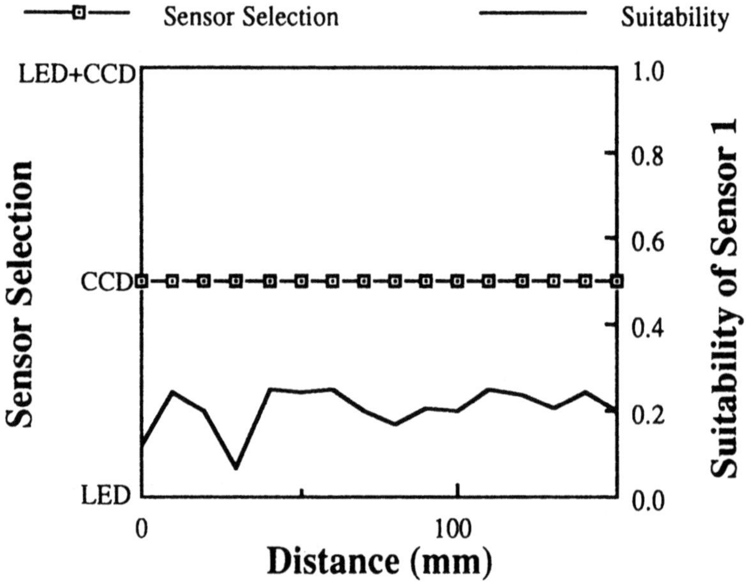

Figure 21 Experimental result (pattern 1, LED sensor is breakdown) sensor selection and estimated suitability.

location including data fusion and error rejection," Proc. of Intelligent autonomous system, pp.886/896, (1989)

[26] H.F. Durrant-Whyte, B.Y.S. Rao, H. Hu., "Toward a fully decentralized architecture for multi-sensor data fusion," Proc. of IEEE Int'l Conf. on Robotics and Automation, pp. 1331/1336, (1990).

[27] T. D'Orazio, M. Ianigro, E. Stella, F.P. Lovergine, A. Distante, "Mobile robot navigation by multi-sensory integration," Proc. of IEEE Int'l Conf. on Robotics and Automation, Vol. 2, pp. 373/379, (1993).

[28] C. Ferrell, "Many sensors, one robot," Proc. of 1993 IEEE/RSJ Int'l Conf. on Intelligent Robots and System, pp 399/406, (1993).

[29] I. Gasparovic, "Integration of the multisensory information system," Proc. of the Second Int'l Symposium on Measurement and Control in Robotics, pp. 125/136, (1992)

[30] J.K. Hackett, M. Shah, "Multi-sensor fusion: a perspective," Proc. of IEEE Int'l Conf. on Robotics and Automation, pp. 1324/1330, (1990).

[31] G.D. Hager, S.P. Engelson, S. Atiya, "On comparing statistical and set-based methods in sensor data fusion," Proc. of IEEE Int'l Conf. on Robotics and Automation, Vol. 2, pp. 352/358, (1993)

[32] L. Hong, T. Scaggs, "Real-time optimal multiresolutional sensor/data fusion," Proc. of IEEE Int'l Conf. on Robotics and Automation, Vol. 2, pp. 117/122, (1993).

[33] K. Hughes, N. Ranganathan, "A model for determining sensor confidence," Proc. of IEEE Int'l Conf. on Robotics and Automation, Vol. 2, pp. 136/141, (1993).

[34] T. Kawashima, T. Nagasaki, Y. Aoki, "Sensor fusion system for model-based object tracking," Proc. of the Second Int'l Symposium on Measurement and Control in Robotics, pp. 265/270, (1992)

[35] A. Kosaka, A.C. Kak, "Data fusion an perception planning for indoor mobile robot navigation," Proc. of the Second Int'l Symposium on Measurement and Control in Robotics, pp. 271/278, (1992)

[36] R.C. Luo, M.G. Kay, W. Gary Lee, Multisensor integration and fusion: issues, approaches, and future trends," Proc. of the IMACS/SICE Int'l Symposium on Robotics, Mechatronics and Manufacturing System'92, pp.1055/1063, (1992)

[37] D. Matusmoto, T. Kimoto, S. Nagata, "A neural network approach to sensorimotor fusion," Proc. of the IMACS/SICE Int'l Symposium on Robotics, Mechatronics and Manufacturing System'92, pp.1105/1110, (1992)

[38] T. Moriizumi, T. Nakamoto, Odor sensing system using neural network pattern recognition," Proc. of 1992 Int'l Conf. on Industrial Electronics Control and Instrumentation, pp.1645/16479, (1992)

[39] T. Moriwaki, V. Mori, "Sensor fusion for in-process identification of cutting process based on neural network approach," Proc. of the

IMACS/SICE Int'l Symposium on Robotics, Mechatronics and Manufacturing System'92, pp.245/250, (1992)

[40] T. Mukai, T. Mori, M. Ishikawa, "A sensor fusion system using mapping learning method," Proc. of 1993 IEEE/RSJ Int'l Conf. on Intelligent Robots and System, pp.391/396, (1993)

[41] R.R. Murphy, "Robust sensor fusion for teleoperations," Proc. of IEEE Int'l Conf. on Robotics and Automation, Vol. 2, pp. 572/577, (1993).

[42] C. Olivier, O. Dessouble, "Heterogeneous sensors cooperation for an advanced perception system," Proc. of Fifth Int'l Conf. on Advanced Robotics, pp.1275/1280, (1991)

[43] T. Oomichi, Y. Fuke, "Hierarchical navigation of legged robot for terrain based on sensor fusion," Proc. of the IMACS/SICE Int'l Symposium on Robotics, Mechatronics and Manufacturing System'92, pp.1071/1076, (1992)

[44] F. Ramparany, "Multisensor data fusion for robotic tasks," Proc. of the IMACS/SICE Int'l Symposium on Robotics, Mechatronics and Manufacturing System'92, pp.1091/1098, (1992)

[45] A. Sabater, F. Thomas, "Set membership approach to the propagation of uncertain geometric information," Proc. of IEEE Int'l Conf. on Robotics and Automation, pp. 2718/2723, (1991).

[46] Y. Sakaguchi, K. Nakano, "Active perception with intentional observation," Proc. of the Second Int'l Symposium on Measurement and Control in Robotics, pp. 241/248, (1992)

[47] K.T. Song, C.C. Chang, "Ultrasonic sensor data fusion for environment recognition," Proc. of 1993 IEEE/RSJ Int'l Conf. on Intelligent Robots and System, pp.384/390, (1993)

[48] A. Takahashi, M. Ishikawa, "Signal processing architecture with bidimensional network topology for flexible sensor data integration system," Proc. of 1993 IEEE/RSJ Int'l Conf. on Intelligent Robots and System, pp 407/413, (1993)

[49] H. Xu., "Effective fusion technique for disparate sensory data," Proc. of 1991 Int'l Conf. on Industrial Electronics, Control and Instrumentation, pp.2535/2540, (1991)

9

ROBOT HAND-EYE COORDINATION BASED ON FUZZY LOGIC

Sukir Kumaresan, Hua Li, and Xing-Min Li*

**Computing Center, The Institute of Textile Engineering*
He-Dong, Tianjin, China
Computer Science Department
College of Engineering, Texas Tech University
Lubbock, TX 79409, USA

ABSTRACT

Hand-eye coordination of a robot system is a classical problem in robotic vision, which mathematically can be described as an ill-posed inverse mapping problem. Different techniques have been developed to solve this inverse mapping problem, but most of the reported work has a common limitation that is the relative position between the given robot and the camera must be fixed once the calibration is finished. However, very often in real-world applications this may not be realistic either due to the random disturbances to the system or due to the need of repositioning the camera or moving the robot base to a new position in order to perform a required task. In this case, usually a lengthy re-calibration process has to be carried out. In this paper, we describe our technique which eliminates the need of re-calibration for the vertical position change of the camera. Our technique is based on the integration of the utilization of a close form mathematical formulation and fuzzy logic. Using the fuzzy logic technique, we are able to deal the unknown amount of vertical changes of the camera position. Using image processing technique to estimate the distance change of the image land marks, we derived a fuzzy reasoning model which allows us to calculate the inverse mapping matrix . The algorithm and fuzzy reasoning system have been developed and implemented in a 6-degree-of-freedom robot system. Experiments were conducted and the result confirmed our design.

1 INTRODUCTION

Hand-eye robotic systems have been increasingly used in industries. In the normal mode of operation, a robotic vision system senses the three-dimensional environment, performs image preprocessing to extract image features of both the environment and work pieces, then calculates the three-dimensional position information, or the depth, of the work piece and the relative position from the image registration points to their position in the three-dimensional space. Once the correspondence between image features and their three-dimensional coordinates is established, the robot can then be programmed to move in to perform various picking, placing tasks.

It is well known that recovering three-dimensional information from a given two-dimensional image is an ill-posed inverse mapping problem in the sense of Hadamard. The ill-posedness was caused by the projection of objects in a three-dimensional space on to a two-dimensional image plane. In order to establish one-to-one correspondence, we have to first convert this ill-posed problem to a well-posed one by regularization process or by imposing additional constraint. Several different techniques have been developed today for this purpose, [3], [18], [19]. However, once the hand-eye calibration process is finished, the relative position between the robot base and the camera pose must be fixed to ensure the inverse mapping relation preserved, so that one-to-one correspondence between the image points and their three-dimensional positions can be recovered. This fixed position has to be maintained all the time during the entire normal operation, which can be disrupted by random noise and limits the operation of moving the camera to a new position for different task.

In many real-world applications, very often the camera position may not be fixed either due to disruption caused by random disturbances or due to the intentional relocation of the camera position for different task. With the current technique, recalibration has to be performed each time this change occurs, which can be slow and lengthy process. Facing this problem, we propose a new technique which allows us to eliminate the recalibration process. In particular our technique works for the situation where the system has been calibrated for two known positions given at the same vertical line in the three dimensional world coordinate system, the camera can be placed anywhere between these two positions.

2 SURVEY OF EXISTING TECHNIQUES

2.1 Conventional Calibration Techniques

The problem of camera calibration has been extensively studied [3]-[4], [10], [16]. Tsai [18], [19], classifies the calibration methods for a camera at a fixed position into five categories:

1. Techniques involving full-scale optimization. This technique allows easy adaptation of calibration accuracy but is computationally intensive and involves nonlinear process.

2. Techniques based on the computation of perspective transformation. Given a set of 3D world coordinates of a large number of points and the corresponding 2D image coordinates, the coefficients in the perspective transformation matrix can be solved by least square solution of a system of linear equations.

3. The two-plane technique which involves solving a set of linear equations. The main disadvantage of this technique is that the transformation between the image features and object 3D coordinates is empirically based only.

4. The geometric technique, which uses a geometric construction to derive a direct solution for the camera location and orientation. None of the camera intrinsic parameters need to be computed. An major drawback is that the lens distortion was not considered.

5. Tsai's method for three dimensional camera calibration, which was developed for efficient computation of camera external position, orientation relative to the object reference coordinate system, effective focal length, and radial lens distortion as well as image scanning parameters.

Until recently, for most vision-based manipulator control systems, hand-eye calibration is carried out before the operation of the system. Liang et al. [11] classified this type of calibration as static batch calibration.

2.2 Calibration Technique With Adaptation Capability

It is desired that a hand-eye system be adaptive and autonomous. Unexpected disturbances to a hand-eye system often occur during or after the operation of robot motion. In such cases, a spatial relation between the camera pose, position and the robot position is not fixed, and the intrinsic and extrinsic parameters are changing. These Changes to the relative position of the camera and the robot have to be taken care of in order to preserve the coordination. This very often demands recalibration process.

A dynamic learning process through an automatic repetition of operation trials has been discussed by Arimoto et al. [1] and Miller [14]. An approach based on a visual-feedback learning process, known as *adaptive calibration*, was first proposed by Liang et al. [12]. In this approach, the hand-eye system calibration is performed in real-time while the system is operating. The hand-eye transformation is learned dynamically through repeated trials until a satisfactory result is obtained. A simple system to intercept a ping-pong ball using a robot and a single camera has been demonstrated by Skaar et al. [17]. Brooks et al. [2] addressed the self-calibration problem of motion and stereo vision for mobile robots obstacle avoidance.

Recursive computation of the transformation between the camera coordinate system and the world coordinate system for a robot hand-eye system using Kalman filtering has been addressed by Izaguirre et al. [6], for calibration based on a stereo vision. Faugeras et al. [3] proposed using a Kalman filter for camera calibration to deal with the uncertainty of the parameters. The calibration was static and carried out using a calibration block before the operation of the system. They used a one-stage extended non-linear Kalman filter to calibrate all the parameters. Since the extended Kalman filter is not optimal in general and numerically unstable, Liang et al. [11] proposed a linear Kalman filter. They used a statistical hypothesis test to detect large parameter changes.

The goal of all the above research was to provide the capability of dynamically calibrating a hand-eye system in real time. General drawbacks of these approaches include (1) intensive computation for parameter identification, (2) high sensitivity to sensor noise, and (3) the selection of a system model. Due to these limitations, several investigators have discussed possibilities for the application of neural networks.

Kuperstein [7] developed an algorithm for visual-motor coordination in multijoint robots using a parallel architecture. Walter and Schulten [20] implemented self-organizing neural networks for the visuo-motor control of an industrial robot. Martinez et al. [13] developed a three-dimensional neural net for learning visuo-motor coordination of a robot arm. There have been successful implementation of neural nets for the calibration of hand-eye systems. However the computationally intensive training process, and the stability concern of the network usually become a obstacle for the wide application of this approach. As an alternative, developing a technique based on fuzzy logic for hand-eye calibration is appealing.

3 MATHEMATICAL FORMULATION

In this section, we will give the mathematical formulation based on the analysis of the hand-eye calibration problem. Given the two-dimensional image positions of the workpiece's features and the pose of the camera, we have to determine the workpiece's position with respect to the world (robot) coordinate system. This problem in nature is an ***inverse mapping problem*** [5], [15].

3.1 Formulation Of The Inverse Mapping

Given the two-dimensional image of the work area, it generally would not be possible to reconstruct the three-dimensional features without the knowledge of the depth information, or additional constraints. The inverse mapping problem can be formulated as follows.

Let's define a camera coordinate system and a world coordinate system. The camera (viewer) coordinate system is defined as the coordinate system of the mounted camera, whose origin coincides with the camera as illustrated in Figure 1 [5]. The world (robot) coordinate system is defined as the coordinate system whose origin coincides with the origin of the robot as in Figure 1.

The notations for describing the coordinates in various coordinate systems are as follows:

1. A point in the world coordinate system is described as (x_i, y_i, z_i).

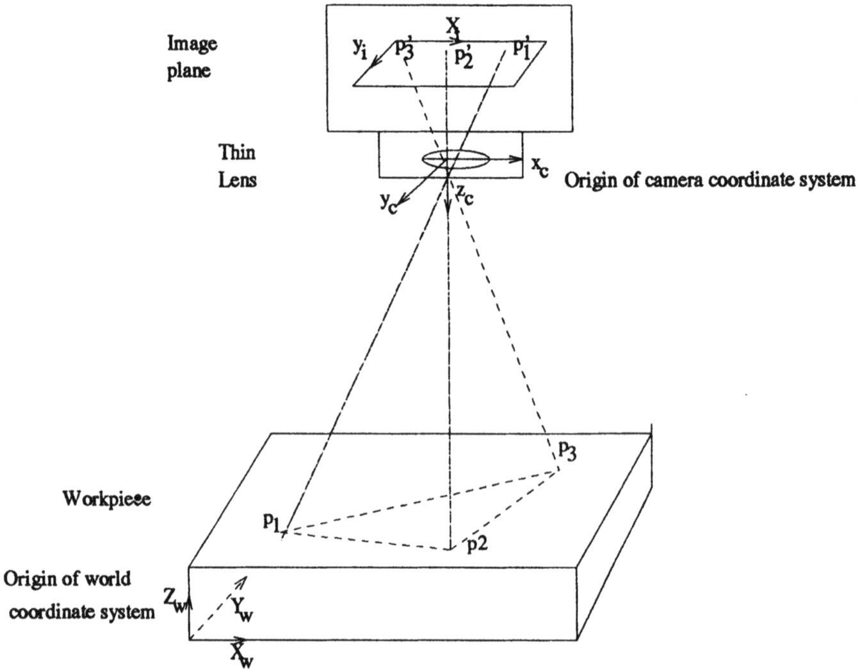

Figure 1 Illustration of the world coordinate system and the viewer coordinate system.

2. The same point, whose world coordinates are (x_i, y_i, z_i) has $(x_i^{'}, y_i^{'}, z_i^{'})$ as its camera coordinates.

3. The perspective projection of $(x_i^{'}, y_i^{'}, z_i^{'})$ on the image plane is $(x_i^{''}, y_i^{''}, z_i^{''})$.

Let $v_i^{'}$ be a point $(x_i, y_i, z_i)^T$ of the workpiece and $v_i^{''}$ be its projection to image plane in the camera coordinate system, which is,

$$v_i^{''} = (x_i^{''}, y_i^{''}, z_i^{''})^T. \tag{9.1}$$

The perspective projection can be described mathematically by

$$v_i^{''} = T_i * v_i^{'}, \tag{9.2}$$

where T_i is nonsingular 3*3 matrix.

Suppose

$$v_i = (x_i, y_i, z_i), \tag{9.3}$$

is the coordinate of $v_i^{'}$ in the world coordinate system, then the transformation from the world coordinate system to viewer coordinate system is given by,

$$v_i^{'} = T_r * v_i + v_s, \tag{9.4}$$

where T_r is 3D rotation matrix and v_s represent shift. Then the relation between its perspective projection and its position in the world coordinate system is

$$v_i^{''} = T_i * (T_r * v_i + v_s) = T_i * T_r * v_i + v_f, \tag{9.5}$$

where $v_f = T_i * v_s$.

Now, given any three known points v_1, v_2 and v_3 in world coordinates, their corresponding points $v_1^{''}$, $v_2^{''}$ and $v_3^{''}$ on image plane can be found by using equation (9.5). So, we have

$$v_j^{''} = T_i * T_r * v_j + v_f, \tag{9.6}$$

where j=1,2,3. By subtraction, we have

$$v_2^{''} - v_1^{''} = T_i * T_r * (v_2 - v_1), \tag{9.7}$$

and

$$v_3^{''} - v_1^{''} = T_i * T_r * (v_3 - v_1), \tag{9.8}$$

or,

$$(v_2^{''} - v_1^{''}, v_3^{''} - v_1^{''}) = T * (v_2 - v_1, v_3, -v_1)$$

where $T = T_i * T_r$ is a 3*3 matrix. Putting the above equations together in the form of x, y, and z components, we have

$$\begin{pmatrix} x_2^{''} - x_1^{''} & x_3^{''} - x_1^{''} \\ y_2^{''} - y_1^{''} & y_3^{''} - y_1^{''} \\ z_2^{''} - z_1^{''} & z_3^{''} - z_1^{''} \end{pmatrix} = \begin{pmatrix} t_{11} & t_{12} & t_{13} \\ t_{21} & t_{22} & t_{23} \\ t_{31} & t_{32} & t_{33} \end{pmatrix} * \begin{pmatrix} x_2 - x_1 & x_3 - x_1 \\ y_2 - y_1 & y_3 - y_1 \\ z_2 - z_1 & z_3 - z_1 \end{pmatrix}. \tag{9.9}$$

Since all projected points have the same z value, $z_2^{''} - z_1^{''} = 0$ and $z_3^{''} - z_1^{''} = 0$. Also assume these objects in the 3D world coordinate system have the same depth. Then, $z_2 - z_1 = 0$ and $z_3 - z_1 = 0$.

Let $v_1'' = 0$ and $v_1 = 0$. Finally, we have

$$T_1 = \begin{pmatrix} t_{11} & t_{12} \\ t_{21} & t_{22} \end{pmatrix}^{-1} = \begin{pmatrix} x_2 & x_3 \\ y_2 & y_3 \end{pmatrix} * \begin{pmatrix} x_2'' & x_3'' \\ y_2'' & y_3'' \end{pmatrix}^{-1} \tag{9.10}$$

Thus, for any point v_i'' in image plane, we can find its corresponding point v_i, by

$$v_i = T_1 * v_i''$$

3.2 The Environment With Random Disturbances

For any change of camera position, there is a corresponding change in the image of the workarea. Depending on the position, the object size in the image may increase or decrease depending on the change of the relative position between the camera and the robot. Given the fact that the position of the work piece in the world coordinate system is not changed but its image position was changed, this leads to the change of the coefficients of the inverse homogeneous transformation matrix.

Using the same notation for defining the coordinate systems as in the previous section, we know the rotation matrix T_r changes to $T_r + \delta T_r$ and v_s changes to $v_s + \delta v_s$. The coordinate of the point in image plane is defined as

$$v_i'' = T_i * (T_r + \delta T_r) * v_i + T_i * (v_s + \delta v_s). \tag{9.11}$$

Three points in the world coordinates are v_1, v_2 and v_3, and their corresponding points on image are v_1'', v_2'' and v_3''. So, we have,

$$v_2'' - v_1'' = T_i * (T_r + \delta T_r) * (v_2 - v_1) \tag{9.12}$$

$$v_3'' - v_1'' = T_i * (T_r + \delta T_r) * (v_3 - v_1) \tag{9.13}$$

or,

$$(v_2'' - v_1'', v_3'' - v_1'') = T'' * (v_2 - v_1, v_3 - v_1) \tag{9.14}$$

where $T'' = T_i * (T_r + \delta T_r)$.

Using the distributive property of matrix multiplication,

$$T'' = T_i * T_r + T_i * \delta T_r = T + T'$$

where $T = T_i * T_r$ and $T' = T_i * \delta T_r$.

The coefficients of T_1 change to the new coefficients of $T_1 + \delta T_1$. Also the image coordinates change to v''_{i1}. So the position of the workpiece in world coordinates, when the camera is moved, is given by

$$v_i = (T_1 + \delta T_1) * v''_{i1}. \tag{9.15}$$

4 ALGORITHM FOR THE COORDINATION BASED ON FUZZY LOGIC

It is difficult to establish a correlation between the change in camera position and change in the value of image coordinates. The change in the coefficients of the inverse homogeneous transformation matrix poses this difficulty. So the hand-eye system has to be recalibrated for every change in camera position. Thus it is important to make the system adaptive to the change of camera position without the need of recalibration. We now describe the algorithm based on fuzzy logic.

We first calibrate the system at two extreme vertical positions. These two positions are termed A and B and are used as the reference points. At A, there is a zero change in the coefficients of T_1. The maximum change in the coefficients of T_1 occurs at B. These two extreme vertical positions form the boundaries within which the camera is moved. By experimentation, the coefficients of significance are t_1 and t_4 corresponding to t_{11} and t_{22} in equation (9.10) respectively. These two coefficients change when the camera position is changed. The changes of these two coefficients have to be characterized and determined. When the camera moves, object size changes in the image. The squared distance between two known points A and B on the image is used to characterize the camera position change. This squared distance is described by linguistic variables and in term can be measured by using fuzzy membership functions. Based on this, we have designed an algorithm for coordination. The algorithm consists of (1) selection of input variables, sampling and fuzzification of the input, (2) design of fuzzy inference rules for reasoning, and (3) defuzzification and computation of the output for reestablishing coordination between camera and the robot.

4.1 Fuzzification

We select the squared distance between two known points A and B in the world coordinate system as landmarks. As the camera displacement from a reference point A increases, the distance between the landmarks on the image plane increases. By the calculation of the distance change on the image plane, we have rather good indication of where the camera moves, and roughly how much the camera moves. The distance d of the landmarks on the image plane is defined as:

$$d = (x_2 - x_1)^2 + (y_2 - y_1)^2, \tag{9.16}$$

where x_1, y_1 are the coordinate of point A on the image plane, while x_2, y_2 are the coordinate of point B on the image plane. They can be determined by image feature extraction algorithm. The values of d at positions A and B are measured during the calibration process. The changes in d correspond to the changes in t_1 and t_4 of the calibration matrix. These changes, denoted as δt_1 and δt_4, have to be determined in order to re-establish the one-to-one correspondence between the object features on the image plane and their position in the world coordinate system. If d is selected as an independent variable, then both t_1 and t_4 are functions of the input d. In this sense, the system is a *single-input-multiple-output* system.

The parameter d, is quantized into five levels. The linguistic variables are used to describe each level of the quantization as: positive large L, positive or negative medium $\pm M$, positive or negative small $\pm S$. For the output parameters δt_1 and δt_4, we quantize them into small S, medium M, and large L. A set of triangular shaped membership functions [8]-[9] are used to describe both the input variable d and the output variables t_1, t_4 respectively. These membership functions are given in Figure 2.

The fuzzification process assigns the input data d with membership values.

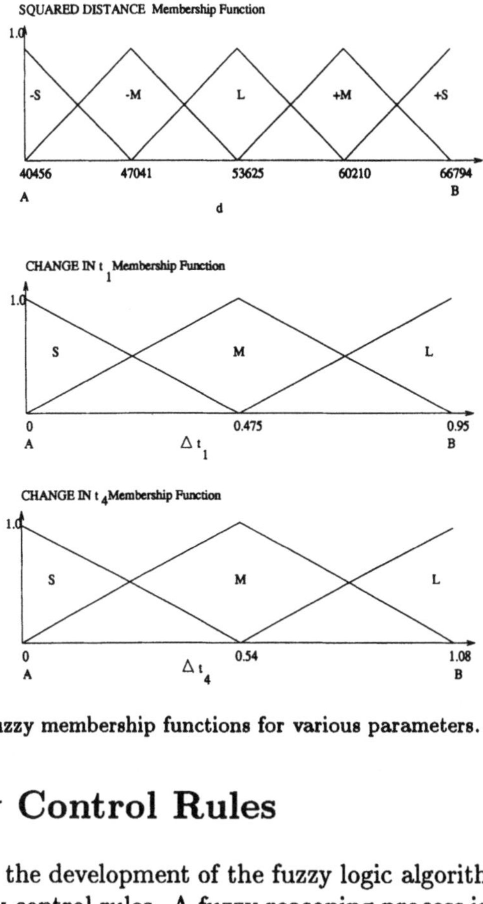

Figure 2 Fuzzy membership functions for various parameters.

4.2 Fuzzy Control Rules

The next stage in the development of the fuzzy logic algorithm is the derivation of fuzzy control rules. A fuzzy reasoning process is characterized by a set of linguistic statements based on expert knowledge which can be framed in the form of "if-then" rules. These rules are conditional statements in which the antecedents are conditions in the application domain and the consequents are control actions. The collection of these rules form the fuzzy control rule base. For the present *single-input-two-output* hand-eye coordination system, we have derived the fuzzy control rules in the following form,

$$R_1\text{:if } x \text{ is } A_1 \text{ then } y \text{ is } B_1,\ z \text{ is } C_1,$$

$$\dots\ \dots\ \dots\ \dots$$

$$R_n\text{:if } x \text{ is } A_n \text{ then } y \text{ is } B_m,\ z \text{ is } C_k,$$

where$x$ is the input variable, and y, z are the functions of it. A_i, and B_i, C_i are linguistic variables associated with x, y, and z. The index $i = 1, 2, ..., n$ defines the number of quantization levels. In particular, for the the hand-eye coordination application, we have $n = 5$ and $\{A_1, ..., A_5\} = \{L, +M, -M, +S, -S\}$, $m = 3$ and$\{B_1, ..., B_3\} = \{S, M, L\}$, $k = 3$ and $\{C_1, ..., C_3\} = \{S, M, L\}$, respectively, as follows,

R_1:if x is A_1 then y is B_2, z is C_2,
R_2:if x is A_2 then y is B_2, z is C_2,
R_3:if x is A_3 then y is B_2, z is C_2,
R_4:if x is A_4 then y is B_3, z is C_3,
R_5:if x is A_5 then y is B_1, z is C_1.

4.3 Defuzzification

Defuzzification process maps the fuzzy control actions into the nonfuzzy control actions. This process produces nonfuzzy control action to best represent the possibility distribution of an inferred fuzzy control value. One of the most widely used methods, the Center of Area Method (COA), [8]-[9], was utilized to generate the center of gravity of the possibility distribution. Hence, a defuzzified crisp control action z_0 is given as

$$z_0 = \frac{\sum_{j=1}^{n} w_j \mu_z(w_j)}{\sum_{j=1}^{n} \mu_z(w_j)}. \tag{9.17}$$

where n is the number of quantization levels of the input and w_j is the support value at which the membership function reaches the maximum value $\mu_z(w_j)$. For the hand-eye system, the outputs are the δt_1 and δt_4. The outputs are computed as linguistic variables for each set of input. Then defuzzification is done by the Center of Area Method as,

$$\delta t_1 = \frac{\sum_{j=1}^{5} w_j \delta t_1}{\sum_{j=1}^{5} w_j}, \tag{9.18}$$

$$\delta t_4 = \frac{\sum_{j=1}^{5} w_j \delta t_4}{\sum_{j=1}^{5} w_j}, \tag{9.19}$$

where w_j is the support value.

5 EXPERIMENTS

The experiments were conducted to test the proposed algorithm for both a stationary environment and the environment with camera displacement and and random disturbances.

5.1 Experimental Setup

A hand-eye robotic manipulator system was employed for the testing of the proposed algorithm. This system consists of a single camera, a six degree-of-freedom robotic manipulator and a workpiece. The physical system setup is given in Figure 3.

A CCD camera for image digitization was mounted on a vertical stand, while its position can be moved up or down on the stand. The relative position from the camera landmark positions (A, and B) to the robot base were measured during the initial calibration process. So the coefficients of the calibration matrices can be calculated using the existing camera calibration technique as described in the previous section. A work piece was placed on the top of the work bench, whose image features were extracted by image processing techniques. These image features together with their known three-dimensional coordinates in the world coordinate system were utilized to set up the calibration. This is illustrated in Figure 4.

A SCORBOT-ER VII, a six degree-of-freedom robot, was used as a part of the hand-eye system. The robot was controlled by a local controller and the 386 host computer for the picking-and-placing task.

First, a stationary environment was chosen to test the initial calibration at upper bound and lower bound of the camera positions respectively. A 256-by-256 grey level image was captured in TGA file format and was then converted to PC/DIP's IMG image format for further processing by the 386 host computer. The image was first preprocessed for better feature extraction. Then feature extraction algorithm was utilized to extract the position information of the workpiece for the later calculation of the robot motion trajectory. The homogeneous transformation algorithm described in

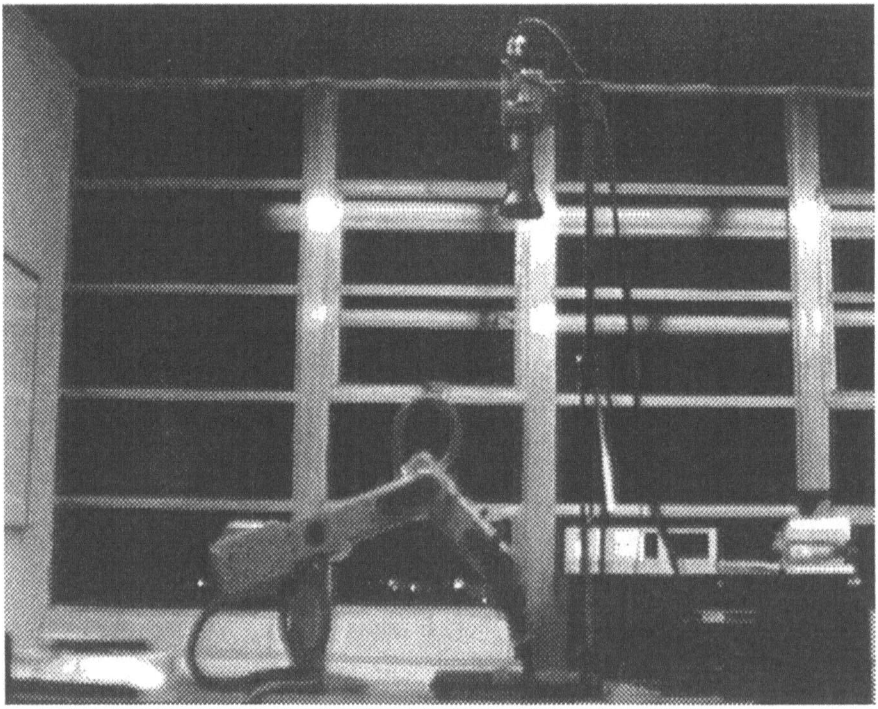

Figure 3 The physical setup of the system.

the previous chapter was utilized to recover the coordinates of the work piece in the three dimensional world coordinate system. This homogeneous transformation algorithm was implemented in Turbo C (Version 2.0) and assembly language.

5.2 Calibration At The Extreme Positions

As the first stage of the proposed hand-eye coordination algorithm, a calibration at two extreme locations (which are the lower and upper bounds of the camera displacement and they are 10 inch apart), A and B, was

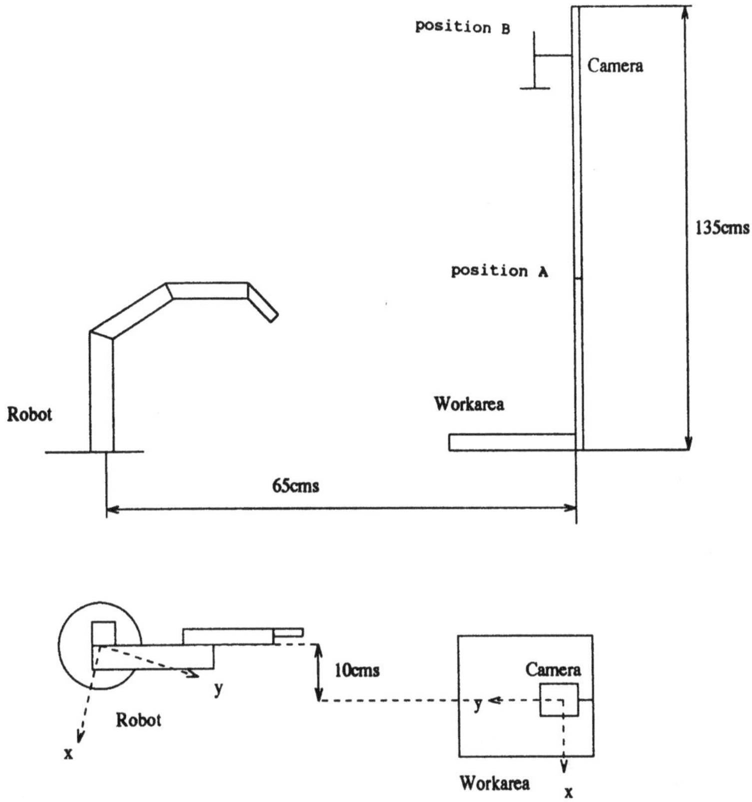

Figure 4 The illustration of the system configuration.

established based on the homogeneous transformation $T_{1,A}$ and $T_{1,B}$ whose coefficients are calculated with the data from the hand measurement. The measured data includes the relative distances between the robot base and the camera, plus the measured coordinate of the work piece in the robot centered world coordinate system. At the lower and the upper bound positions, A and B, images were digitized and processed to extract features. Figures 5 and 6 give two images digitized at lower and upper bounds respectively.

Then the distances for these two different images, d_A and d_B, were computed and saved as the reference value to set up the relation between transformation matrics $T_{1,A}$, $T_{1,B}$ respectively. The coefficients of these matrices are used to define the ranges upon which the membership functions will be defined.

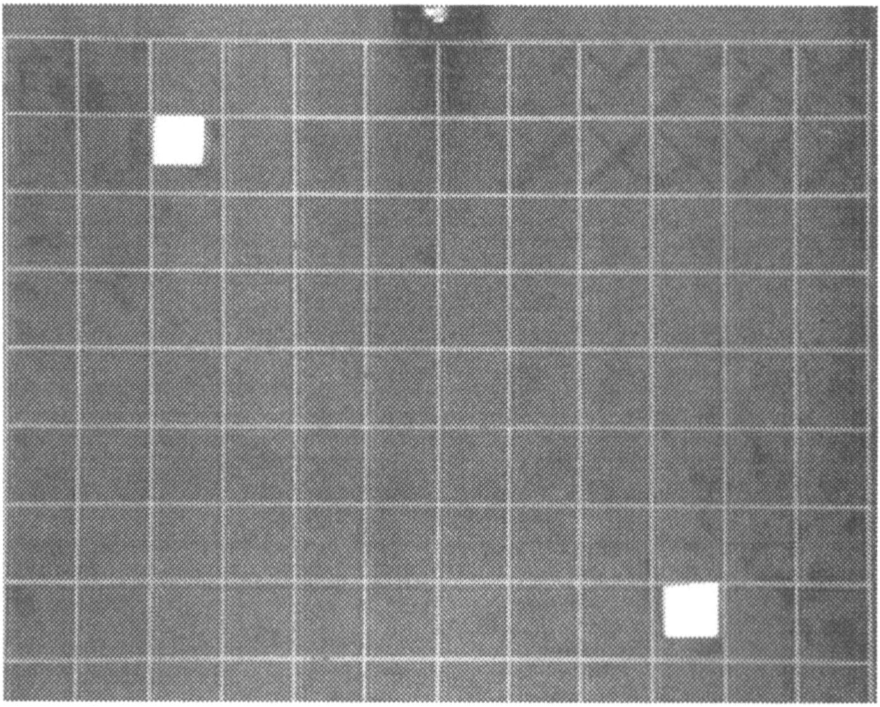

Figure 5 An image obtained when the CCD camera was placed at the upper bound position.

6 OPERATION UNDER CAMERA DISPLACEMENT

The second stage of our work is the implementation and testing of the fuzzy logic calibration algorithm for the environment with camera displacement. The amount of the displacement was not known prior to the experiment, nor is the direction of the displacement, which may be from the random disturbances or from specifically required task. The only indication of the camera displacement was obtained from the digitized images form the CCD

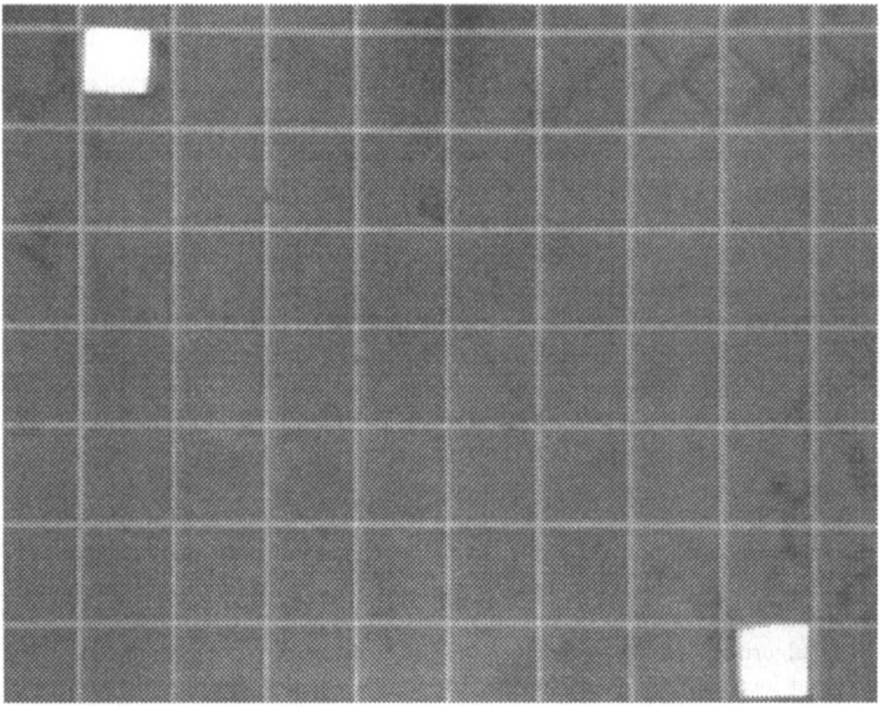

Figure 6 An image obtained when the CCD camera was placed at the lower bound position.

camera input. With this input, and the initial calibration at the landmark positions, we are now ready to move to fuzzy logic algorithm.

Using the digitized image data, preprocessing of the image removes undesired features, then image segmentation was performed. As the result, a binary image with only two different grey levels was generated. With the foreground as the potential candidate of the work piece. Then random noise removal was performed by ereasing the forground patterns, if any of these patterns do not fit into a predetermined size range. In other words, two thresholds T_{low} and T_{high} are chosen to define a size range $[T_{low}, T_{high}]$, if the size of a binary pattern $U_i < T_{low}$, or $U_i > T_{high}$, this pattern should be removed. A region labeling algorithm was utilized for this purpose to label each region with different index. Then the removal of the region of particular index was performed. After this removal process, pattern recognition technique of finding the curvatures, the orientations, and the high order moments were used to identify the landmarks and the work piece.

Next the squared distance d between two landmarks was changed due to the displacement of the camera. The changed distance was then recalculated. In addition, the position of the work piece on the image plane was calculated for the later use of calculating its position in the three dimensional, robot-centered world coordinate system. The fuzzification was then performed to fuzzifies the newly calculated distance d. Depending on its value, d was assigned to *all* different quantization groups, $\{L, \pm M, \pm S\}$ with corresponding membership values. Then the fuzzy reasoning was executed, for each rule, there was a corresponding control action with different membership value. Finally, the defuzzification process was executed to give the desired value of δt_1 and δt_4 respectively. From δt_1 and δt_4, changes in t_1 and t_4 were calculated. This then redefined the transformation matrix T_1 for the new camera position.

The algorithm was tested by displacing the camera at both regularly spaced positions between the lower and the upper bounds, and randomly chosed positions. it was working reasonably well. To be able to analyze the accuracy of the algorithm, we now describe the test of using 14 regularly spaced, different locations. At lower bound, the image was processed, and the transformation matrix was computed, then the location of the work piece in the three-dimensional world coordinate system was computed by using the transformation matrix. The robot was moved to this location to pick the object. The end effector's position was recorded. Then the camera is moved to the next position, which is 0.75 inche up towards the upper bound. The same computation was carried out and the data was recorded. This process was repeated for total 14 regularly spaced positions.

The coordinate of the workpiece at each different camera location is used as the reference point, denoted as (x_0, y_0), while the actual end effector's location is denoted as (x, y). The error is defined as

$$E = {\delta_x}^2 + {\delta_y}^2, \tag{9.20}$$

where $\delta_x = x - x_0$ is the error in the X direction and $\delta_y = y - y_0$ is the error in the Y direction. Table 1 gives the experimental result.

Table 1 Results of experimentation using fuzzy logic algorithm

Position	Squared Distance d	X	Y	δ_x	δ_y	E
A	40500	-53.7	-589.0	0.0	0.0	0.00
1	42800	-49.5	-587.0	-4.2	-2.0	21.64
2	44800	-50.0	-590.0	-3.7	+1.0	14.69
3	47000	-48.0	-592.0	-5.7	+3.0	41.49
4	49000	-51.7	-587.5	-2.0	-1.5	6.25
5	50100	-53.9	-535.0	-0.1	-4.0	16.01
6	51400	-58.0	-583.0	+4.3	-6.0	54.49
7	52700	-54.3	-585.9	+0.6	-3.1	9.97
8	54000	-55.0	-582.8	+1.3	-6.2	40.13
9	55300	-52.6	-585.2	-0.7	-3.8	14.93
10	56800	-54.8	-584.5	+1.1	-4.5	21.46
11	58280	-52.1	-586.4	-1.6	-2.6	9.32
12	60600	-53.9	-586.0	+0.2	-3.0	9.04
13	61800	-53.5	-587.0	-0.2	-2.0	4.04
14	63800	-58.0	-585.0	+4.3	-4.0	34.49
B	66790	-53.7	-589.0	+0.0	+0.0	0.00

Value of X and Y is in millimeters.
value of d is in square of pixels.
Cumulative Error for Fuzzy Logic E_c = 267.94

Note that at both the upper and lower bounds of the camera position, there is no error because of the initial calibration was performed at these two extreme points. This set of experimentation results was also given in Figure 7.

From the results, it has been observed that the maximum error occurred around the center of the entire displacement region.

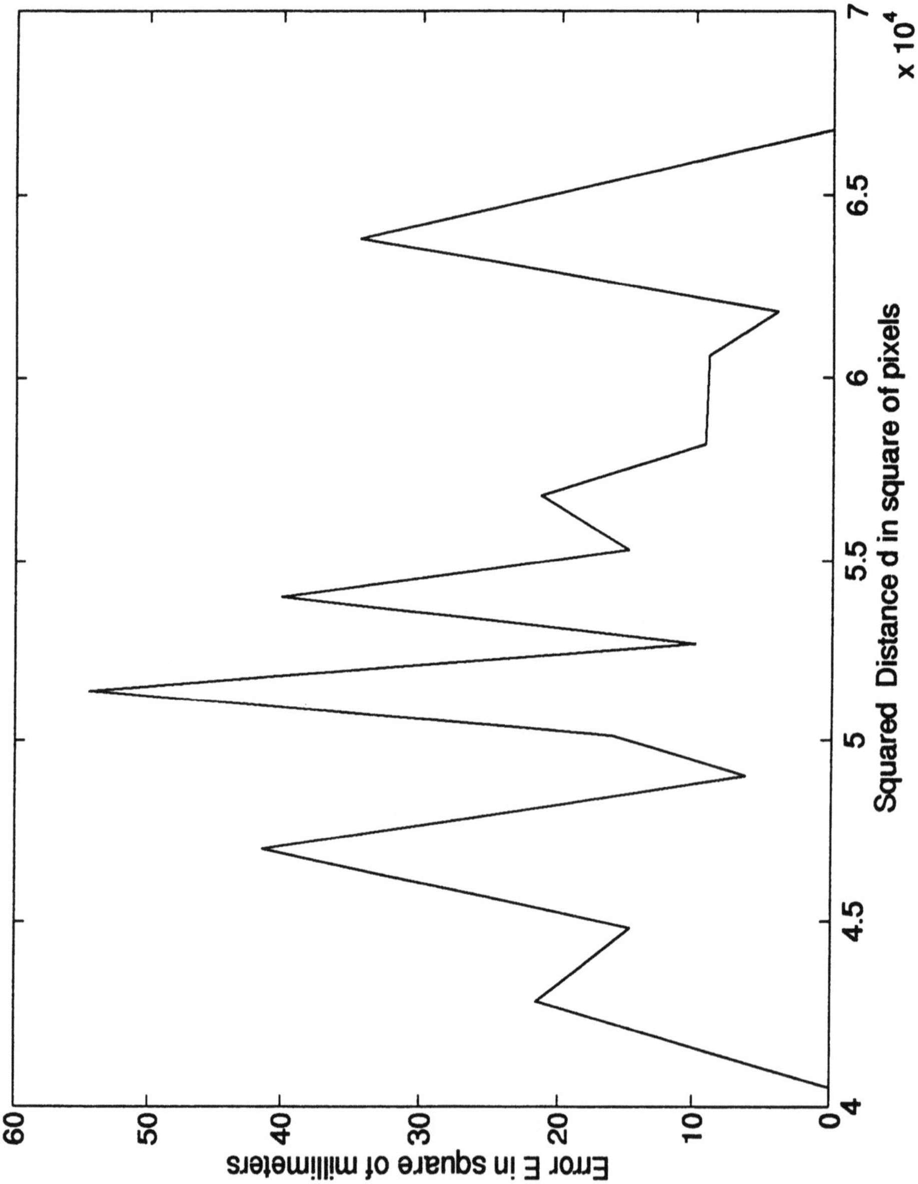

Figure 7 The plot of the experimental data based on fuzzy logic algorithm.

7 COMPARISON

To evaluate the proposed algorithm, we have also developed linear interpolation algorithm, which is given as,

$$t_1 = t_{1A} + \frac{(t_{1B} - t_{1A})}{(d_B - d_A)} * (d - d_A), \tag{9.21}$$

$$t_4 = t_{4A} + \frac{(t_{4B} - t_{4A})}{(d_B - d_A)} * (d - d_A), \tag{9.22}$$

where A and B are the two extreme points, lower and upper bounds of the camera displacement range. The linear interpolation algorithm was tested for the same set of experiment conditions as that of the fuzzy logic algorithm. Table 2 gives the results of the experiments.

Table 2 Results of experimentation with linear interpolation

Position	Squared Distance d	X	Y	δ_x	δ_y	E
A	40500	-53.7	-589.0	0.0	0.0	0.00
1	42800	-50.0	-587.0	-3.7	-2.0	17.69
2	44800	-52.0	-588.0	-1.7	-1.0	3.09
3	47000	-56.4	-587.0	+2.7	-2.0	11.29
4	49000	-57.7	-583.0	+4.0	-6.0	52.0
5	50100	-56.2	-534.0	+2.5	-5.0	31.25
6	51400	-61.0	-583.0	+7.3	-6.0	89.29
7	52700	-59.8	-584.0	+6.1	-5.0	62.21
8	54000	-61.0	-583.2	+7.3	-5.8	86.93
9	55300	-58.5	-583.0	+4.8	-6.0	59.20
10	56800	-63.1	-584.0	+9.4	-5.0	113.36
11	58280	-59.5	-585.0	-5.8	-4.0	49.64
12	60600	-59.6	-587.0	+5.9	-2.0	38.81
13	61800	-55.1	-586.0	+1.4	-3.0	10.96
14	63800	-56.7	-588.0	+3.0	-1.0	10.0
B	66790	-53.7	-589.0	+0.0	+0.0	0.00

Value of X and Y is in millimeters.
value of d is in square of pixels.
Cumulative Error for Linear Interpolation $E_c = 632.63$

Figure 8 gives the experiment result of both fuzzy logic and linear interpolation algorithms. It is found that the linear interpolation algorithm is generally more erroneous than that of the fuzzy logic algorithm due to the fact that the relationship between the changes of d and the changes of the coefficients of the transformation matrix is nonlinear.

We defineded an accumulative error E_c as the total error over the entire range of the camera displacement,

$$E_c = \sum_{j=1}^{n} E_j. \tag{9.23}$$

where E_j is the error occurring in the X and Y coordinates for any j position and n is the number of positions where the camera was placed. E_c for the fuzzy logic algorithm was found to be 267.94 and for linear interpolation algorithm was found to be 632.63. From these values, it can be observed that the fuzzy logic algorithm gives a smaller cumulative error than that of the linear interpolation algorithm.

It was also found that for our particular applications, the error introduced by the fuzzy logic algorithm lies within the tolerable range for the robot to perform pick-and-place operation. In the case of linear interpolation algorithm, at certain places, the error was beyond the tolerable range which cause the failure of the pick-and-place operation.

8 CONCLUSION

In this study, we have analyzed the effect of the change in the camera position and its impact on the calibration of the hand-eye system. We have described a fuzzy logic algorithm for taking care of the camera displacement either due to the random disturbances or due to prior unknown task requirement. The fuzzy logic algorithm is able to maintain the hand-eye coordination without the need of recalibration within a certain limits. This algorithm saves time since there is no need of calibrating the hand-eye system every time the camera is moved. This algorithm can be used in the places where different tasks require the camera be placed at different position while still maintaining hand-eye coordination. As the future research, we feel there is a need to address the camera position change in more general situations, to include pan, tilt operations. Also there is a need to further improve the accuracy of the present algorithm.

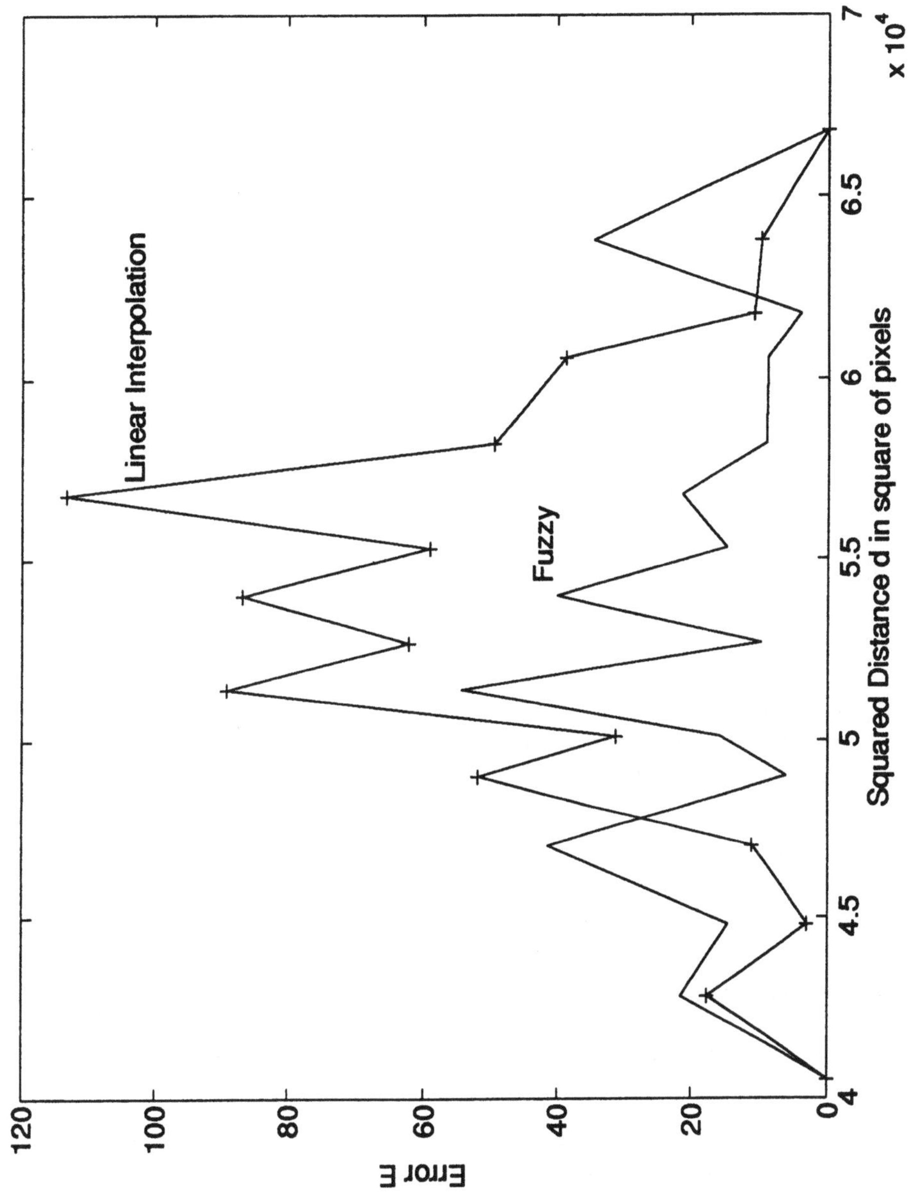

Figure 8 The comparison of the fuzzy logic and linear interpolation algorithms.

REFERENCES

[1] S. Arimoto, S. Kawamura, and F. Miyazaki, Can mechanical robots learn

by themselves?," *Robotics Research: 2nd Int. Symp.* Cambridge, MA: MIT Press, 1985, pp. 127-134.

[2] R.A. Brooks, A.M. Flynn, and T. Marill, "Self calibration of motion and stereo vision for mobile robots," *Proc. 4th Int. Symp. Robotics Research*, Santa Cruz, CA, Aug. 1987.

[3] O.D. Faugeras and G. Toscani, "Camera Calibration for 3-D computer vision," *Proc. Int. Workshop Industrial Applications of Machine Vision and Machine Intelligence*, Tokyo, Japan, Feb. 1987, pp. 240-247.

[4] J.T. Feddema, G. Lee, and R. Mitchell, "Model-Based Visual Feedback Control for a Hand-Eye Coordinated Robotic System," *IEEE Transactions on Computers*, page 21-29, Aug 1992.

[5] M.A. Fischer and R.C. Bolles, "Random Sample Consensus: A Paradigm for Model Fitting with Applications to Image Analysis and Automated Cartography," *Comm. ACM*, Vol. 24, No. 6, June 1981, pp. 381-395.

[6] A. Izaguirre, P. Pu, and J. Summers, "A new development in camera calibration: Calibrating a pair of mobile cameras," *Int. J. Robotics Res.*, vol 6, no.3, pp. 104-116, 1987.

[7] M. Kuperstein, "Adaptive visual-motor coordination in multijoint robots using parallel architecture," *Proc. 1987 IEEE Int. Conf. Robotics and Automation*, Raleigh, NC, 1987, pp. 1595-1602.

[8] C.C. Lee, "Fuzzy Logic in Control Systems: Fuzzy Logic Controller-Part I," IEEE Transactions on Systems, Man, Cybernetics, Vol 20, No 2, March/April 1990, pp 404-418.

[9] C.C. Lee, "Fuzzy Logic in Control Systems: Fuzzy Logic Controller -Part II," *IEEE Transactions on Systems, Man, Cybernetics*, Vol 20 , No 2, March/April 1990, pp 419-434.

[10] R.K. Lenz and R.Y. Tsai, "Techniques for calibration of the scale factor and image center for high accuracy 3-D machine vision metrology," *Proc. IEEE Int. Conf. Robotics and Automation*, 1987, pp. 68-75.

[11] P. Liang, Y.L. Chang, and S. Hackwood, "Adaptive self-calibration of vision-based robot systems," *IEEE Trans. on Systems, Man, and Cybernetics*, vol 19, no. 4, July 1989, pp. 811-824.

[12] P. Liang, J. F. Lee, and S. Hackwood, "A general framework for robot hand-eye coordination," *Proc. IEEE Int. Conf. on Robotics and Automation*, 1988.

[13] T.M. Martinez, H.J. Ritter, and K.J. Schulten, "Three-dimensional Neural Net for Learning Visuomotor Coordination of a Robot Arm," *IEEE transactions on neural networks*, Mar 01 1990, v 1, n 1, Page 131.

[14] W.T. Miller III, "Sensor-based control of robotic manipulators using general learning algorithm," *IEEE J. Robotics Automation*, Vol. RA-3, pp. 157-165, Apr. 1987.

[15] T. Poggio and C. Koch, "Ill-posed Problems in early vision: From computational theory to analog networks," *Proc. of Royal Society, London* B-226, pp 303-323, 1985.

[16] A.C. Sanderson and L. E. Weiss, "Image-based servo control of Robots," *Proc. 26th Ann. SPIE Technical Symp.*, Aug. 1987.

[17] S.B. Skaar, W.H. Brockman, and R. Hanson, "Camera-space calibration," *Int. J. Robotics Res.*, vol 6, no. 4, pp. 20-32, 1987.

[18] R.Y. Tsai, "An efficient and accurate camera calibration technique for 3D machine vision," *Proc. IEEE Computer Vision and Pattern Recognition*, 1986, pp. 364-374.

[19] R.Y. Tsai and R.K. Lenz, "A new technique for fully autonomous and efficient 3D robotics hand/eye calibration," *Proc. 4th Int. Symp. Robotics Research*, Santa Cruz, CA, Aug. 1987.

[20] J.A. Walter and K.J. Schulten, "Implementation of Self-Organizing Neural Networks for Visuo-Motor Control of an Industrial Robot," *IEEE transactions on neural networks*, Jan 01 1993 v 4 n 1 page 86.

10

USING FPGA TECHNIQUE FOR DESIGN AND IMPLEMENTATION OF A FUZZY INFERENCE SYSTEM

Donald Hung

Department of Electrical Engineering
Gannon University
Erie, PA 16541, USA

1 INTRODUCTION

Recently, there has been increased use of fuzzy logic in control applications. The kernel of a fuzzy logic controller (FLC) is a fuzzy inference engine which includes the knowledge base, the decision making logic, and the defuzzification unit, as shown in Figure 1. For most of the reported applications, fuzzy inference algorithms of the FLCs were implemented in software and executed on a standard Von Neumann type processor or microcontroller. Although software-based FLCs are in general more economical and flexible, they often have difficulty in dealing with control systems that require very high processing and I/O handling speeds. For this reason, in the recent years, a number of hardware fuzzy inference systems have been reported or proposed [1-5, 7, 9-16] that reflect the diversity in fuzzy inference methods and in technologies related to hardware design. In general, analog hardware is relatively simple but lacks accuracy and reliability; on the other hand, digital hardware is accurate and reliable but is more complex and lacks speed if an iterative algorithm is unavoidable. In recent years, rapid advances in digital technology allow system designers to design custom computing machines based on a variety of technologies: full custom application-specific IC (ASIC), standard-cell, gate array, FPGA, programmable logic device (PLD), standard IC, etc.. Among these, FPGA is especially suitable for fast implementation and quick hardware verification. Besides technology alternatives, design decisions have to be made based on tradeoffs on performance, hardware resource and flexibility. In general, higher performance can be achieved at the expense of more hardware resource and less flexibility.

In this chapter, the design and implementation of a digital hardware fuzzy logic controller is discussed. Since the guideline for the design was to pursue high performance, the designed FLC is algorithm- specific, i.e., instead of trying to support a large number of possible fuzzy inference methods, the FLC is dedicated to the algorithm which is based on the most widely adopted fuzzy inference mechanism, namely, GMP-based inference with the "mini" operator for fuzzy implication and the "max-min" operator for fuzzy composition. To achieve high performance, a system architecture with high degree of parallelism was chosen. For the defuzzification algorithm, the design adopted the most favorable COG method. Traditionally, digital hardware designs for COG-based defuzzification employ iterative multiplier and divider circuitry. As will be shown in the chapter, this limits the overall throughput of the system and becomes the computational bottleneck of the fuzzy inference process. To increase processing speeds, a table-lookup approach for implementing the COG function was proposed which eliminated the need for multiplier and divider circuitry and significantly improved the system's throughput. For on-line adaptation, the FLC must be able to update its knowledge base. The idea in designing the FLC is to let the system support the maximal amount of rules; therefore updating the knowledge can be done by just updating the contents of the storage modules. With this in mind, the designed FLC uses two memory banks for its storage modules. Both memory banks can be accessed by a host processor. When the FLC is operating based on the knowledge based stored in one of the memory bank, the host processor can update the knowledge based stored in the second memory bank at the same time. For hardware verification, a two-input (antecedents), one-output (consequent) FLC based on the general design consideration was implemented with a Xilinx XC4008-6 FPGA and a Cypress CY7C251 EPROM. As expected, the system achieved a very high processing speed. Details of the system's functional blocks and tested performance data are provided and some remaining issues are discussed as the conclusion of the chapter.

2 THE FUZZY INFERENCE ALGORITHM

About forty different fuzzy inference methods have been proposed in the literature [6,8], as the results of different fuzzy inference rules and different fuzzy implication, composition and defuzzification methods used in the inference process. In practice, most of the fuzzy applications adopt an inference mechanism which is based on the combination of the following:

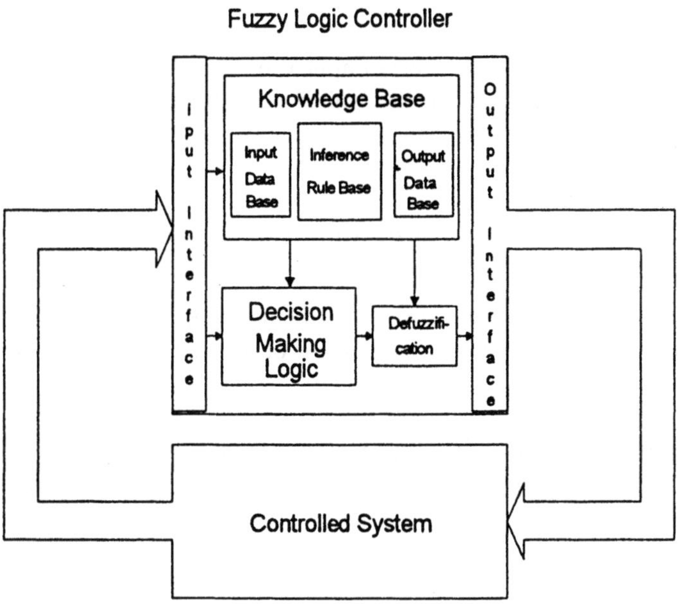

Figure 1 A fuzzy logic controller in a control loop

- Inference rule: Generalized Modus Ponens (GMP)
- Implication operator: Mini
- Composition operator: Max-Min
- Defuzzification method: Center of Gravity (COG)

Our discussion on hardware design is based on the fuzzy inference mechanism listed above.

The GMP fuzzy inference rule is stated as below:

Premise 1:	If input is A then output is B
Premise 2:	Input is A'
Consequence:	Output is B'

where A, B, A', B' are fuzzy sets.

Premise 1 is described by the fuzzy implication function $\mathbf{R} = \mathbf{A} \rightarrow \mathbf{B}$, where **R** is a fuzzy relation and, when the "mini" operator is used, its membership function values are computed as:

$$\mu_R(u, v) = \min(\mu_A(u), \mu_B(v)); u \in U, v \in V. \tag{10.1}$$

In (10.1), $\mu_R(u,v), \mu_A(u)$ and $\mu_B(v)$ are membership functions of **R, A and B** respectively, and **U** and **V** are universes of discourse of **A** and **B**, respectively.

The *Consequence* is determined by the fuzzy composition $\mathbf{B}' = \mathbf{A}' \circ \mathbf{R}$ and, when the "max-min" operator is used, the membership function value of $\mathbf{B}'$ is computed as:

$$\mu_{B'}(v) = \max \min(\mu_{A'}(u), \mu_R(u, v)) \tag{10.2}$$

where $\mu_{B'}(v)$ and $\mu_{A'}(u)$ are membership functions of $\mathbf{B}'$ and $\mathbf{A}'$, respectively.

The above fuzzy inference procedure corresponds to a SISO (single-input single-output) FLC. GMP- based multiple-input multiple-output (MIMO) FLCs can be partitioned into a set of multiple-input single- output (MISO) subsystems with respect to each output. Further, when the "mini" fuzzy implication function and the "max-min" fuzzy composition rule are used, the following relation holds [6]:

$$\begin{aligned} B' &= (A'_1, A'_2, \ldots, A'_n) \circ (A_1 \times A_2 \times \ldots \times A_n \rightarrow B) \\ &= \min[(A'_1 \circ (A_1 \rightarrow B), A'_2 \circ (A_2 \rightarrow B), \ldots, A'_n \circ (A_n \rightarrow B)] \end{aligned} \tag{10.3}$$

and the membership function value of $\mathbf{B}'$ can be computed as

$$\mu_{B'} = \min\{[\min(\max\min(\mu_{A'i}, \mu_{Ai}))], \mu_B\}, \; for\, i = 1, 2, \ldots, n. \tag{10.4}$$

In applications, inputs and outputs of a FLC are fuzzy linguistic variables, each of which is characterized by usually more than one fuzzy linguistic values (fuzzy sets). As a consequence, with a sample of observed input values, multiple inference rules can be activated simultaneously. Since a crisp output (decision) is required by most applications, FLCs normally adopt certain defuzzification strategies to determine a crisp value in the output space. The computation required by the widely used COG-based defuzzification method is

$$z_0 = \frac{\sum w_i z_i}{\sum w_i}, \tag{10.5}$$

where w_i is the weight (membership function value) of the i th activated rule and z_i is the correspondent crisp value in the universe of discourse of the output space.

A graphic interpretation of the computation described by (4) through (6) is given in Figure 2, where the simple fuzzy inference system has two inputs and one output with two inference rules: 1) if $\mathbf{A_1}$ and $\mathbf{B_1}$ then $\mathbf{C_1}$, 2) if $\mathbf{A_2}$ and $\mathbf{B_2}$ then $\mathbf{C_2}$ and the observed inputs $\mathbf{A}'$ and $\mathbf{B}'$ are assumed to be crisp.

With the help of Figure 2, the hardware algorithm for fuzzy inference systems under discussion can be summarized as follows:

Hardware Algorithm

1. Perform the "maximum of minimum" operation between each observed input linguistic value (a fuzzy set in general) and the system's input membership functions in the correspondent input space.

2. Perform the "minimum" operation among values obtained from *Step 1* and relate to the same inference rule.

3. For all weights obtained from *Step 2*, perform

   (a) the "summation" operation $\sum w_i$,

   (b) the "multiplication" operation $w_i z_i$.

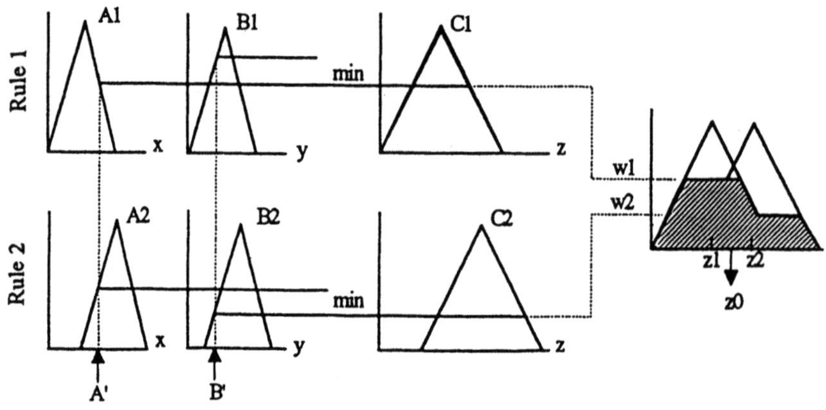

Figure 2 A graphical interpretation of the fuzzy inference computation.

4. Perform the "summation" operation $\sum w_i z_i$ for all $w_i z_i$ obtained from *Step 3b*.

5. Perform the "division" operation $\frac{\sum w_i z_i}{\sum w_i}$ using the results obtained from *Step 3a* and *Step 4*.

The five steps listed above must be executed sequentially, but operations inside each step can be executed concurrently. From a hardware point of view, data flow from *Step 1* to *Step 2* clearly bears a parallel architecture, and *Step 3a, 3b* can be pipelined. *Step 3* through *Step 5* are for the COG-based defuzzification. In this part, multiplication and division are encountered. Since their correspondent arithmetic algorithms are complicated and require a significant amount of iterative operations, hardware fuzzy inference systems that employ multiplier and divider circuitry in their COG design suffer from greater complexity and lower overall processing speed.

3 FPGA DESIGN CONSIDERATIONS

A major concern in this design is to keep the system architecture with high degree of parallelism and eliminate the iterative multiplication and division operations in ***Step 3b*** and ***Step 5*** of the hardware algorithm discussed in ***Section 2***. Along with the hardware algorithm, the following constraints have been applied to the design:

Design Constraints

1. The observed inputs of the FLC are crisp and digitized.

2. For each input variable, the overlapping degree of its membership functions is limited to 2 (i.e., no more than two input membership functions can be overlapped in the same input space) and the maximum overlap of two adjacent membership functions is 0.5.

3. Output membership functions are symmetric in shape.

The first two constraints are based on practical considerations and experience gained from previous fuzzy applications. The third constraint is basically for simplicity in implementation. These constraints relate to the following hardware design considerations:

Design Considerations

1. With ***Design Constraint 1***, the "maximum of minimum" operation in the hardware algorithm (***Step 1***) discussed before can be simplified as a table search operation. Values of a input membership function can be prestored in a lookup table and the digitized crisp values of the observed input serve as the addresses for accessing the table.

2. ***Design Constraint 2*** means that for a FLC with n inputs, given a sample of observed input values, at most 2^n inference rules can be activated. This constraint also reduces the estimated lengths of the binary sum $\sum w_i$ and $\sum w_i z_i$ in the hardware algorithm (***Steps 3a, 4 and 5***).

3. ***Design Constraint 3*** significantly reduces the storage and computation required by the COG-based defuzzification by allowing the usage of a

simplified version of the COG algorithm in which the z_i's are considered the center values in the universe of discourse of output membership functions. The ith output membership function therefore can be treated as a fuzzy singleton situated at z_i, in the universe of discourse of the output space. This approach is known as the *truth value fuzzy inference* (TVFI). The simplicity of the TVFI leads to the next two alternative design considerations that eliminate the need of iterative multiplication and division hardware according to the hardware algorithm.

4. Since the input and output membership function values have limited resolution (scales), for a particular fuzzy output singleton situated at z_i, all possible $w_i z_i$ values can be precalculated and stored in a lookup table which has values of w_i as the addresses for accessing the table.

5. Since both the $\sum w_i$ and $\sum w_i z_i$ in the hardware algorithm have limited length, all possible values as the results of the division $\frac{\sum w_i z_i}{\sum w_i}$ can be precalculated and stored in a lookup table, and values of the concatenated $\sum w_i z_i$ and $\sum w_i$ serve as the addresses for accessing the table.

The overall block diagram of a MISO FLC based on the above discussion is shown in Figure 3. In the data section of the FLC, the *Input* lookup table is for *Step 1* of the hardware algorithm; the *Min* module performs the "minimum" operation described in *Step 2* of the hardware algorithm; the *Max* module selects the activated inference rules and produces their weights (w_i), based on the current sample of the observed input values; the *Product* lookup table produces the products (values of $w_i z_i$) correspondent to the weights. To reduce hardware resources, the *Sum* module is shared by the w_i inputs and the $w_i z_i$ inputs and a partial pipeline is formed here, i.e., while the module is adding up the $\sum w_i$, values of $w_i z_i$'s are being fetched from their lookup tables. The $\sum w_i$ and $\sum w_i z_i$ are finally latched to form the address bus to the *Division* lookup table. The depth of the table is 2^m, where m is the total length of the concatenated binary data $\sum w_i$ and $\sum w_i z_i$. Note that the *Division* lookup table only depends on the lengths of $\sum w_i$ and $\sum w_i z_i$. As long as its address bus is wide enough to cover the length of the concatenated $\sum w_i$ and $\sum w_i z_i$, the *Division* lookup table is independent of changes in the rest parts (the fuzzification strategy, the choices of the system's data base and rule base, etc.) of the FLC. The registers (R1 through R7) are inserted into the data path to facilitate the implementation of pipelines.

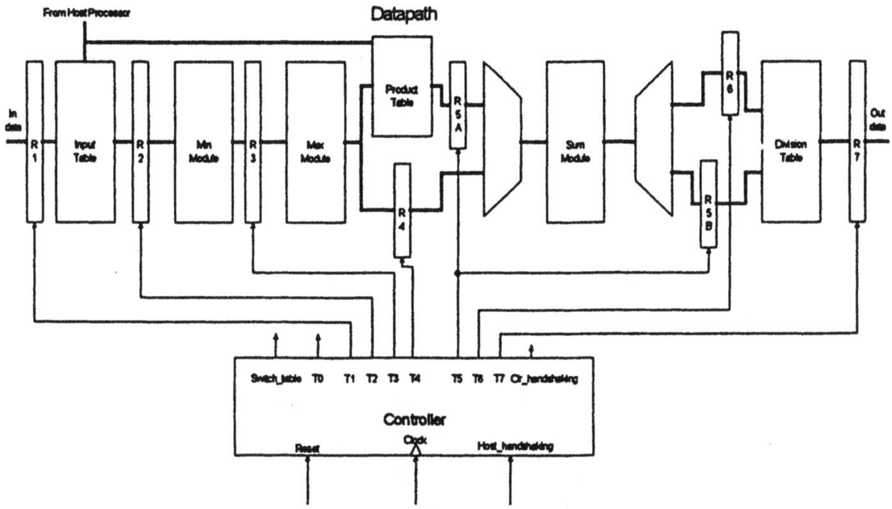

Figure 3 Block diagram of the fuzzy logic controller.

The control section of the FLC is a finite state machine (FSM) which controls operations of the FLC's data path by generating the following basic commands:

- latch the system inputs (control signal T1)
- latch the *Min* module inputs (control signal T2)
- latch the *Max* module inputs (control signal T3)
- latch the w_i inputs to the *Sum* module (control signal T4)
- latch the $\sum w_i$ outputs from the *Sum* module (control signal T5)
- latch the $w_i z_i$ inputs to the *Sum* module (control signal T5)
- latch the $\sum w_i z_i$ outputs from the *Sum* module (control signal T6)

- latch the *Division* lookup table outputs (control signal T7)

The control logic also controls the FLC's adaptation action by communicating with the host processor through handshaking signals 'host_handshaking' and 'clear_handshaking', and generates the 'switch_table' signal accordingly. The basic configuration of the FSM can be easily modified to pipeline the above operations with different overlapping stages.

4 DESIGN METHODOLOGY AND IMPLEMENTATION

Based on discussions in the previous section, a two-input (fuzzy linguistic variables **E** and **D**), single output (fuzzy linguistic variable **C**) FLC has been implemented. The observed input values are assumed to be digitized into six bits. Each input variable of the system has three membership functions which implies up to nine inference rules can be adopted. All membership function values are represented by four bits of binary which gives sixteen levels of resolution. A maximum of nine distinct fuzzy singletons are equally distributed in the universe of discourse of the output space. The defuzzified output values are represented by eight bits of binary numbers. The fuzzy output singletons are separated sixteen units apart from each other therefore only four bits are needed to represent their locations.

The overall design was managed in a layered hierarchy, and the top-down approach was adopted. The design started at the top level where the overall system was specified by behavioral level VHDL description. The correctness of the overall design concept was verified by comparing the test results generated by a VHDL test bench and those computed by a C++ program. The design and verification process then proceeded to the second level where the individual modules of the datapath and the control logic of the FLC were specified in behavioral level VHDL code and then simulated using separate VHDL test benches. When this step succeeded, the overall system was constructed by wiring all the modules in the datapath and the control logic together through structural level VHDL description. Since the system was targeted at the Xilinx 4000 family FPGAs, the third-level design, i.e., the hardware details of each individual module and the control logic, were entered by using the Viewlogic schematic capture tool Viewdraw, based on macros and primitives available in the XC4000 library. Building blocks of the main modules in the FLC's datapath are introduced below:

4.1 The Input Lookup Table

This module contains twelve 16 x 4 bits SRAMS (XC4000 macro RAM16 x 4) in parallel to store the input membership function values. Six of them for normal operation; another six belong to the secondary memory bank for on-line adaptation. Each of the 16 x 4 SRAM can be accessed by both the related sensor input and the host processor, and switch between the memory banks are controlled by the FLC's control logic, as shown in Figure 4.

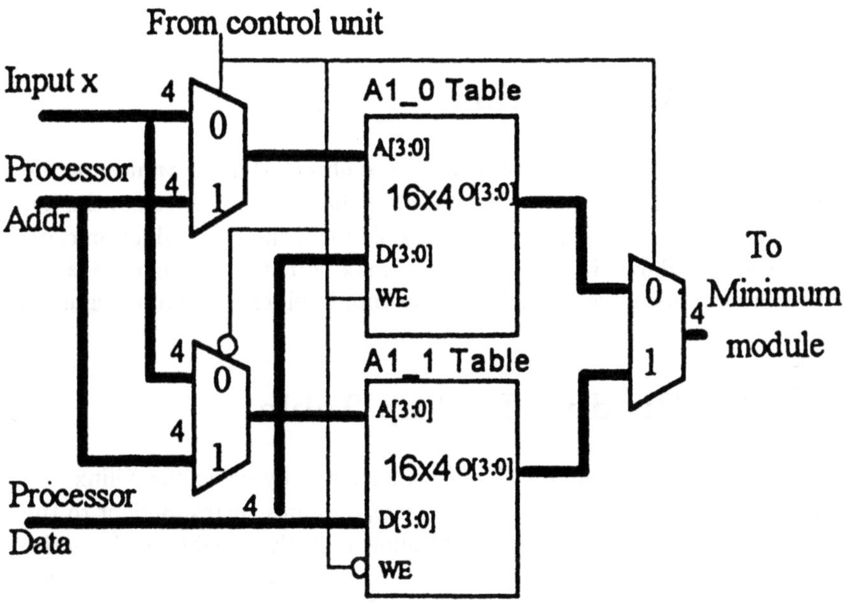

Figure 4 The input lookup table.

4.2 The Min Module

This module contains nine identical "mini" blocks (constructed by using XC4000 macros COM4 and X74-257) that execute in parallel. Each of the "mini" block selects the minimum from its two 4-bit inputs, as shown in Figure 5.

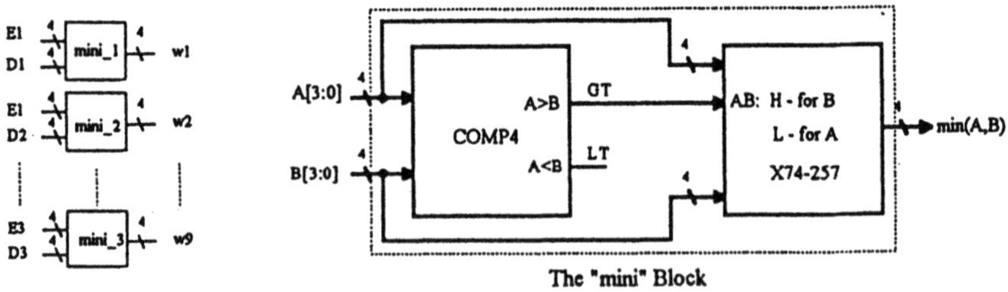

Figure 5 Building block of the Min module.

4.3 The Max Module

Due to *Design Constraint 2* , for a sample of observed input values, no more than four inference rules can be activated in this system. To select the four active inference rules among a set of possible nine, the rule reduction criterion was derived and shown in Figure 6. The "max" function in Figure 6 is implemented in a way very similar to the "mini" blocks shown in Figure 5.

4.4 The Product Lookup Table

This module includes memory blocks (constructed by using the Xilinx FPGAs' on-chip resources) that store the $w_i z_i$ values correspondent to the weights, and some decoding logics that allow only the weights of the activated inference rules to access the table. Figure 7 shows part of the data flow within the *Max Module* and the *Product Lookup Table* which illustrates how these two modules work together. For simplicity, the second memory block for adaptation is not shown.

4.5 The Sum Module

This module produces the two sums $\sum w_i$ and $\sum w_i z_i$. It consists two levels of adders (XC4000 macros ADD8 and ADD12) in a tree structure, as shown in Figure 8.

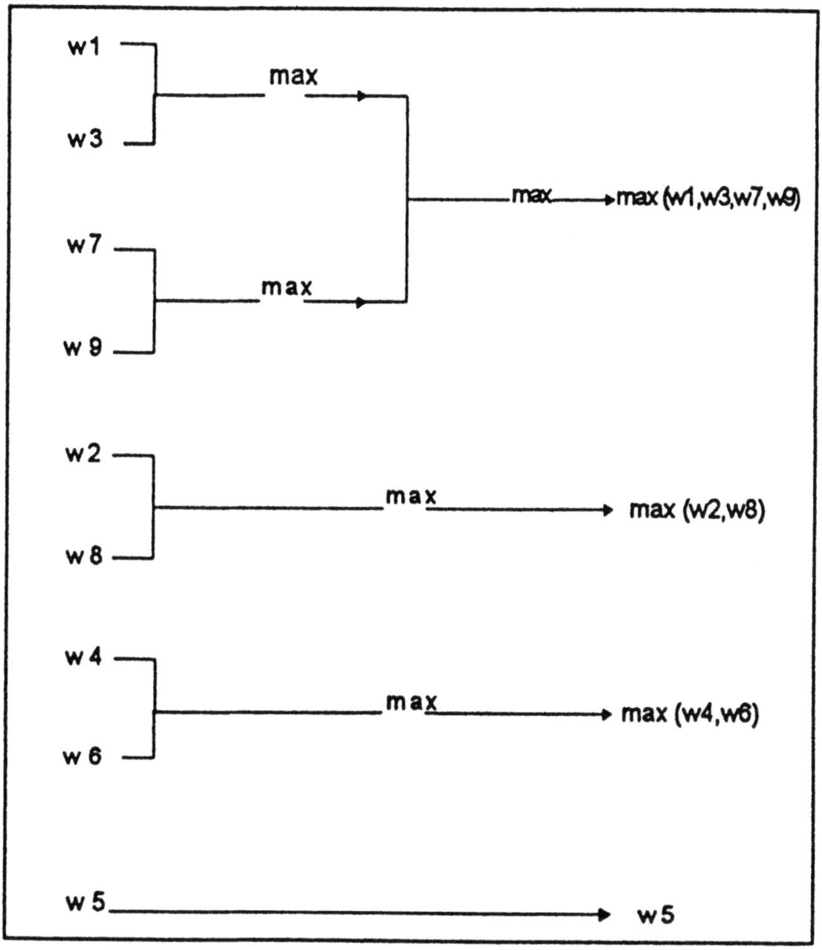

Figure 6 The rule reduction criterion.

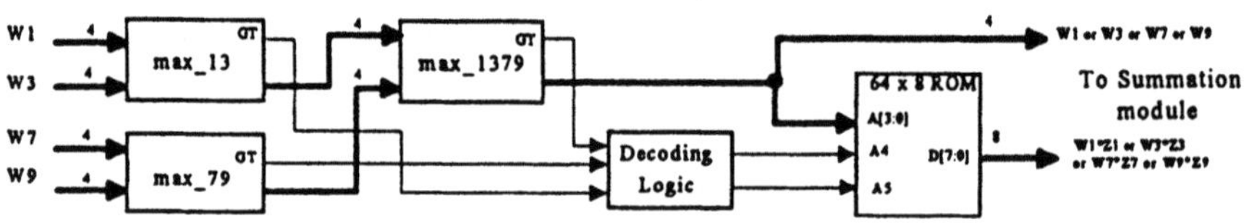

Figure 7 Part of the connected Max module and the product lookup table.

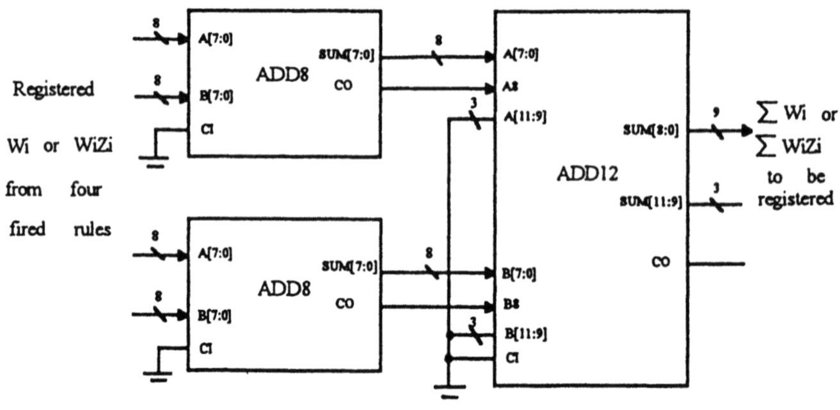

Figure 8 The sum module.

4.6 The Division Lookup Table

With all the given information, the reader can verify that 1) the $w_i z_i$ values can be stored in eight-bit binary format, 2) the sum $\sum w_i$ can be represented by a five-bit binary number, and 3) the sum $\sum w_i z_i$ can be represented by a nine-bit binary number. The *Division* lookup table was implemented by using a separate Cypress CY7C251 16K x 8 EPROM with 50 ns access time. Again, this module is independent of the rest of the FLC.

The FLC with hardware details (targeted to the Xilinx 4008-6 FPGA) was simulated with the Viewlogic simulation tool Viewsim, and then implemented by using the Xilinx placement and routing tools. The datapath of the FLC was then "locked" to avoid unpredictable delays as the results of later placement and routing processes. At this time, no overall pipelining was considered in the control logic design. After the post-layout timing information of the "locked" datapath was collected and carefully analyzed, the delays times were inserted into the second level VHDL specifications. With detailed timing information on different segments of the datapath, the FLC's control logic was modified to improve the system's over all throughput.

5 TESTING RESULTS

The FLC (except its *Division* module) was implemented on a Xilinx XC4008-6 FPGA with 6 ns delay per configurable block (CLB). Performance of the FLC with three different versions of the control logic is listed below:

System Clock	Pipelining	Control Actions/Second
10 Mhz	None	1.43 M
40 Mhz (15 wait states)	T1, T7	1.90 M
10 Mhz	T1, T2, T3, T4, T7	3.30 M

By using a faster system clock and inserting wait states, throughput of the fully pipelined FLC can be further increased.

6 CONCLUSION AND DISCUSSION

The design considerations of a hardware fuzzy inference system in general, and the implementation of a two-input, one-output system as a special case, have been described in previous sections. As a conclusion of this chapter, some remaining issues are briefly discussed below.

Performance. Due to the high degree of parallelism of the system's architecture and the simplicity of its computational modules, successful performance results (measured by control actions or fuzzy inference per second) were obtained compared to those available from other reported systems [1, 9, 10, 12, 13, 14]. Considering that the implemented system was targeted at a Xilinx XC4000 FPGA with 6 ns delay per CLB, system performance may be further improved by targeting the design to other technologies or to the same device with a better speed grade.

Rule Aggregation. In the rule base of a certain fuzzy inference system, if more than one inference rules that relate to the same output membership function are activated by the same inputs, e.g., if rules $\mathbf{A_1} \times \mathbf{B_1} \rightarrow \mathbf{C_1}$ and $\mathbf{A_2} \times \mathbf{B_2} \rightarrow \mathbf{C_1}$ are activated simultaneously for a given input pair $(\mathbf{A'}, \mathbf{B'})$, theoretically the two inferred $\mathbf{C_1'}$'s need to be aggregated before entering the defuzzification stage. (The aggregation operator can be either disjunctive, e.g., the "max" operator or conjunctive, e.g., the "min" operator, depending on which one is more appropriate for a specific application). Based on our experience, when the COG method is used in defuzzification, the rule aggregation stage can be skipped without causing noticeable differences on the final defuzzified results. For this reason hardware for rule aggregation function was not considered in the implemented system discussed in *Section 4* , although the inclusion of this function may not necessarily increase the complexity of the overall system.

Extendibility As stated before, a MIMO FLC based on GMP can be decomposed into a number of MISO subsystems in parallel. Theoretically, the concept and architecture of the implemented two-input, one-output FLC can be extended to handle any number of inputs. Due to the architecture's high degree of parallelism, the extension should not cause deterioration in overall performance. The real constraint, however, comes from the availability of hardware resource. For a MISO FLC with n inputs, assuming that each input is characterized by m membership functions in the input data base, the FLC may have up to m^n different inference rules in its rule base. With *Design Constraint 2* mentioned in *Section 3* , up to 2^n inference rules may be activated by a given input sample. These observations show that the required hardware resource increases exponentially with the increase of the system's input. Although the rule reduction circuitry based on *Design Constraint 2* will eliminate a considerable portion of the required hardware resource, the complexity of the rule reduction circuitry itself may increase remarkably. To cope with large number of inputs by using the two-input, one-output FLC discussed in *Section 4* as a basic module, one possible approach is to decompose an inference rule with a long list of antecedents (input linguistic variables) into a set of two-antecedent inference rules that are linked by some appropriate connectors. For instance, decompose $\mathbf{A} \times \mathbf{B} \times \mathbf{C} \times \mathbf{D} \rightarrow \mathbf{E}$ into

$(\mathbf{A} \times \mathbf{B} \rightarrow \mathbf{E})$ & $(\mathbf{C} \times \mathbf{D} \rightarrow \mathbf{E})$ where **'&'** is the connector, and $\mathbf{A}, \mathbf{B}, \mathbf{C}, \mathbf{D}, \mathbf{E}$ are linguistic variables.

All subjects discussed in this section are under further study.

REFERENCES

[1] W.D. Detloff, K.E. Yount and H. Watanabe, "A fuzzy logic controller with reconfigurable, cascadable architecture," Proceedings of 1989 IEEE International Conference on Computer Design: VLSI in Computers and Processors, pp.474-478, 1989.

[2] M.A. Eshera and S.C. Barash, "Parallel rule-based fuzzy inference on mesh-connected systolic arrays," IEEE EXPERT, Winter 1989, pp.27-35.

[3] D.L. Hung and W.F. Zajak, "Implementing a fuzzy inference engine using field programmable gate array," Proceedings of the IEEE International ASIC Conference, pp349-352, 1993.

[4] D.L. Hung, "Custom design of a hardware fuzzy logic controller" Proceedings of the 3rd IEEE International Conference on Fuzzy Systems, June, 1994.

[5] C. Isik, "Inference hardware for fuzzy rule-based systems," FUZZY COMPUTING, M.M. Gupta and T. Yamakawa Ed., ELSEVIER Science Publishers B.V. (North Holland), 1988, pp.185-194.

[6] C.C. Lee, "Fuzzy logic in control systems: Fuzzy logic controller, Part II," IEEE Transactions on Systems, Man, and Cybernetics, Vol.20, No. 20, pp.419-435, March/April 1990.

[7] M-H. Lim and Y. Takefuji, "Implementing fuzzy rule-based systems on silicon chips," IEEE EXPERT, February 1990, pp.31-45.

[8] M. Mizumoto and H. Zimmermann, "Comparison of fuzzy reasoning methods," FUZZY SETS AND SYSTEMS, Vol.18, pp.253-283, 1982.

[9] Sujal Shah and Ralph Horvath, "A hardware digital fuzzy inference engine using standard integrated circuits," Proceedings of the 1st International Conference on Fuzzy Theory and Technology, pp.109-114. Durham, 1992.

[10] Masaki Togai and Hiroyuki Watanabe, "Expert system on a chip: an engine for real-time approximate reasoning," IEEE EXPERT, Fall 1986, pp.56-58.

[11] Masaki Togai and S. Chiu, "A fuzzy logic chip and a fuzzy inference accelerator for real-time approximate reasoning," Proceedings of the 17th International Symposium on Multiple-Valued Logic, pp.25-29, 1987.

[12] A.P. Ungering, K. Thuener and K. Goser, "Architecture of a PDM VLSI fuzzy logic controller with pipelining and optimized chip area," Proceedings of the 2nd IEEE International Conference on Fuzzy Systems, 1993, Vol. 1, pp.447-452.

[13] H. Watanabe, W.D. Detloff, K.E. Yount, "A VLSI fuzzy logic inference engine for real-time process control," IEEE Journal of Solid-State Circuits, Vol. 25, No.2, pp.376-382, 1990.

[14] H. Watanabe, W.D. Detloff, K.E. Yount, "VLSI fuzzy chip and inference accelerator board systems," Proceedings of the 21th International Symposium on multiple-valued logic, pp.120-127, 1991.

[15] T. Yamakawa, "High-speed fuzzy controller hardware system: the Mega FIPS machine," INFORMATION SCIENCES, Vol.42, 1988.

[16] T., Yamakawa, "Fuzzy microprocessors - rule chip and defuzzifier chip," Proceedings of the International Workshop on Fuzzy System Applications, Iizuka, Japan, August 1988, pp.51-52.

11

AN EMPIRICAL ANALYSIS OF ONE TYPE OF DIRECT ADAPTIVE FUZZY CONTROL

Hugues Bersini and Vittorio Gorrini

IRDIA - CP 194/6, Universite Libre de Bruxelles
50, av. Franklin Roosevelt, 1050 Bruxelles, Belgium

ABSTRACT

This chapter will address, both in an analytical and experimental ways, various issues related to the automatic construction and on-line adaptation of fuzzy controllers. First we will show a DAFC (Direct Adaptive Fuzzy Control), i.e., an adaptive control methodology requiring a minimal knowledge of the process to be coupled with, can be derived in a way very reminiscent of neurocontrol methods. Indeed a main point to be argued and illustrated in this chapter is the case to import methods and ideas emerging in t he connectionist community for control applications as soon as the fuzzy controller is supplied with a gradient method for the automatic tunning of its parameters (such as the membership functions) akin to the well known backpropagation for multilayer neural nets. Since fuzzy PID is one of the most popular fields of investigation in the fuzzy control community with researchers trying to understand better the kind of non-linear extrapolation the fuzzyfication of classical PID can provide, we will show how to extend DAFC to fuzzy PID. An adaptive fuzzy satisfies both objectives to make the resulting control and to offer a method for automatic discovery as well as mechanisms of adaptation for processes in varying environments. Besides, it has been recently shown that radial-basis neural networks were nearly equivalent to Sugeno's type of fuzzy systems (the only type we are using) making any fuzzy-neural comparison and merging often very redundant and confusing. We will finally attempt to clarify what is alike and what is different between a Sugeno's fuzzy system and a radial-basis neural net.

1 INTRODUCTION

Since neural nets and fuzzy systems are easy to implement, robust intrisicaly parallel, lend themselves to a dedicated material realization and can approximate to any degree every non-linear mapping, their use for adaptive control (mainly for non-linear process) is rapidly expanding. the main objective of this chapter is to illustrate how fuzzy systems, provided a gradient method for automatic tunning of their parameters (for instance the membership functions) akin to the well known backpropagation for multilayer neural nets, can benefit from ideas and methods appeared these last years when using neural nets for adaptive control.

In the next section, we present the gradient based method we developed (slightly improving on a method developed by Nomura et al, and described in [11]) for the automatic adjustment of membership functions of Sugeno type fuzzy systems. The third section presents a simple DAFC derived from a previous work involving one of the authers but based on neural nets instead [13]. Some results obtained when applying the method to the well known cart-pole problem are discussed. The fourth section shows the same DAFC exploited for the adaptive adjustment of fuzzy PID. PID controllers are still one of the frequently met methods for adaptive control and their fuzzyfication might increase the range of non-linearity they are able to handle. An adaptive fuzzy PID satisfies both objectives to make the resulting controller to be grounded on the firm strategy of control and to offer a method for automatic discovery as well as mechanisms of adaptation for processes in varying environments. Furthermore we will show how to maintain the linear structure of the PID in its fuzzy form, as a preliminary controller to be subsequently adjusted by the gradient descent. Experimental results indicating the benefits gained by coupling an adaptive mechanism on the fuzzy PID will be shown.

Besides, it has been recently shown [15] that Sugeno's type of fuzzy systems provided certain restrictions on the architecture and the operators used by the system, were equivalent to radial-basis neural networks (RBF) (i.e. based on local non-monotonous activation function like Gaussian, rather than global monotonous ones like the sigmoid). In the last section, we will discuss this equivalence showing that some of the restrictions to be imposed on fuzzy systems so as to obtain a RBF can be seen as improvement and in some circumstances their presence is worth to be preserved. However and due to this equivalence, any form of fuzzy-neural comparison or merging often turn out to be very confusing and, in the future, should require a very precise description to help understand better where the benefits of this hydridization, if any, come from.

2 A GRADIENT METHOD FOR SUGENO'S FUZZY SYSTEM

In this section we will concentrate on the simplest member of Sugeno's fuzzy systems given by the following type of mapping of R^n in R, $y = f(x_j)$:

$$\text{IF } x_1 \text{ IS } A_1^a \text{ AND } x_2 \text{ IS } A_2^b \text{ AND } x_3 \text{ ......, THEN } y = F^d \qquad (Rule\alpha)$$

where $j = 1, 2...$, is the index of the variables, α is the rule (we suppose nr rules) and a, b, c, d ... are the types of linguistic terms. For reasons of homogeneity, the output y although crisp can only has a fixed number of discrete values corresponding to large, small, very large, very small, ... The linguistic terms A_j^a are fuzzy sets characterized by isosceles triangular shape given by:

$$A_j^a(x_j) = \max\left(1 - \frac{2\left|x_j - a_j^a\right|}{b_j^a}, 0\right),$$

then each linguistic term for each variable A_j^a is described by two parameters a_j^a and b_j^a that a gradient method will try to optimize. The crisp value F^d for the output will also be subject to an optimization process.

If $\mu_\alpha = \prod_j A_j^a$ (for the A_j^a that appears in rule α) , the final output y is given by:

$$y = \begin{cases} \dfrac{\sum_{\alpha=1}^{nr} \mu^\alpha F^\alpha}{\sum_{\alpha=1}^{nr} \mu^\alpha} & \text{if } \sum_{\alpha=1}^{nr} \mu^\alpha \neq 0 \\ 0 & \text{if } \sum_{\alpha=1}^{nr} \mu^\alpha = 0 \end{cases}$$

The central difference (and we believe improvement) of the algorithm we propose with respect to Nomura et al. work is that we don't really try to optimize the rules but the linguistic terms instead. Briefly, we try to adjust a_j^a and not at all $a_j^{\alpha,a}$. At the end of the optimization process, the linguistic term "small" for the variable x_j will always mean the same independently on the rule i which it appears. This alternative method presents various beneficial consequences: mainly it accelerates the optimization process and it keeps the final fuzzy system more coherent (respecting the linguistic interface inherent to fuzzy system), easier to understand and to interface with. If K_a, K_b and K_f are the learning parameters associated to the position of the triangle a, its base b and the output of the rule F, three expressions can easily be derived:

$$\Delta F^d = -K_f \frac{1}{\sum_{\alpha=1}^{nr} \mu^\alpha} \sum_{\substack{all rules \alpha \\ in which \\ F^d appears}} \mu^\alpha,$$

$$\Delta a_j^a = -K_a \frac{1}{\sum_{\alpha=1}^{nr} \mu^\alpha} \left(\sum_{\substack{all rules \alpha \\ in which \\ A_j^a appears}} \mu^\alpha F^\alpha - y \sum_{\substack{all rules \alpha \\ in which \\ A_j^a appears}} \mu^\alpha \right) sgn(x_j - a_j^a) \frac{2}{b_j^a A_j^a(x_j)},$$

$$\Delta b_j^a = -K_b \frac{1}{\sum_{\alpha=1}^{nr} \mu^\alpha} \left(\sum_{\substack{allrules\alpha \\ inwhich \\ A_j^a appears}} \mu^\alpha F^\alpha - y \sum_{\substack{allrules\alpha \\ inwhich \\ A_j^a appears}} \mu^\alpha \right) \frac{1 - A_j^a(x_j)}{A_j^a(x_j)} \frac{1}{b_j^a}.$$

The slight difference with [11] is clear since the sum is computed only on the rules in which a specific linguistic term associated to a specific variable, the one to be adjusted, appears.

3 A SIMPLE DIRECT ADAPTIVE FUZZY CONTROL

Suppose in general a nonlinear mapping from input I to output O (we will call it a FUZZY system (a mixture of Fuzzy and NN)). This mapping (Fig. 1) is characterized by a set of parameters symbolically designated by W hence $O = F_W(I)$.

Figure 1 A FUNNY system

The partial derivatives $\partial O/\partial W$ will be assumed to be known. We computed them for the Sugeno's mapping tackled in the previous section. Once

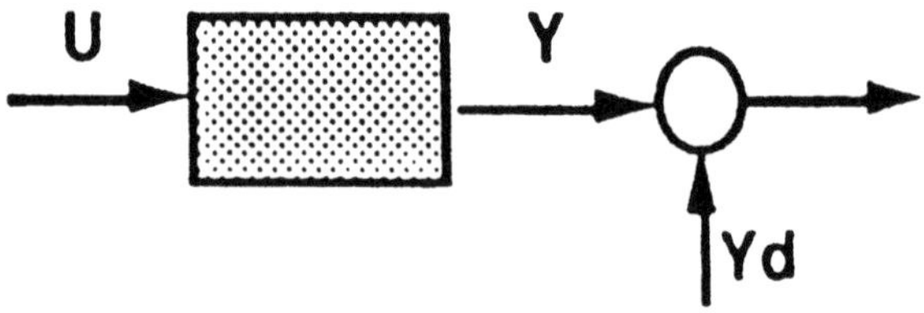

Figure 2 The process to control.

furnished with a gradient method for the parametric tuning of whatever fuzzy system ($\Delta W = -\eta \partial O / \partial W$), every use of neural nets in a control problem can be similarly contemplated using this fuzzy system. As Werbos [22] points out, it is quite immediate to substitute the neural net box by the fuzzy system box in any neural control application. Suppose the process to control indicated in Fig. 2, the objective being to tune the control parameter U to an unknown value Ud such to drive the process output Y to a desired value Yd.

A first control approach amounts to teach FUZZY to reproduce the human operator successful actions. One observes this operator and collects a set of efficient actions he executed on the process. These actions are expressed as pairs $Y \rightarrow Ud$. Then FUNNY must encode through the gradient method this set of pairs: $\Delta W = -\eta \partial (O - Ud)^2 / \partial W$. A further approach called "general learning" and represented in Fig. 3 presents a prior training stage which consists in using the process to produce a set of input-output pairs $U \rightarrow Y$. These pairs are then used as patterns for training FUNNY. During the training stage, the control parameters U are chosen randomly to scan over a certain working range. Those parameters are then injected into the process which supplies output values Y. The associated gradient method is used to train FUNNY to relate its output U to its input Y. After this learning stage, The FUNNY controller is able to provide the correct control parameter to reach any target Yd. This method has several drawbacks: first it raises enormous problems in case of non-invertible process; second, learning must be performed off-line and it is hard to call it an adaptive control, finally FUNNY cannot limit its working range to the Y that are actually relevant.

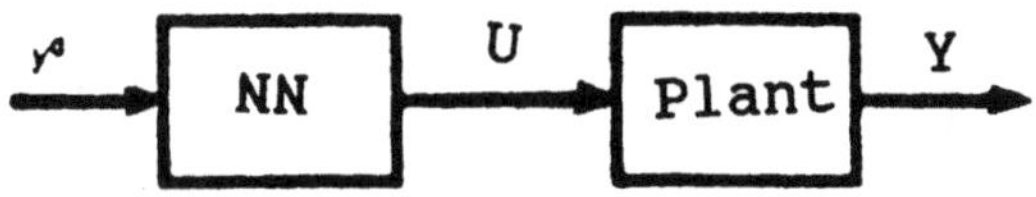

Figure 3 General learning architecture.

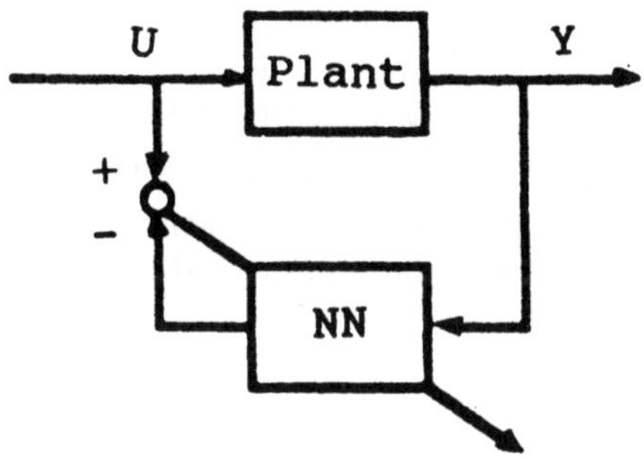

Figure 4 Indirect learning.

Another possible approach is called the "iterative inversion technique" in which first FUNNY learns the forward model of the process in terms of pairs: $U \rightarrow Y$. Then the control results from the inversion algorithm which backpropagates the error signals $(Y - Yd)^2$ down to the input layer to update

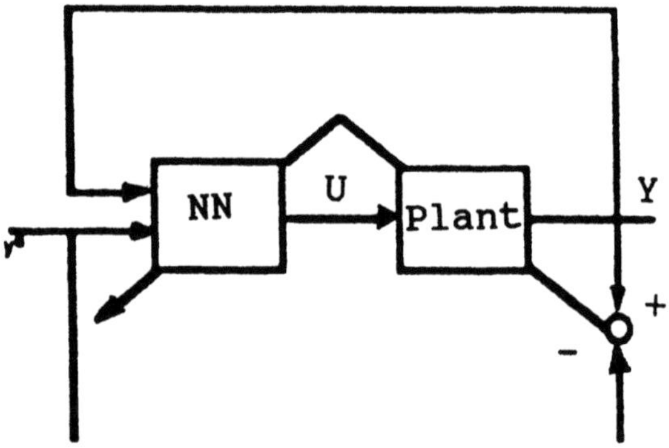

Figure 5 Specialized learning.

the activation values of input (and no more W) so that the output error is decreased. In the indirect learning architecture (Fig. 4), the desired target Yd propagates through FUNNY producing the process input $U1$. This same input is tried on the process which responds by an output Y distinct, at the beginning, from the desired target. The Y is injected as input of FUNNY and gives $U2$ as output. The learning phase immediately follows aiming at minimizing the error $(U1 - U2)^2$ obtained when presenting Y as input of FUNNY. However simulations show that FUNNY can settle to a solution that maps all the desired target to a single plan input, which gives zero training error, but obviously a non-zero total error. Learning is indirect because the difference between the output of the process and the desired output are not directly reduced.

The final approach, the one that we have adopted , has been called specialized learning in the NN literature (Fig. 5). Specialized learning differs from general and indirect learning by the fact that the FUNNY controller learns no longer from input-output pairs but from a direct evaluation of FUNNY accuracy with respect to the output of the process. FUNNY uses the difference between the actual output of the plant Y and the desired output Yd to adapt its parameters W. Specialized learning avoids several drawbacks of general learning: there is no longer a specific training stage during which the controller is not operational, and FUNNY learns directly on the domain of relevant Y. It learns continually and is therefore adaptive. In fact, the specialized learning approach fits in perfectly with the classical adaptive control attitude. FUNNY controls and learns to control simultaneously. However, to make specialized learning possible, we need some prior knowledge on the way the plant reacts to slight control modification, i.e., the Jacobian of the plant. The reason for this requirement is obvious when computing the gradient no more on basis of the error obtained at output of FUNNY: $(O - Od)^2$ but at the output of the process instead: $(Y - Yd)^2$. Od is all the less unknown since it is what is looking for.

$$\Delta W = -\eta \frac{\partial (Y - Y_d)^2}{\partial W} = -2\eta(Y - Y_d)\frac{\partial Y}{\partial W} = -2\eta(Y - Y_d)\frac{\partial Y}{\partial O}\frac{\partial O}{\partial W}$$

$\partial O/\partial W$ corresponds to the gradient method akin to the backpropagation for multilayer neural nets and derived in the previous section in case of Sugeno's fuzzy system. What is unknown in the absence of an analytical model of the the process is the Jacobian $\partial Y/\partial O$. One possible strategy consists in approximating the partial derivatives by plotting the process reactions to slight control modifications at the operation points. We applied this method with very good results for the control of a static robotics arm [4]. In terms of prior knowledge of the process, it is less requiring approach. This application revealed a further interest of specialized type of approach as compared with general learning approach: its adequacy for controlling non-invertible processes such as a robotics arm by selecting among multiple solutions one and only one of them.

Another recent and sophisticated technique has been proposed by Jordan and Rumelhart [6] which incorporates the default prior knowledge in another FUNNY system and links the FUNNY controller to this FUNNY emulator of the plant - a fuzzy or neural system that identifies the process. It is easily shown that the Jacobian of the process can be derived by a mechanism of backpropagation applied not on the weights but on the input of the FUNNY emulator (like for the iterative invertion technique mentioned before). The most salient application of such double FUNNY techniques has been, in the case of neural nets, the control and backing up of a truck-and-trailer [10]. A similar double fuzzy system is easy to conceive. Interestingly enough, Kosko [7] has turned to be onbe of the lead advocates of the use of fuzzy logic for control while generating a fuzzy controller which compares favorably with the neural controller in terms of black-box development effort, black-box computational load, smoothness of truck trajectories, and robustness. The resulting control neither requires any automatic parametric tuning nor a fuzzy emulator. The rules were based on common-sense and gradually improved through manual trial-and-error. Nothing is really surprising in this double absence since the emulator is a direct consequence of the automatic tuning method and the non-necessity of one entails the non-necessity of the second.

The double FUNNY method corresponds to the classical indirect control approach where the parameters of the process are estimated at any instant and the parameters of the controller are adjusted assuming that the estimated values of the process represent the true values. This technique makes the control quite flexible for indeterminate learning problems, allowing generic constraints to be expressed separately at the control and at the behavioral level. However it demands, in compensation, either a preceding learning stage (the identification of the process) or a "neurally or fuzzily" expressed prior knowledge of the dynamic of the process.

The specialized learning technique that we will present in the next section makes the bet that in many circumstances a satisfactory control can be attained even on basis of the least knowledge of the process, namely the sign of its Jacobian. Indeed the gradient method indicates that the adjustment of W follows the sign and amplitude given by $\partial Y/\partial O.\partial O/\partial W$. Since for single output process, substituting $\partial Y/\partial O$ by its sign ± 1 just alters the amplitude not the direction of the variation, the gradient method can still give a adequate value of the control parameters.

Obviously although this method is straightforward in case of single output process with Jacobian of constant sign, its extension for multiple output processes demands some slight adjustments which are described in [16]. Reducing to such minimum the knowl edge of the process necessary to its control makes the resulting control still more "direct". The process can remain a black-box with just few holes allowing to observe qualitatively the way it reacts to increases and decreases of its input. This type of knowledge is quite common and certainly the first shallow type of knowledge a user or observer of the process is aware of and able to communicate. In [13], we have derived the extension of this direct method (either double funny ro sign-based) for processes characterized by arbitrary delay no longer restricted to one time step. We also have explained and justified the kind of simplification called the "non-recurrence-in-learning" which leads to this method and which consists in restricting the dependency of the FUNNY parameters to adjust on the most recent output.

In [12][13] a simple Direct Adaptive Neural Controller (DANC) has been achieved and tested among other toy problems on the catt-pole. One and only one multilayer neural net was necessary, the sign of the cart-pole Jacobian was known, and this suffices to achieve a very satisfactory control with respect to the numerous other neuralcontrols we were aware of and with respect to more conventional, like MRAC, approaches. In this section we will describe in more details the equivalent simple Direct Adaptive Neural Controller (DANC) similarly applied to cart-pole problem (a problem frequently met in NN and fuzzy control literature, see [2]). In our application the objective of the control is to bring the cart-pole, whatever its initial conditions, to a stable equilibrium position with the four variables describing the system: x, v, θ, ω driven to 0. The fuzzy control is one of the Sugeno's basic type with four variables in the IF part and only one variable (given by a "fuzzy set" restricted to a single values of the force to excert on the cart , in the THEN part. An example of rule is:

IF x is large and v is very small THEN f is large (=10N).

The parameters to be adjusted are the memberships functions coding the linguistic terms appearing in the premise of the rules and the different possible values of the force appearing in the sequent part. The gradient method is the one described in section 2. With regard to the Jacobian, the signs of the four partial derivatives: $\partial x/\partial F, \partial v/\partial F, \partial\theta/\partial F, \partial x/\partial F$ are all negative. The error is given by:

$$Err = 0.5(a_1 x^2(t) + a_2 v^2(t) + a_3\theta^2(t) + a_4\omega^2(t)),$$

where a_i weights the relative contribution of each variable. The global gradient method immediately follows:

$$\Delta W = -\eta\left(a_1 x\frac{\partial x}{\partial F_{(t-1)}} + a_2 v\frac{\partial v}{\partial F_{(t)}} + a_3\theta\frac{\partial\theta}{\partial F_{(t-1)}} + a_4\omega\frac{\partial\omega}{\partial F_{(t)}}\right)\frac{\partial F}{\partial W},$$

with $\partial F/\partial W$ given by the method presented in section 3 and $\partial Err/\partial F$ easily obtained by substituting $\partial x/\partial F, \partial v/\partial F, \partial\theta/\partial F, \partial x/\partial F$ all by -1. The reason for both t and t-1 is linked to the slower reaction a delay equal 2 rather than 1) of x and θ in comparison with v and ω. The difficulties raised by delay superior to 1 are discussed in details in [13].

The results were very impressive. For a large range of initial conditions even drastic, the equilibrium position was reached after an average of 5 falls and an average of 2500 learning steps. For these experiments and in order to make possible the comparison with the results of our neural controller for the same problem, the rules were generated, like a synaptic matrix at random. More precisely, any initial set of rules contains 30 rules. In each rule, the probability of appearance on one of the four variables is 0.3 (and no rule can be generated without variables), each variable is associated to a linguistic terms taken randomly among the seven classical possibilities. On the whole the learning was faster than for the neural equivalent and, confirming results exposed by Kong and Kosko [7] for another application, the final controller tolerated more initial conditions and results more robust than neural controller. However as stated previously, since a radial-basis NN is nearly equivalent to the kind of Sugeno's mapping we used for the control, it is more informative to compare NN with fuzzy system without stating with precision what kind of NN and Fuzzy systems are under study. In our case, the NN architecture we tested was sigmoid-based with one hidden layer and then the reasons for the fuzzy superior performance can just be due to the use of non-monotonous activations functions instead of sigmoid. Similar performances could have been reached by using RBF.

4 AN ADAPTIVE FUZZY PID

Because fuzzy PID is one of the most popular fields of investigation in the fuzzy control community with researchers trying to understand better the kind of non-linear extropolation the fuzzyfication of a classical PID can provide [5], we decided to extend our adaptive mechanism to a fuzzy PID .

When defining:

$$\begin{aligned} e(t) &= y(t) - y^o(t) \text{ (with } y^o(t) \text{ the desired output for the process)} , \\ \delta e(t) &= e(t) - e(t-1) = y(t) - y(t-1) - y^o(t) + y^o(t-1), \\ \delta\delta e(t) &= \delta e(t) - \delta e(t-1) = y(t) - 2y(t-1) + y(t-2) - y^o(t) \\ &\quad + 2y^o(t-1) - y^o(t-2), \\ u(t) &= u(t-1) + \delta u(t). \end{aligned}$$

A fuzzy PID is particular member of Sugeno's systems composed of a bunch of fuzzy rules similar to

If e is NL and δe is PM and $\delta\delta e$ is ZE then δu=NL .

Although the controller is different from the previous ones due to the nature of its input (error and error derivatives) and output (increment of control), the adaptive mechanism is basically the same:

$$\Delta w = -\eta \partial E/\partial w = -\eta(y - y^o)\partial y/\partial w = -\eta(y - y^o)\partial y/\partial u \; \partial\delta u/\partial w, \quad (11.1)$$

the last term is still given by our gradient method applied on the fuzzy PID. We tested this adaptive method on the process given below:

$$y(t) = ay(t-1)\sin(y(t-2)) + \cos(ku(t-1))$$

with a and k initially equal to 1.

Our fuzzy controller contains 27 initial rules generated in a way that tries to keep the linearity and proportionality inherent to PID structure (very similar to what has been done in [5]). This is achieved by means of integers associated to the linguistic terms partitioning the variable domain of variation. For instance, 0 is associated to the fuzzy set "zero", 1 to the fuzzy set "medium", 2 to "big", 3 to "very big", and so on. The construction of the fuzzy PID rules results from elementary arithmetic, f or instance:

if e is medium (1) and δe is big (2)
then δu is very big (3) .

Applied on this pseudo-linear fuzzy structure, the results of the gradient tuning are illustrated in Fig. 6. For such process, the fuzzy PID is not easy to tune manually and an adaptive method allows its on-line automatic discovery, As you can see in the last box where the same simulation is run again with the final fuzzy PID found by the optimization algorithm, htis on-line optimization leads to an efficient controller.

Next, we perturbed the process by increasing the time constant a by 20%($1 \to 1.2$) and we can see in Fig. 7 that the fuzzy PID previously found is no more adapted to handle this new situation. The second and third boxes below show how the on-line adaptation m echanism leads to discovery of a new PID which brings back the process to nominal conditions. These two simple experiments enable to show the two advantages gained by the addition of such on-line optimization algorithm: automatic discovery of an efficient fuzzy PID structure and good robustness in varying environments by continuous adaptation.

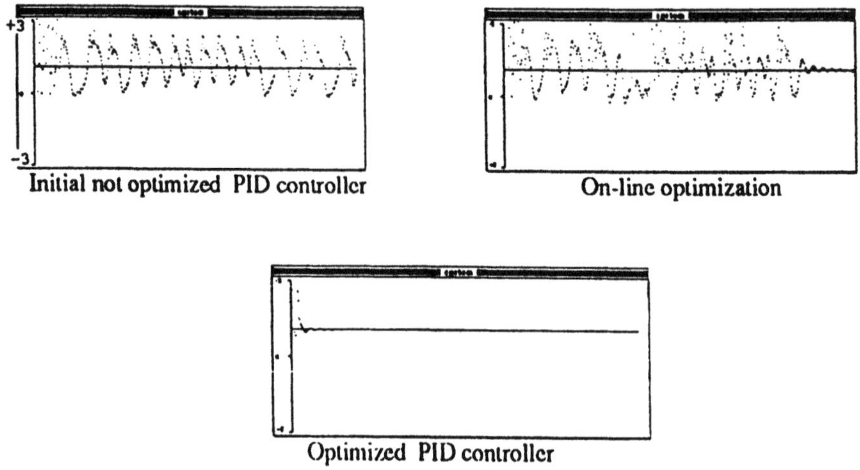

Figure 6 Automatic discovery of an efficient fuzzy PID.

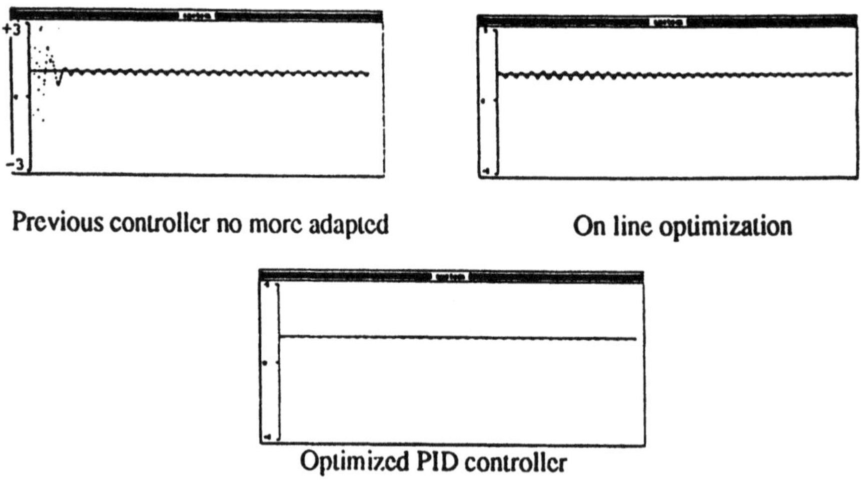

Figure 7 Illustration of the adaptive capacities of the fuzzy PID.

5 ON THE EQUIVALENCE BETWEEN RBF AND FUZZY SYSTEMS

As discussed by Roger Jang and Sun [15], five conditions are necessary for establishing the equivalence between Sugeno;s system and the RBF:

1. The number of RBF receptive fields is equal to the number of fuzzy if-then rules.

2. The output of each fuzzy if-then rule boils dowm to a constant.

3. The membership functions within each rule are chosen as Gaussian functions with the same variance (however multiple variances, one by Gaussian, can easily be achieved in RBF.)

4. The T-norm operator is multiplication (which together with the presence of exponential makes the equivalence possible.)

5. Both the RBF and the fuzzy inference system use the same method to derive their overall output. We believe it is necessary to add a sixth condition,

6. All the input variables (associated to RBF input to the RBF input units) appear in all the rules.

While these six conditions are necessary to turn a fuzzy system into RBF in terms of the input/output mapping, four of them deserve a deeper discussion. With respect to the second condition, clearly a fuzzy system presents more freedom than a usual neural net for the fixing of its output. In RBF the counterpart of the fuzzy constant output are the output synaptic weights and, within a near classical perspective, it would appear very odd to transform these synaptic weights into functions of the input variables. Moreover a classical neural architecture is feedforward and all the direct dependency with respect to the input variables is given one for all in the hidden layer (in the RBF case the Gaussian units). In Sugeno's mapping, this dependency can app ear two times, one to establish the membership function and the second to compute the local approximator i.e. whatever function of the input prevailing in the fuzzy regions traced by the input fuzzy sets. Indeed, although both RBF and fuzzy system are universal approximator, due to its knowledge-based origin and in contrast to neural net type of approaches, the way a fuzzy system tries to approximate any non-linear mapping is more transparent, thus more easy to grasp and eventually more elegant. For each fuzzy subpart of the input space, the mapping is approximate by a certain function of the input (not just a constant). In between these parts (given by the fuzzy sets intersection), the fuzzy system realizes a smooth interpolation between the local approximato rs. This smooth interpolation is based on the respective overlap of the state to be treated with the intersecting fuzzy partitions. The more the state belongs to one fuzzy part, the more the local approximator associated to it will prevail in the resulting combination.

This indeed relates to the fifth condition. It is easy to understand why a weighted average imposes itself more naturally on fuzzy mapping than a weighted sum. When there is no intersection between the fuzzy partitions of the input space, the output of th e fuzzy system should just amount to the approximator associated to each particular partition. The approximator shape will only be modified as a consequence of the intersection of the fuzzy partitions of the input space. By contrast, connectionists have n ever been in favor of activation function accounting for events not directly connected with the neuron, accordingly they will prefer a weighted sum where neural contribution remains local. In consequence , in the neural case and for non-intersecting input partitions, the output will be the product of the approximator and the Gaussian partitions. On account of its attempts at respecting biological structure, the RBF way to approximate non-linear mapping looses in generality, transparency and elegance.

Related to condition three, an important issue to address is the efficiency of local Gaussian type of activation function as compared with global sigmoid type. In the connectionist literature, it is generally accepted that during the learning phase a RBF learns faster to approximate a function than a sigmoid-based NN. There is a reasonable explaination for that. In RBF only the approximator associated to the counter part of the function to match will be adjusted by gradient descent, the other local approximators will remain intact (then no degradation of what they learned in previous phases will occur). In adaptive control, speed is a crucial issue then largely favoring the use of RBF at the expense of the use of monotonous activation function. For other possible uses of NN, like for classification tasks, the NN generation capability becomes a more important quality than the learning speed. Monotonous activation functions divide the input space into regions separated by hyper planes and then could help the neural net to generalize better than by a lot of small bounded regions distributed in the input space.

The sixth condition again makes fuzzy system more flexible than RBF for function approximation. Suppose to know that when one of the input variable takes a certain range of value, just its contribution needs to betaken into account for computing the output. In such case, it is very helpful to posses fuzzy rules whose intrinsic structure allows to change the number of input variables to appear in each rule. This restriction is harder to achieve in a non tricky way in RBF.

Finally, in their paper Roger Jang and Sun have restricted their comparison between the fuzzy systems and RBF to the way output data map to input data leaving apart the important chapter of how and what both systems learn (although learning is a key aspect of neural nets). Since in fuzzy systems, the intrinsic linguistic nature of the system fundamental (completely absent in RBF), serious attempts to graft a learning algorithm onto the fuzzy system will try to preserve the linguistic interface, an aspect which does not preoccupy connectionists. For instance a classical learning algorithm for NN will independently modify all parameters without respecting any linguistic significance. In contrast, fuzzy practioners will try as far as possible to keep the linguistic significance in the after learning system. Among these attempts, we can include what we did and presented in the second section, that is learning the linguistic terms instead of all the parameters independently We can also mention the Werbo's elastic logic [23] aiming at a similar preservation of the linguistic nature of the fuzzy system. Also during the learning phase, fuzzy practioners, when using as fuzzy sets limited shapes like triangle and trapezoid (not Gaussians), generally try to preserve the fuzzy overlapping either by a subtitle selection of the parameter to optimize or performing optimization under constraint. Nevertheless, apart from this linguistic safeguard, it would be a welcome contribution to test experimentally to what extent such constraints will result in profit in terms of learning speed or quality and robustness of the solution obtained by the optimization process.

6 STABILITY AND CONCLUSIONS

The main point developed in this chapter is the ease to import methods and ideas appeared in the connectionist community for control problems as soon as the fuzzy controller is supplied with a gradient-based automatic tuning of its parameters. We have pr esented two different DAFC, the second one coupling a fuzzy PID with an automatic adjustment mechanism, and shown the good performances attained when applied these two DAFC on simple control applications. A further claim is to describe with more accuracy current or future comparisons between NN and fuzzy control systems on any kind of problems: function approximation, classification, control, ... since it has been shown that although following different histories these two technique have turned out to be close in terms of the input/output mapping they generate .

A large part of the algorithms for adaptive fuzzy and neurocontrol developed so far are gradient-based [3] [11] [12] [13] [16] and do not provide any information on stability of the overall process. Stability is a fundamental requirement in adaptive control theory [1] [8], and certainly a crucial issue for the future development and application of fuzzy adaptive controllers. Only recently, researchers tried to design adaptation laws that ensure the stability of the closed-loop process [21]. Inspired by the work already done in neurocontrol (by Renders [12] and Saerens, Renders and Bersini [16]), the same authors gave applied the hyperstability formalism and Lyapunov theory [8] [17] [20] for single-input single-output (SISO) discrete processes in order to prove the stability of the resulting closed-loop system, provided that the initial parameters of the fuzzy system are not too far from their optimal values [15]. This extends the work, of Wang [21], where only processes linear in the input signal were considered. When restricting the gradient-based tuning to the consequent part of the rules, linear with respect to the parameters to adapt, the stability results are strictly valid; that is the parameters values do not have to be initialized around the perfectly tuned values. For this reason, we propose to link the fuzzy controller with a linear regulator, so that the fuzzy controller only takes the nonlinear part of the control law and the parameters can beconsidered as close to their optimal values.

On the whole, fuzzy controller requires two types of tuning which Lee [9] designates as structural and parametric tuning. the first one concerns the structure of the rules: the variables to account for, for each variable the partition of the universe of discourse, the number of rules and the conjunctions which constitute them. We are aware of very interesting works done by Sugeno's team, based mainly on clustering techniques [18] for the automatic tuning of the structure. On account of the highest sensitivity to structural aspects than to membership shapes, structural adjustment is without doubt a more promising way of automatic discovery of fuzzy system than parametric adjustment. However in order to be applied, this technique needs, from the very big inning, all the input/output pairs which will be necessary to these structural adjustments. If such requirement is likely to be fulfilled in case of process modeling in which a simple observation of the process behavior can generate this whole initial set of data, it is no more the case in direct adaptive control where the data upon which the learning is based are supplied on-line, while the control proceeds and in a continuous way. When interested in on-line adaptive control, an optimization technique, allowing only parametric adjustment and although leading to a non-minimal structure, appears to be the only alternative.

ACKNOWLEDGEMENTS: This work was partially supported by the ARC 92/97-160 (BELON) project from the "Communauté Française de Belgique", and the FALCON (6017) Basic Research ESPRIT project from the European

Communities. We would like to thank Marco Saerens and Jean-Michel Renders for their precious collaboration.

REFERENCES

[1] Åströ̈m K.J. & Wittenmark B., 1989, "Adaptive control." Addison-Weskey Publishing Company.

[2] Barto, A., Sutton, R. AND C. Anderson, 1983, "Neuronlike adaptive elements that can solve difficult learning control problems" - IEEE Transactions on Systens, Man and Cybernetics, 13(5).

[3] Bersini H., Notdvik, J-P. & Bonaribi, A., 1993, "A Simple Direct Adaptive Fuzzy Controller Derived from its Neural Equivalent." Proceedings of the Second IEEE International Conference on Fuzzy Systems - pp. 345 - 350.

[4] Bersini, H, 1993, "Comparison des régulateurs neuronaux pour la commande d'un bras de robot" - IRIDIA Internal Report

[5] Foulloy, L. AND S. Galichet, 1992, "Controlleurs Flous: représentation, équivalences et études comparatives" - Rapport LAMII 92-4.

[6] Jordan,M.I. AND D. Rumelgart, 1991, "Internal world models and supervised learning" In Proceedings of the eight International Workshop on Machine Learning, pp 70 -74

[7] Kong S.G. AND B. Kosko, 1992, "Adaptive Fuzzy Systems for Backing up a Truck-and-Trailer" In IEEE transactions on Neural Networks, Vol. 3, No 2, March 1992, pp 211 - 223

[8] Landau Y., 1979, "Adaptive control. The model refernce approach." Marcel Dekker.

[9] Lee, C.C. 1990, "Fuzzy logic in control systems: "Fuzzy logic controller -Parts I, II" IEEE Trans. Syst. Man Cybern., vol 20, no2, pp 404-435

[10] Nguyen D. AND B. Widrow, 1989, "The truck backer-upper: an example of self-learning on neural networks, Washington, Vol. II, pp 347 - 353.

[11] Nomura H., Hayashi I. & Wakwmi N.,1992, "A learning method of fuzzy inference rules by descent method." Proceedings of the IEEE International Conference on Fuzzy Systems, pp. 203-210.

[12] Renders J.-M., 1993, "Biological metaphors for process control." Ph.D Thesis, Université Libre de Bruxelles, Faculté Polytechnique, Belgium.

[13] Renders J.-M., Bersini H. & Saerens M., 1993, "Adaptive neurocontrol: How black-box and simple can it be ?" Proceedings of the tenth International Workshop on Machine Learning, Amherst, pp. 260-267.

[14] Renders,J.-M., Saerens, M. AND H. Bersini, 1993, "On the Stability of Direct Adaptive Fuzzy Controllers" - Submitted in IEEE Transactions on Fuzzy Sets

[15] Roger Jang J.-S. & Sun C.-T., 1993, "Functional equivalence between radial basis function networks and fuzzy inference systems." IEEE Transactions on Neural Networks,4(1), pp. 156-159.

[16] Saesens M., Renders J.-M. & Bersini H., 1993, "Neural controllers based on backpropagation algorithm." In the "IEEE Press Book on Intelligent Control" by the Gupta M. & Sinha N. (editors), under press. IEEE Press.

[17] Slotine J.-J. & Li W., 1991, "Applied nonlinear control." Prentice-Hall

[18] Sugeno M. & Tanaka K., 1991, "Successive identification of fuzzy model and its application to prediction of complex system." Fuzzy Sets and Systems, 42, pp. 315-344.

[19] Tanaka K. & Sugeno M., 1992 "Stability analysis and design of fuzzy control systems." Fuzzy Sets and Systems, 45, pp. 135-136.

[20] Vidyasagar M., 1993, "Nonlinear system analysis, second edition." Prentice-Hall

[21] Wang L.-X., 1993, "Stable adaptive fuzzy control of nonlinear systems". IEEE Transactions on Fuzzy Systems, 1 (2), pp. 146-155.

[22] Werbos, P.J. 1992, "Neuralcontrol and Fuzzy Logic: Connections and Designs." In international Journal of Approximate Reasoning; 6; pp. 185-219.

[23] Werbos, P.J. 1993, "Elastic Fuzzy Logic: A better Way to Combine Neural and Fuzzy Capacities" - in Proceeedings of the World Congress on Neural Networks, Vol II - pp. 623-626

12

AUTOMATIC OPTIMAL DESIGN OF FUZZY SYSTEMS BASED ON UNIVERSAL APPROXIMATION AND EVOLUTIONARY PROGRAMMING

Mattias Nyberg and Yoh-Han Pao

Department of Electrical Engineering and Applied Physics
Case Western Reserve University
Cleveland, OH 44106, USA

1 INTRODUCTION

The theory of fuzzy sets was initiated in 1965 by Zadeh [1]. Since then, it has been expanded and has found its way into a large number of applications, for example in the areas of expert systems and control (Mamdani 1975 was the pioneer in control [2]). For several application areas, there are many examples of solutions based on fuzzy sets that outperform traditional solutions, examples in control are for control of cement kiln [3], car parking [4] and automatic train operation [5]. The reason for this is its unique properties of handling nonlinearity as well as uncertainty. An advantage of fuzzy systems is that they are based on linguistic rules, and therefore it is often easy to understand the underlying functionality of the systems. This is of great help when these systems are designed, because expert knowledge and common sense can often contribute in a natural way. However sometimes, sufficient knowledge is not available, which makes the design phase difficult. Also the process of designing fuzzy systems is usually heuristic, and systematic, approaches that lead to optimal systems do not exist. This chapter describes progress towards an universal systematic design strategy for automatically producing optimal fuzzy systems. In this strategy, the fuzzy system is viewed as an universal approximator [10] and optimized by using Guided Evolutionary Simulated Annealing (GESA) [23]. This design strategy is tested on two applications: general function approximation and control.

2 FUZZY SETS AND SYSTEMS

2.1 Fuzzy Sets

The concept of fuzzy sets was introduced as a generalization of the classic set theory. In the real world, many classes of objects are not as well defined as regular set theory would suggest. For example the set of all tall persons includes of course all persons taller than 1.9m, but what about persons of length 1.8m. It should be clear that the concept of a crisp set could be very hard to deal with sometimes. The definition of a fuzzy set have what is needed to define a set that has inexact boundaries, i.e. objects with vague membership. For example the set A can be defined to be the set of tall people. A concept associated with fuzzy sets is membership function. A fuzzy set's membership function takes an object and returns the degree of membership in that fuzzy set. In the example with tall persons, all persons are tall to a certain degree defined by the value returned by the membership function of the fuzzy set "tall persons". Usually the membership function has a range from 0 (is not a member), to 1 (is undoubtedly a full member). As in the regular set theory, a linguistic term is usually associated with each set. The membership function can be interpreted as a measurement of the compatibility between the linguistic term and an object.

Because fuzzy set theory is a generalization of the classic set theory, it can not be defined by the notation used in classic set theory. However the membership function is a normal functional mapping and with its help, a fuzzy set A in X has been defined in [30] and [31] as a set of ordered pairs

$$A = \{(x, y)|x \in X, y = \chi_A(x) \in [0,1]\} \tag{12.1}$$

where $\chi_A(x)$ is the membership function. In the example of tall persons, A would be the fuzzy set "tall persons" and X the set of all persons. This definition says that a fuzzy set is a function with a range from 0 to 1, and this function is the fuzzy set's membership function. (In other words, this definition implies that a fuzzy set and its membership function is the same thing!) In the special case of crisp sets, $\chi_A(x)$ can only take the value 0 or 1.

Fuzzy sets are defined in a crisp domain, e.g. temperature. An example of three fuzzy sets and their membership functions in the domain temperature, is shown figure 1. The three fuzzy sets are named with the linguistic terms 'cold', 'medium' and 'warm'.

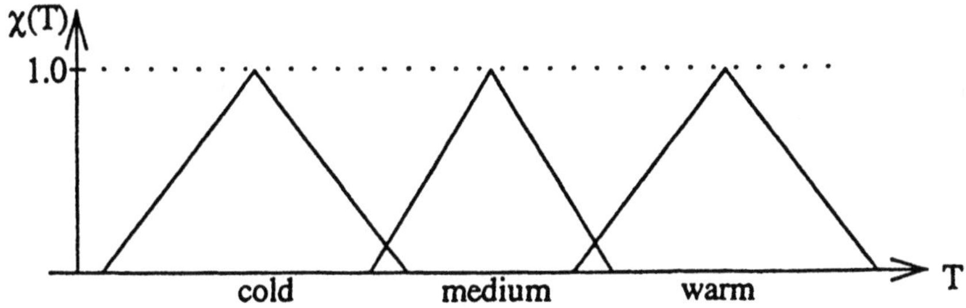

Figure 1 The crisp physical domain temperature in which three fuzzy sets have been defined.

2.2 The Fuzzy System

In this chapter, a system based on fuzzy sets theory is called a *fuzzy system*. Such a system has an input and an output. The input is usually a vector and for the sake of simplicity, the output is taken to be a scalar. However the results can be generalized to a system with vector outputs. The principal components of a Fuzzy System are shown in figure 2. The crisp inputs are fuzzified by the Fuzzification Interface. Based on the knowledge in the Rule Base, the Decision Making Logic infers decisions using rules of fuzzy inference. The Defuzzification Interface produces a crisp output from the inferred fuzzy decisions.

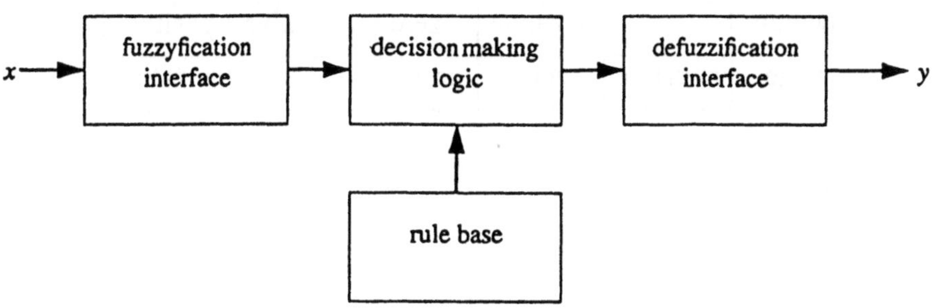

Figure 2 The Principal Components of a General Fuzzy System

An example of a rule in the rule base is

IF x_1 is A_2 AND x_2 is A_2 AND x_3 is A_3 THEN z is C.

where x_i is the system input variables, z is the system output, and A_i and C are fuzzy sets. The interpretation of a rule is that the consequent (the part after THEN) is implied if all antecedents (the parts before THEN, between the ANDs) are true. There is usually one antecedent for each system input component. The rule base can be represented by a list of linguistic rules or by a N-dimensional matrix, where N is the number of antecedents. In the matrix case, each rule is represented as an matrix entry which contains the consequent. See figure 3 for an example of a 2-dimensional rule matrix. Note that not all entries need to contain a rule. For example, one of the rules in the rule base represented by the rule matrix in figure 3 is

IF x_1 is positive AND x_2 is small THEN z is zero .

2.3 Different Types of Fuzzy Systems

In the design of fuzzy systems, several issues have to be considered. First it has to be determined what type of fuzzy system that is going to be used. For most fuzzy systems, the type is defined by the following two design issues:

x_2 \ x_1	very negative	negative	zero	positive	very positive
small				zero	zero
medium		negative	zero	positive	positive
large	negative	zero	positive	zero	negative

Figure 3 The rules of a fuzzy system represented as a 2-dimensional rule matrix.

- mechanism of inference
- defuzzification method

The decision regarding the mechanism of inferring consists of determining the function of the rules. That is to determine the exact meaning of the sentence connective AND and the fuzzy implication function, i.e. the meaning of THEN. Usually AND, i.e. x_1 is A AND x_2 is B, is implemented as

$$\alpha = \min\{\chi_A(x_1), \chi_B(x_2)\}, \tag{12.2}$$

or

$$\alpha = \chi_A(x_1) \cdot \chi_B(x_2). \tag{12.3}$$

where $\chi_A(x_1)$ and $\chi_B(x_2)$ are the membership functions of the fuzzy sets A and B respectively. x_1 and x_2 are the crisp input variables [7].

Almost 40 distinct fuzzy implication functions have been described in literature but the two most common are Larsen's product operation and Mamdani's minimum operation [7]. The inferred membership function $\chi_{C_w}(z)$ in each case is

$$\chi_{C_w}(z) = \alpha \cdot \chi_C(z) \quad \text{product operation} \tag{12.4}$$

$$\chi_{C_w}(z) = \alpha \wedge \chi_C(z) \quad \text{minimum operation} \tag{12.5}$$

where α is the value (crisp) resulting from the combination, using the implementation of AND, of the antecedents of the rule. $\chi_C(z)$ is the membership function of the fuzzy set C belonging to the consequent, $\cdot$ is the algebraic product and the meaning of $\wedge$ is $\min\{\alpha, \chi_C(z)\}$.

At least 6 defuzzification methods have been described in literature. Of these the Center of Area method (COA) also called the Centroid method is the most common. It determines the center of the area below the membership function of the combined fuzzy set,

$$C_{comb} = \bigcup_k C_w^k, \tag{12.6}$$

where k is the rule index, and $\cup$ is the fuzzy union defined by $\chi_{A\cup B}(x) = \max\{\chi_A(x), \chi_B(x)\}$. This is illustrated in figure 4. The formal expression of the Centroid calculated output is:

$$y = \frac{\int z \cdot \max_k \left\{\chi_{C_w^k}(z)\right\} dz}{\int \max_k \left\{\chi_{C_w^k}(z)\right\} dz}. \tag{12.7}$$

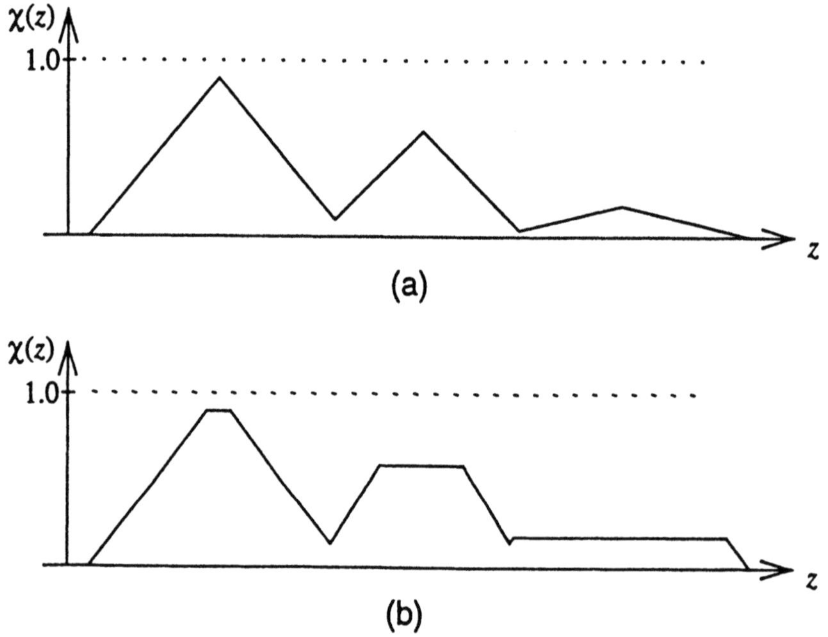

Figure 4 The combined output membership function for which the center of area is calculated with the COA method. The fuzzy system has 3 fuzzy sets defined in the output space. (a) The implication function is Larsen's product operation. (b) The implication function is Mamdani's minimi operation.

In addition to these fuzzy systems, other systems that are not separable, in Decision Making Logic and Defuzzification Interface, have been described in literature. An example of such a system is the Takagi-Sugeno-Kang (TSK) fuzzy system [15], in which each rule produces a crisp output value as a function of the inputs. The system output is then the weighted average of the outputs from the rules.

Some research has been done trying to compare different fuzzy systems. For example different implication rules have been compared in [9] and different defuzzification methods have been compared in [8] and [13]. However the results are often application specific. Therefore it is difficult to state any general recommendation for the choice of type of fuzzy system.

2.4 The Parameters of a Fuzzy System

When the type of fuzzy system is given, there are still parameters left to be determined. These parameters are:

- the number of fuzzy sets
- membership functions
- the rules

Few general investigations have been done regarding these issues. However, most common is to use 3 to 7 fuzzy sets in each universe of discourse. This is the same as saying that the *fuzzy partition* is 3 to 7 in each universe of discourse. The most common kind of membership functions is triangular shaped. When determining the rules, expert knowledge and common sense can be used to some extent. It is clear that an universal systematic design method does not exist. Therefore engineering skills and 'trial and error' are in most cases the only used methods. Also because of this 'ad hoc' approach it is very difficult to predict the performance of the system. Even if good results have been obtained, there is nothing that guarantees that an optimal system has been found. Sometimes the designed fuzzy system can be very close to an optimal system. It might only require some small changes of the system, e.g. in the position of the membership functions, to achieve optimality. However even these small changes can be hard to perform because the lack of systematic optimization techniques.

3 AUTOMATIC DESIGN OF FUZZY SYSTEMS

The function of fuzzy systems have the advantage of being relatively easy to understand , but there are always design parameters that can not be determined with expert knowledge or common sense. For instance it might be possible to let an expert determine most of the rules, but not exactly all. Also the approximate positions and shape of the membership functions can be determined by common sense but the **exact** (optimal) position and shape is unknown. Therefore there is a need for different degrees of automatic design or tuning of fuzzy systems. The term *automatic design* is used when considerable features of the fuzzy system are determined by an automatic process. The term *automatic tuning* is used when only smaller changes of an already existing system are done automatically. The borderline of the two is somewhat floating. Several techniques have been proposed and some of them are reviewed in the sections below. Usually these techniques don't treat all the stages of the design procedure. Often only one parameter is treated, e.g. the position of the membership functions. Therefore, although these techniques produce systems with improved performance, they may suboptimal. A thrilling property of these techniques is their capability of self discovery and for the teaching of humans. An automatically designed fuzzy system can sort out important features from unimportant and also discover relationships. Because of the fact that fuzzy systems are easy to understand, a person studying at the system can then learn these discoveries.

x_2 \ x_1	-3	-2	-1	0	1	2	3
-3	-6	-5	-4	-3	-2	-1	0
-2	-5	-4	-3	-2	-1	0	1
-1	-4	-3	-2	-1	0	1	2
0	-3	-2	-1	0	1	2	3
1	-2	-1	0	1	2	3	4
2	-1	0	1	2	3	4	5
3	0	1	2	3	4	5	6

Figure 5 An example of a performance table of a SOC.

3.1 SOC

One of the earliest techniques for automatic fuzzy system tuning is the Self-Organizing Controller (SOC) [16]. It was originally implemented by Procyk [28] 1977 and it is basically a technique for tuning the consequent of the rules. The idea is to try to match an ideal trajectory with the fuzzy control system. It is assumed that this ideal trajectory lies along a straight line in the state space. Any deviation from the ideal trajectory is corrected by a change of the rule or rules that were responsible for the undesirable process behavior. This is implemented by having a 'performance table' which is a table of the same shape as the state space. Each entry in the table contains a change of rules. These changes are applied to the consequent of the rules so that the process trajectory is guided back to the ideal trajectory. An example of a performance table is shown in figure 5. The diagonal of zeros represents the ideal trajectory. For example, if a deviation from this trajectory occurs and the system reach the state $x_1 = -1, x_2 = 0$, the consequent of the responsible rule is changed one "step" to the preceding fuzzy set. Experience with the SOC has concluded that it is difficult to analyze mathematically. Therefore use of the system requires a lot of trial and error. Also its performance is very much dependent on the application.

3.2 Neural Network Approach

Inspired by neural networks, researchers [32, 33, 34] have set up connective models of a fuzzy system. Common optimization techniques in neural network theory, such as gradient search (also called backpropagation) and conjugate search have been used. An example of this approach is [32] which use a TSK fuzzy system with gradient search optimization. As in neural networks, the problem is that these optimization techniques only guarantees a convergence to a local minima, not a global minima. Also the convergence rate can be vary slow. Another drawback with the neural network approach, is that in its original form, it can only handle continuous variables. This means that the technique is limited in use, to types of fuzzy systems with continuous rule representation, e.g. the TSK model.

3.3 Input-Output Clustering

Kosko [29] has proposed clustering of pairs of input and output data as a way of designing the rule base of a fuzzy system. The method consists of sampling the behavior of a master system, a human or an automated system. The sampling data is clustered with arbitrary clustering algorithm into prototypes. Each prototype is classified to belong to one rule. The number of prototypes in each rule determines the weight of that rule. Only rules with a weight greater than zero become rules of the final system. The method has been used with satisfactory results in, for example, control of an inverted pendulum. The limitation of this method is that it is really only a method for modeling an existing system with a fuzzy system.

3.4 Evolution Algorithms

Recently the optimization concept of evolution algorithms[1], usually in the version of Genetic Algorithms, has been used for automatic design and tuning of fuzzy systems. (See section 5.2 for a detailed description of evolution algorithms.) This is the area in which this chapter is positioned. Most research has concentrated on optimizing only the membership functions [42] or only the rule base [43, 44]. The drawback of these techniques is their inflexibility of being designed for optimization of either membership function or the rule base. However some research has been concerned with the problem of optimizing membership functions and the rule base at the same time [38, 45]. Usually these chapters claim that the techniques that fix one part of the fuzzy system and only optimize the other part, are suboptimal. This is true but a closer look at the techniques that claims to be optimal 'full design' techniques, reveals that they are often also suboptimal in the sense that they for example optimize only one parameter of the membership function or fix the fuzzy partition of each subspace. The meaning of the term 'optimal' is also a bit provocative because evolution algorithms in practice never guarantee an optimal solution and all of the proposed techniques fix the type of fuzzy system. Because almost all people use the Genetic Algorithms version of the evolution algorithms paradigm, it is difficult to perform continuous optimization (see section 5.2 for details).

There are no chapters reporting use of 'full design' techniques, i.e. techniques that deals with all parts of a fuzzy system, for anything else than small fuzzy systems. This can be interpreted as meaning that the use of evolution algorithms for 'full design' may be a theoretically promising technique but in practice not that useful because of the size and complexity of the search space. However, even if a 'full design' technique is not able to produce an optimal fuzzy system it may nevertheless produce a system that is sufficiently good. Also the techniques can be used for different degrees of system tuning. These two application areas are interesting enough to motivate further research in the use of evolution algorithms for fuzzy system design.

[1] This term is used here as the collection name for the paradigms of genetic algorithms, evolutionary programming and simulated annealing.

3.5 Other Techniques

This review of automatic design and tuning techniques doesn't claim to be complete in any sense. Only some of the most common techniques are covered. To briefly mention some more, there are optimizing with Box's complex algorithm [47], rule base design based on the cell state space algorithm [46].

4 UNIVERSAL APPROXIMATION WITH FUZZY SYSTEMS

Neural networks have been proved to be universal approximators, i.e. capable of approximating any real continues function on a compact set to arbitrary accuracy. Although this is a theoretical result, these existence theorems are an important strength even in the practical use of neural networks. Of the same reason it is desirable to have existence theorems for fuzzy systems as well. A couple of theorems, concerning some special types of fuzzy system's ability of universal approximation, have been published.

In [11], it is shown how a fuzzy system with a finite number of fuzzy sets can approximate any function from R^1 to R^1. The key is that the membership functions are tailor made to exactly fit the function that is approximated.

A system with the Center-of-Sums defuzzification (see the next section below) is proved to be an universal approximator in [12]. Also this proof deals only with functions from R^1 to R^1. Another constraint is that the membership functions have to be symmetric and have a finite width.

Finally the fuzzy system that is used in this chapter is shown to be an universal approximator in [10]. This system is not limited in use for functions from R^1 to R^1, it can approximate a function between domains of arbitrary dimensions. The system and the proof is explained in detail below.

4.1 Theory

Recently it has been proved by Wang [10] that a special type of fuzzy system is a universal approximator. The result can be interpreted as an existence theorem of an optimal fuzzy system. The system uses Gaussian membership functions in the input space. That means that the membership functions are defined over the range $[-\infty, \infty]$. In the output space, the membership function's shape is redundant as long as they have equal **height** and **shape**, which implies that they have equal area. This means that they can be represented by fuzzy singletons with any finite area. The meaning of the sentence connective AND is implemented as a product (see formula (4)), and the fuzzy implication function is Larsen's product operation (see formula (5)). The defuzzification method is a special version of the COA called the Center-of-Sums (COS) defuzzification [14]. A fuzzy system with these properties is referred to as Wang's Fuzzy System (WFS).

The COS method consider distribution of the area of the inferred membership functions from each rule individually. This means that if the inferred membership functions are overlapping, their areas are counted more than once. Compared to formula (12.7), instead of fuzzy union, the algebraic sum defines the combination of individual consequent fuzzy sets. The COS is formally given by

$$y = \frac{\int z \cdot \sum_k \chi_{C_w^k}(z)dz}{,} \Big/ \int \sum_k \chi_{C_w^k}(z)dz \tag{12.8}$$

where $\chi_{C_w^k}(z)$ is the membership function inferred by the kth rule (see section 2.3). By first changing place of the integral and the sum and then replacing the integral with a constant s^k times the membership function's height, the defuzzification can be expressed as

$$y = \frac{\sum_k \int z \cdot \chi_{C_w^k}(z)dz}{\sum_k \int \chi_{C_w^k}(z)dz} = \frac{\sum_k z^k \cdot s^k \cdot \chi_{C_w^k}(z^k)}{\sum_k s^k \cdot \chi_{C_w^k}(z^k)dz}, \tag{12.9}$$

where z^k is the center of $\chi_{C_w^k}(z)$ (and $\chi_{C^k}(z)$), and $\chi_{C_w^k}(z^k)$ is the height. This simplification is the reason why COS is much less computationally intensive than COA. Note that the constant s^k is not needed if the shapes are equal, which is actually the case in WFS. To completely and rigorously describe the WSF, some more notation is needed:

$f : U \subseteq R^N \to R^1$,
subspace i is the projection of U into the ith coordinate of R^N ,
K is the number of rules ,
N is the dimension of the system input, i.e. the number of input subspaces ,
A_i^k is the fuzzy set in rule k, belonging to the antecedent of subspace i ,
$a \in [0,1]$ is the height ,
m is the center ,
$\sigma > 0$ is the width ,
x_i is the ith coordinate of the system input x ,
$\chi_{A_i^k}(x_i) = a_i^k \exp\left[-0.5 \cdot \left(\frac{x_i - m_i^k}{\sigma_i^k}\right)^2\right]$
is the membership function of the fuzzy set A_i^k ,
C^k is the fuzzy set of the consequent in rule k ,
z^k is (as before) the center of the membership function belonging to the fuzzy set C^k ,
rule k: IF x_1 is A_1^k AND x_2 is A_2^k AND ... AND x_N is A_N^k THEN z is C^k.

By using this notation and in formula (12.9) use the substitution

beginequation $\chi_{C_w^k}(z^k) = \prod_{i=1}^N \chi_{A_i^k}(x_i) \cdot \chi_{C^k}(z^k)(12.10)$

which is the product implementation of AND together with the product operation as fuzzy implication function (see section 2.3), the following expression is obtained

$$y = \frac{\sum_k z^k \cdot s^k \cdot \prod_{i=1}^N \chi_{A_i^k}(x_i) \cdot \chi_{C^k}(z^k)}{\sum_k s^k \cdot \prod_{i=1}^N \chi_{A_i^k}(x_i) \cdot \chi_{C^k}(z^k)} \tag{12.11}$$

The shapes are all Gaussian, i.e. equal, and the heights are also equal. This means that the expression can be reduced by eliminating s^k and $\chi_{C_w^k}(z^k)$. The WFS can now be defined as follows.

Definition: WFS is the set Y of all fuzzy systems with the properties stated above and consists of all functions of the form

$$f(x) = \frac{\sum_{k=1}^K z^k \prod_{i=1}^N \chi_{A_i^k}(x_i)}{\sum_{k=1}^K \prod_{i=1}^N \chi_{A_i^k}(x_i)}. \tag{12.12}$$

The output from the rules in a system with 4 rules and 4 different output fuzzy sets, one for each rule, can be drawn as in figure 6. Each membership function is shaped as a Dirac pulse and scaled by its rule. The crisp system output is calculated by taking the normalized mean of the scaled rule outputs in accordance with the Center-of-Sums defuzzification method.

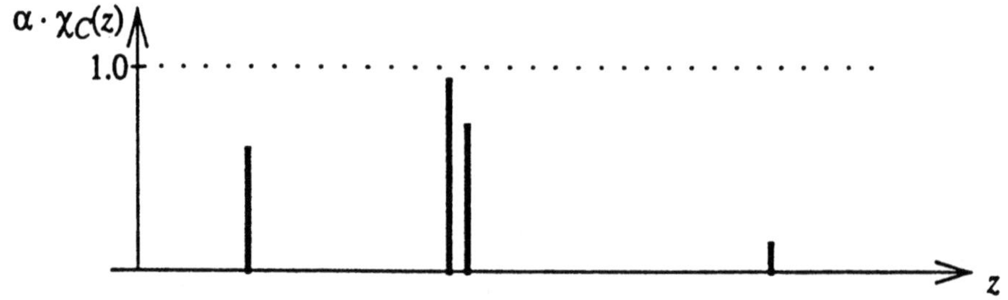

Figure 6 The output from each rule in a WFS with 4 rules. Here the output membership function are represented as Dirac pulses.

To prove that WFS, i.e. all members of Y, is an universal approximator, a metric space has to be defined. Let $d_\infty(f_1, f_2)$ be the sup-metric defined by

$$d_\infty(f_1, f_2) = \sup_{x \in U} \{|g(x) - f(x)|\}, \tag{12.13}$$

then (Y, d_∞) is a metric space. It can be proved [10] that Y is non empty and (Y, d_∞) is well-defined. By using Stone-Weierstrass Theorem to prove that (Y, d_∞) is dense in $(C[U], d_\infty)$, where $C[U]$ denotes the set of all continues functions defined on the compact set $U, f \in Y$ has been shown to be an universal approximator.

Stone-Weierstrass Theorem : Let Z be a set of real continues functions on a compact set U. If:

1. Z is an algebra, i.e. the set Z is closed under addition, multiplication, and scalar multiplication,

2. Z separates points on U, i.e. for every $x, y \in U, x \neq y$, there exists $f \in Z$ such that $f(x) \neq f(y)$,

3. Z vanishes at no point of U, i.e. for each $x \in U$ there exists $f \in Z$ such that $f(x) \neq 0$, then the uniform closure of Z consists of all real continues functions on U, i.e. (Z, d_∞) is dense in $(C[U], d_\infty)$.

Wang proves [10] that Y is an algebra, Y separates points on U and Y vanishes at no point of U. Now the main result can be stated:

Wang's Theorem: For any given real continues function g on the compact set $U \in R^N$ and arbitrary $\varepsilon > 0$, there exists $f \in Y$ (see the definition with equation (12.2)) such that

$$\sup_{x \in U} \{|g(x) - f(x)|\} < \varepsilon \tag{12.14}$$

Proof: Due to formula (12.12), Y is obviously a set of continues functions on U. Wang's Theorem is therefore a direct consequence of the Stone-Weierstrass Theorem.

4.2 Limitations in Practice

Wang's theorem provides the mathematical basis for the use of WFS as an universal approximator. However it is assumed that K can take any value and that each rule has its own fuzzy sets. Also the heights of the input membership functions are assumed to be an arbitrary value between 0 and 1. In practice the fuzzy sets in the rules are picked from a predefined set of fuzzy sets for each subspace. As said above, the number of fuzzy sets in each subspace is usually between 3 and 7. Also the maximum of the membership functions is usually 1, because the linguistic property of the fuzzy set is totally true for at least one value in the universe of discourse.

In spite of these somewhat unrealistic assumptions, it is still a strength to know that the WFS in theory can approximate any function. This has not been proved for any other type of fuzzy system with multi dimensional input vector. To make WFS usable in practice, some restriction have to be introduced. In this chapter, the chosen restrictions are determined by the following definitions:

n_i is the number of fuzzy sets in the ith subspace ,
B_{ji} is the jth fuzzy set in the ith subspace ,
the height a^{ji} is always set to 1 ,
m_{ji} is the center ,
σ_{ji} is the width ,
$A_i^k \in \{Bji|j = 1 \ldots n_i\}$,
n_z is the number of fuzzy sets in the output space ,
z_j is the center of the membership function of the ,
j^{th} fuzzy set in the output space ,
$z_k \in \{z_j|j = 1 \ldots n_z\}$.

4.3 Increased Freedom and Understanding of WFS

Wang has in his chapter assumed that the membership functions in the output space have equal height and shape. Therefore the only thing that counts is the membership functions' position of the center. However Wang's Theorem can be generalized to be valid for output membership functions with different height and shape. In formula (12.11), $\chi_{C^k}(z^k)$ together with s^k defines the areas of the consequence' membership functions of the kth rule. For each k, these constants can be included in any of the parameters a_i^k. If a_1^k is chosen, then the new a_1^k gets the value

$$a_{new1}^k = s^k \cdot \chi_{C^k}(z^k) \cdot a_1^k. \tag{12.15}$$

With this substitution, the function of the system can again be written as in formula (14) and Wang's theorem is therefore valid.

All input membership functions in WFS are Gaussian with height 1. If the output membership functions also are chosen like that, the function of the system can be written as:

$$f(x) = \frac{\sum_{k=1}^{K} z^k \cdot s^k \cdot \prod_{i=1}^{N} \chi_{A_i^k}(x_i)}{\sum_{k=1}^{K} s^k \cdot \prod_{i=1}^{N} \chi_{A_i^k}(x_i)} \tag{12.16}$$

where , $s^k = \sqrt{2\pi} \cdot \sigma^k$ which is only dependent on the width-parameter of the membership function. If this property is introduced, the freedom is increased compared to the restrictions stated in the definitions of section 4.2. Also it is easier to associate the function of this fuzzy system with the established concept of fuzzy sets, where variables are fuzzy (have different distribution) and not crisp (singletons). It should be noted that only the **relations** between the s^k values (and analogous σ^k) are important. Because of the normalizing division in (12.16), any common factor can be multiplied to the s^k values and (19) would still evaluate to the same result. These extra parameters s^k, or σ^k, are used in this chapter as a way of increasing the freedom and to make the system more intuitively understandable (see section 6.1 for more details).

5 GUIDED EVOLUTIONARY SIMULATED ANNEALING

Classical optimization techniques are usually not capable of optimizing complex systems such as a fuzzy system. A class of algorithms, with its origin in natural processes, has become widely spread the last couple of years. Based on these approaches the algorithm GESA has been developed recently and that is the algorithm that is used in this chapter.

5.1 Classical Optimization

To find the optimal parameters of a fuzzy system is a non linear optimization problem. Traditional techniques for non linear optimization are:

- Newton Raphson
- gradient search
- conjugate search
- stochastic search

Common to the first three of these techniques is that they require an analytical description (or an implementation with good numerical accuracy) of the problem. Gradient search and conjugate search have been successfully applied to complex problems such as perceptron- like neural networks [17]. They are guaranteed to converge to a minima. This minima can be a global minima but in most cases, it is more likely to be a local minima. Stochastic search have the advantage of not getting trapped into a local minima that easily, but the convergence rate is very slow because of the lack of guidance in the search.

5.2 Nature Inspired Approaches to Optimization

By studying nature, three new classes of optimization algorithms have been developed. The first has its origin in biological evolution and the second comes from the study of annealing, i.e. the process of heating followed by cooling of solids so that they crystallizes into a perfect lattice. The third is the Hopfield Neural Network [22] that can be used for optimization. Because this chapter uses an optimization algorithm with its origin in the evolution and annealing algorithms, a brief overview of these two paradigms is given below.

Goldberg [18] has popularized Genetic Algorithms (GA) that is based on the mechanisms of natural selection and genetics. Each solution, described as a parent or child, is coded as a binary vector (string). Fogel [19] is associated with a paradigm that is called Evolutionary Programming (EP). It is based on Darwin's evolution theory. These two paradigms are basically the same and the basic algorithm is shown in figure 7.

The conceptual difference between GA and EP is the way children are generated. In GA a child is generated by combining **two** parents (crossover) and then applying a random change (mutation). In EP a child is generated from **one** parent by a random change. In addition to this, GA has fixed on the idea of representing a solution as a binary string. This has the disadvantage that the representation is discrete, which means that GA's primary application area is that of combinatorial optimization. Therefore in continuous optimization, a somewhat strained implementation is enforced. It has been argued whether GA or EP is best, in other words if crossover is good or bad. No definitive answer has been given. The question of which one to choose seems to be problem and implementation dependent.

By applying the Monte Carlo's simulation procedure [20] to annealing, Kirkpatrick, Gelatt and Vecchi proposed the Simulated Annealing (SA) [21] technique for optimization. The algorithm is shown in figure 8. Note that a

```
generate N number of initial parents
repeat
      generate M children from the parents
      evaluate all N + M solutions
      select the N best solutions as parents for next generation
until solution found
```

Figure 7 The basic GA/EP algorithm.

```
set initial temperature t
generate randomly a best solution
evaluate the solution -> y_best
repeat
      repeat k(t) times
            generate a new solution from the current best solution
            evaluate the new solution -> y_new
            accept the new solution as current best solution if
                  exp[-(y_new - y_best) / t] > p
      decrease t
until solution found
```

Notations:
t is the temperature
y_{new}, y_{best} are the objective values of the new and current best solutions respectively.
p is a uniformly distributed random number between 0 and 1.

Figure 8 The basic SA algorithm.

lower objective value, in this example, is a better one. The condition for checking if a new solution is going to be accepted as the best solution is

$$e^{-\frac{y_{new}-y_{best}}{t}} > p \in [0,1]. \tag{12.17}$$

The purpose is to always accept a new solution if it is better than the best current one, and with a probability proportional to how good it is, also accept it even if it is not as good as the current best solution.

These two paradigms, GA/EP and SA, are very similar and are the basis of the GESA algorithm. Before it is explained, a comparison of the two paradigms is presented. A good optimization technique should be guided, it should have the ability to escape from local minima and the ability to converge to a solution with arbitrary accuracy. The comparison is based on these three criteria and the similarities and differences are explained.

- Ability to escape from local minimas.

  GA/EP: Doesn't have any special mechanism but parallelism decreases the probability of getting trapped in a local minima.

  SA: It is not parallel but because of the formula (12.17), a trial solution may be examined even if it is not as good as the current best solution.

- Guidance - how new solutions are guided to be generated in the most promising region.

  GA/EP: Only the best solutions become parents. In GA the property of guidance can be damped or absent depending on the string representation.

  SA: A better solution has larger probability of becoming the 'parent' for next generation, see formula (12.17), and k increases when the temperature is decreased.

- Convergence - how the algorithm can converge to a solution with arbitrary accuracy.

  GA/EP: No special mechanism.

  SA: Its ability of convergence is achieved to some extent by the temperature in formula (12.17) together with the increasing k.

  Both paradigms: Children can be generated with smaller and smaller change of the parents as the execution proceeds. This can be implemented with a temperature dependent mechanism so that the amount of change is proportional to the decreasing temperature.

5.3 The GESA algorithm

Many algorithms have originated from the concepts of GA/EP and SA. Yip and Pao proposed 1993 the GESA algorithm [23] for artificial neural networks training, but the use is not limited to neural networks and it has been applied to a variety of optimization problems, see for example [24] and [25]. The philosophy of GESA is to pick and combine the best properties of GA/EP and SA. The properties that are used is:

- parallelism.
- the simulating annealing formula (20) for accepting children as parents for next generation so that the best child can be accepted even if it is not as good as the parent.
- only the best N children are accepted as parents for the next generation
- the number of children generated from one parent is proportional to the average quality of the family, i.e. the parent and its children.
- because of a new concept with families, the same effect as increasing k in SA is obtained children are generated with smaller and smaller change of the parents as the execution proceeds so that convergence can be obtained.

A family is a parent together with its children. Because of the concept of families, two kinds of competitions exist. Local competition is the competition, between the children within a family, of who is going to be the next generation's parent. Global competition is the competition of the distribution of children between the families. The algorithm is shown in figure 9. In practice, the temperatures differ only by a constant coefficient. Also the decrease of temperature is usually implemented as tnew = c told, where c is a constant less than 1.

The algorithm is very suitable for implementation in parallel computers of the type SIMD (Single Instruction Multiple Data) because the total number of children is always constant. In that case the evaluation of the object function is done in parallel with one processing unit handling each child. The complexity is very much dependent on the evaluation of the children because in a typical application, the most time is spent in this step. GESA has been

```
main algorithm:
    set initial temperatures t1, t2 and t3
    generate randomly N parents
    evaluate these parents
    repeat
        for each family do
            generate children* from the parent by a random change that is proportional to t3
            evaluate these children
            find the best child
            accept this child as the parent for next generation if
                exp[-(y_new - y_best) / t1] > p
        find the number of children that will be generated in each family in the next generation
                by calling the subroutine
        decrease the temperature coefficients
    until solution found

* the number of children is M the first time

subroutine:
for each family i do
    acc_i = 0
    for each child in family i do
        if exp[-(y_child - y_best) / t2] > p
            then acci = acci + 1
sum_acc = Σ_i acc_i
for each family i do
    the number of children in next generation is
        M · N · acc_i / sum_acc

Notation:
    t1, t2 and t3 are the temperatures
    N is the number of children
    M is the avarage number of children in each family
    y_new, y_best and y_child are objective values (lower is better)
    y_best is the objective value for the globally best solution found so far
    acc_i is the number of accapted children in family i
    sum_acc is the total number of accepted children
```

Figure 9 The GESA algorithm.

experimentally compared to other optimization techniques in combinatorial [24] as well as continues [25] optimization. The conclusion of these benchmarks is that it can compete well with GA, EP, SA and Hopfield Nets in both continues and combinatorial problems.

5.4 Asymptotic Convergence Theorem

In [26], a proof for asymptotic convergence of the SA algorithm is given. This proof can be applied to the GESA algorithm because GESA can be thought of as a number of parallel SA processes. The proof is based on the theory of finite Markov chains and the SA process is modeled as a Markov chain. Before the convergence theorem can be stated, the three matrices G_{ij}, A_{ij} and P_{ij} have to be defined:

Definition: Generation probability G_{ij}

$$G_{ij} = \begin{cases} 0 & \text{if } j \ni S_i \\ \frac{1}{|S_i|} & \text{if } j \in S_i \end{cases} \tag{12.18}$$

where S_i is the set of solutions in the neighborhood of solution i. This is the probability of generating solution j if the current best solution is i.

Definition: Acceptance probability A_{ij}

$$A_{ij}(c_k) = \begin{cases} \exp(-\frac{f(j)-f(i)}{c_k}) & \text{if } f(j) - f(i) > 0 \\ 1 & \text{otherwise} \end{cases} \tag{12.19}$$

where c_k is the temperature at step k, and $f(i)$ and $f(j)$ are the objective values of the solutions i and j. This is the probability of accepting solution j as the new solution if i is the current best solution.

Definition: Transition probability P_{ij}

$$P_{ij} = \begin{cases} G_{ij}A_{ij} & \text{if } i \neq j \\ 1 - \sum_{l \neq i} P_{il} & \text{if } i = j \end{cases} \tag{12.20}$$

This is the probability of generating and accepting j as new solution if i is the current best solution. With these definitions the following theorem about stationary distribution can be stated:

Theorem [26]: Let (S, f) denote an instance of a combinatorial optimization problem and $P(c)$ denote the transition matrix associated with the SA algorithm defined by a Markov chain and (12.18), (12.19) and (12.20). Furthermore, let the following condition be satisfied:

$$\forall i, j \in S \exists p \geq l, \exists l_0, l_1, ..., l_p \in S, \text{ with } l_0 = i, l_p = j, \text{ and} \tag{12.21}$$

$$G_{l_k l_{k+1}} > 0, k = 0, 1, \ldots, p-1 \tag{12.22}$$

Then the Markov chain has a *stationary distribution* $q(c)$, whose components are given by

$$q_i(c) = \frac{1}{N_0(c)} \exp\left(-\frac{f(i)}{c}\right) \text{ for all } i \in S \tag{12.23}$$

where

$$N_0(c) = \sum_{j \in S} \exp\left(-\frac{f(j)}{c}\right) \tag{12.24}$$

Aart proves this theorem by first showing that $P(c)$ is *irreducible* and *aperiodic* and then he uses Feller's [27] theorem to show that the Markov chain has a stationary distribution. The main theorem is given here without a proof but Aart shows that it can be proved by a straight forward rewriting of expression (12.23).

Theorem [26]: Given an instance (S, f) of a combinatorial optimization problem with a stationary distribution given by (12.23), then

$$\lim_{c \to 0} q_i(c) = \begin{cases} 0 & \text{if } c \ni S_{opt} \\ \frac{1}{|S_{opt}|} & \text{if } c \in S_{opt} \end{cases} \tag{12.25}$$

where S_{opt} denotes the set of globally optimal solutions.

This result says that the SA algorithm is guaranteed to find an optimal solution asymptotically. It is true for combinatorial optimization problems but can be generalized to continuous problems where the optimization goal is a value with finite accuracy.

Name	Symbol	Type	Range	Number
center	m_{ji}	continuous	$-\infty - \infty$	n_{tot}
width	s_{ji}	continuous	$0 - \infty$	n_{tot}
consequent	C^k	discrete	$1 - n_z$	K

Table 1 The parameters of RWFS.

6 WANG'S FUZZY SYSTEM AND GESA

The idea in this chapter is to use GESA to find the parameters of WFS so that an universal approximator can be constructed. The mathematical basis is covered by the previous sections and it can be summarized:

- Wang's theorem says that WFS is capable of approximating any function to arbitrary accuracy
- Aarts theorem says that GESA is capable of optimizing any function, including finding the parameters of WFS.

This is two very powerful tools but also very computationally intensive. To be able to treat real world problems, some restrictions have to be made.

6.1 Optimized Parameters

The restrictions proposed in this chapter is the one listed in the definition in section 4.2 with the extra freedom of the width in the output membership function discussed in section 4.3. WFS with these restrictions is referred to as the Restricted Wang's Fuzzy System (RWFS). The problem of designing and optimizing the (RWFS) can be viewed as a mathematical optimization problem with both continues and discrete variables. These variables, i.e. the parameters of the RWFS, are shown in table 1 for a RWFS that has N input subspaces, K rules and $n_{tot} = \sum_{j=1}^{N} n_j + n_z$[2].

[2] The index here is not a numerical index. It refers to the output space.

Note that $m_{j,z}$ is the same as z_j, and that the C parameters' datatype is the fuzzy sets in the output space, but implemented as integer values. The total number of parameters are therefore $2n_{tot} + K$.

This optimization problem is non linear and in most cases very complex. Due to the discussion in section 5.1, regular optimization techniques are not capable of solving this kind of problem. Therefore the algorithm GESA is used. Each solution, parent or child, is represented as a vector of m, σ and C element:

$$\{m_{11}, m_{21}, \ldots, m_{n_z N}, m_{1z}, \ldots, m_{n_z z}, \sigma_{11}, \sigma_{21}, \ldots, \sigma_{n_z z}, C^1, C^2, \ldots, C^k\}$$

All elements are continues except for the last K ones. The random change, by which children are generated in GESA (see figure 9) consists of first interpreting each element as continues and then adding a Gaussian distributed value (because the width of σ is always greater than zero, the distribution of the added random value is in this case a half Gaussian, with mean equal to 0, to each element. The variance is depending on the application and the kind of the element, i.e. m, σ or C. Because the C elements have to be integers, they are then rounded off to the closest valid integer value. As stated in the definition of GESA, the variance decreases with the temperature t_3.

6.2 Proposed Design Strategy

In spite of the restrictions introduced in the RWFS, for large fuzzy systems the search space can still be very large. To be able to solve these problems within reasonable computing time, some guidance or help has to be provided to GESA. This can consist of:

- recognizing symmetry in the problem
- constraints on the centers m_{ij}
- predetermine and temporarily fix some parameters

The exact design strategy is of course problem dependent, but can in general terms be stated as:

1. Use expert knowledge and engineering skills to predetermine:
   (a) the number of variables involved, i.e. the number of subspaces
   (b) fuzzy partition of each subspace
   (c) the number of rules and the antecedents of the rules
   (d) as many as possible of the other design parameters, i.e. rules and membership functions.
2. With these fixed parameters, use GESA to determine the rest of the rules and membership functions.
3. If necessary, tune and optimize the whole system, i.e. previously fixed parameters and by GESA determined parameters.

Step 2 can be said to deal with automatic fuzzy system design while step three deals with automatic fuzzy system tuning, if the terminology that was introduced in section 3 is used. To investigate how this technique works in practice, it has been applied to two real problems, a function estimation problem and a control problem. These are discussed in the following two sections.

7 FUNCTION ESTIMATION: THE GENERALIZED PARITY-2 PROBLEM

Neural networks are generally very good at function estimation. Here it is examined if the proposed technique for fuzzy system design , in practice, can produce a fuzzy system with a similar capability of function estimation. A classical problem in the area of neural network training is the parity-2 problem. This problem in a generalized version is approached with the fuzzy system technique.

7.1 The Parity-2 Problem

The original version is to classify four two dimensional patterns into two classes. In this implementation the patterns are all combinations of -1 and 1

Pattern	Class
(-1, -1)	1
(-1, 1)	-1
(1, -1)	-1
(1, 1)	1

Table 2 The original parity-2 problem

and the classes are defined by the functional values -1 and 1. The parity-2 problem is given by table 2. The classification can be viewed as a continues function and if the pattern (0, 0) with function value 0 is added, a generalized version of the problem is obtained.

Neural networks trained with these patterns usually produces a smooth functional surface. The problem here is to see if the fuzzy system can be designed to approximate a similar surface. In order to more strictly define the surface, some constraints about the gradient are added. The original constraints together with the gradient constraints define the generalized parity-2 problem and are as follows:

$$\begin{aligned} f(-1,-1) &= 1 \\ f(-1,1) &= -1 \\ f(1,-1) &= -1 \\ f(1,1) &= 1 \\ \nabla f(-1,-1) &= 0 \\ \nabla f(-1,1) &= 0 \\ \nabla f(1,-1) &= 0 \\ \nabla f(1,1) &= 0 \\ \nabla f(0,0) &= 0 \end{aligned}$$

Many functions fulfill these constraints, for example some 2-dimensional polynomials. Because of the symmetry, this must be a polynomial with even grade. A 2-dimensional polynomial is obviously not enough so the simplest

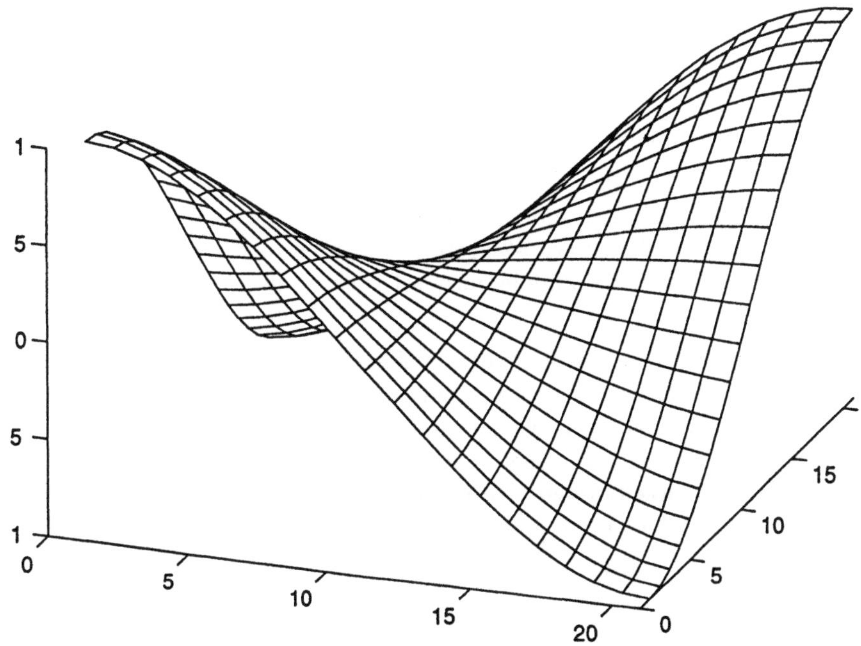

Figure 10 The generalized parity-2 problem viewed as a continuous function.

polynomial is of the fourth degree. The function that is used in this example is given in the following equation and illustrated in figure 10.

$$f(x,y) = -0.5xy^3 - 0.5x^3y + 2xy. \tag{12.26}$$

7.2 Design Of The Fuzzy System

One issue of interest is how the fuzzy partition of each subspace affect the performance of the fuzzy system, that is the accuracy in the function estimation. Therefore three different partitions are investigated: 2, 3, and 5. Note that for each case, the same partition is used in all subspaces. The design strategy proposed in section 6.2 is followed:

1. Predetermined parameters:

   (a) The numbers of input variables is 2, i.e. $N = 2$.

   (b) Three different fuzzy partitions of each subspace are used, i.e. $n_i = 2, n_i = 3$ and $n_i = 5$, where $i = 1, 2, z$.

   (c) The maximum number of rules with all possible antecedents is chosen. That means $K = n_1 \cdot n_2 = n_i^2 = 4, 9$ and 25.

   (d) The center of the first and the last fuzzy sets in each subspace is, due to the properties of the parity-2 problem, prefixed to -1 and 1 respectively. Also the center of the middle fuzzy set is set to 0.

2. Also the center and the width of the membership functions are symmetric around the crisp value 0. To get an easier optimization problem when the fuzzy partition is 5 and there are 25 rules, it is observed that, because of symmetry, the rules in one corner of the rule matrix determine all rules. These rules are marked in figure 15(c). This means that the vector that is optimized by GESA is, in each of the three examples are

$$\{\sigma_{1x}, \sigma_{1y}, \sigma_{1z}, C_1, C_2, C_3, C_4\} \qquad n_i = 2$$

$$\{\sigma_{1x}, \sigma_{2x}, \sigma_{1y}, \sigma_{2y}, \sigma_{1z}, \sigma_{2z}, C_1, C_2, \ldots, C_9\} \qquad n_i = 3$$

$$\{c_{2x}, c_{2y}, c_{2y}, \sigma_{1x}, \sigma_{2x}, \sigma_{3x}, \sigma_{1y}, \sigma_{2y}, \sigma_{3y}, \sigma_{1z}, \sigma_{2z}, \sigma_{3z}, C_1, C_2, C_6, C_7\} \quad n_i = 5$$

   The continues parameters that are optimized are illustrated in figure 11. As objective function, the sum of the quadratic estimation error over 121 equally distributed sample points (training patterns) is used:

$$err_{tot} = \sum_{k=1}^{121} \left(\tilde{f}(x_k, y_k) - z_k \right)^2 \tag{12.27}$$

   where x_k, y_k are the position in input space of the training patterns and z_k is the correct function value due to formula (12.26).

3. To tune the system, all parameters are at this stage optimized. This means that in addition to all parameters tuned before, the centers of the first and last fuzzy sets in each subspace, are also tuned.

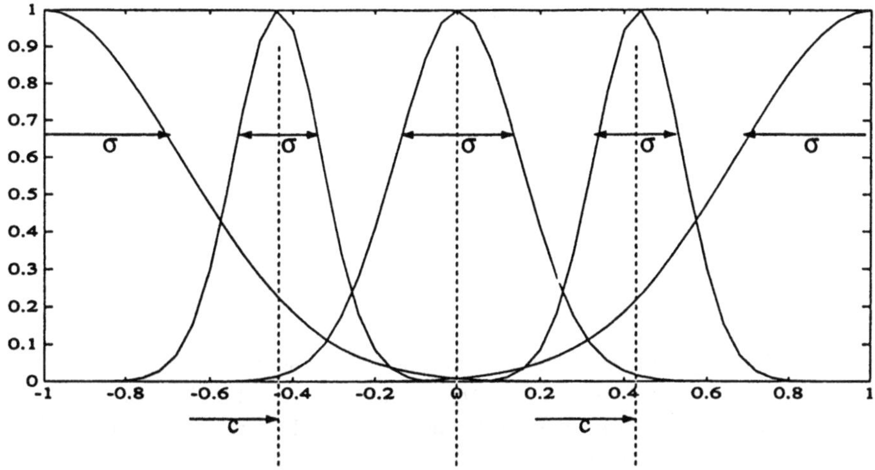

Figure 11 For the case when the fuzzy partition is 5, the 7 continues parameters that are optimized in the step 2 are illustrated by arrows.

7.3 Results

The final RMS error of the 121 sample points for the three examples are shown in table 3. The first two columns show the result after step 2 (see section above) and the third column shows the result after extra tuning is done. In the first column, the result using the same shape (width) of all output membership functions is shown. In the second column, the output membership functions are allowed to have different shapes, due to section 4.3. In the third column, the result from the second column is further optimized including also the outmost centers (see step 3 in the section above).

In tables table 4, 5, and 6, the centers and widths of all membership functions for each example respectively, are shown. The numbers within parentheses are the values after step 2 and the other numbers are the values after step 3. As expected, because of symmetry the values for x and y are identical. The same data is shown graphically in figure 12, 13 and 14. The resulting rule base is shown in figure 15 for each of the three cases.

fuzzy partition	same shape of the output space membership functions	different shape of the output space membership functions	tuned, after step 3
2	0.051	N/A	0.013
3	0.043	0.017	0.0047
5	0.013	0.0059	0.0022

Table 3 RMS error of the 121 sample points for the three examples. The two first result columns shows the result after step 2 and the last column shows the result after step 3.

The fact that the centers of the first and last fuzzy sets in each subspace are allowed to go beyond 1 produces better results due to table 3. However this does not have a natural intuitive interpretation, and therefore it is arguable if the release of this constraint can be allowed. The explaination to this phenomena is to be find in the properties of the COA and COS defuzzification procedures in combination with Gaussian membership functions. Because the center of the area is calculated and Gaussian function have an infinite range where it is greater than zero, all membership functions within a subspace are always going to contribute to the center of the area. This means that COA/COS can not possibly give any of the end positions (first or last centers) as a result. Therefore the first and last centers have to be a bit beyond -1 and 1, if the system is going to be able to give -1 and 1 as outputs.

In figure 15, 16 and 17, the 3-dimensional surfaces of the absolute error compared to the correct surface in figure 10 are shown. It is easy to see that the error decreases when the fuzzy partition increases. Also the error after step 3 is much less.

8 ON FUZZY LOGIC CONTROL: THE INVERTED PENDULUM

The inverted pendulum is a classical control problem, that has become a benchmark problem for non linear control. This problem originated from a need to keep a space launch vehicle under control as it left the launch pad and also to keep it balanced as it is rolled out to the launch pad on a railway cart.

subspace	param.	negative	positive
x	center	-1 (-0.93)	1 (0.93)
	width	0.75 (0.86)	0.75 (0.86)
y	center	-1 (-0.93)	1 (0.93)
	width	0.75 (0.86)	0.75 (0.86)
z	center	-1 (-1.44)	1 (1.44)
	width	N/A	N/A

Table 4 The parameters that specifies the membership functions of the fuzzy system when the fuzzy partion is 2 in each subspace. The first values in each cell are the values obtained after step 2. Values within parentheses are the values obtained after step 3, i.e. the finial fine tuning.

subspace	param.	negative	zero	positive
x	center	-1 (-1.08)	0 (0)	1 (1.08)
	width	0.57 (0.68)	0.38 (0.46)	0.57 (0.68)
y	center	-1 (-1.08)	0 (0)	1 (1.08)
	width	0.57 (0.68)	0.38 (0.46)	0.57 (0.68)
z	center	-1 (-1.13)	0 (0)	1 (1.13)
	width	0.41 (0.56)	0.21 (0.21)	0.41 (0.56)

Table 5 The parameters thet specifies the membership functions of the fuzzy system when the fuzzy partion is 3 in each subspace. The first values in each cell are the values obtained after step 2. Values within parentheses are the values obtained after step 3, i.e. the finial fine tuning.

subspace	param.	very neg.	negative	zero	positive	very pos.
x	center	-1 (-1.06)	-0.44 (-0.34)	0 (0)	0.44 (0.34)	1 (1.06)
	width	0.41 (0.44)	0.22 (0.34)	0.28 (0.36)	0.22 (0.34)	0.41 (0.44)
y	center	-1 (-1.06)	-0.44 (-0.34)	0 (0)	0.44 (0.34)	1 (1.06)
	width	0.41 (0.44)	0.22 (0.34)	0.28 (0.36)	0.22 (0.34)	0.41 (0.44)
z	center	-1 (-1.07)	-0.44 (-0.70)	0 (0)	0.44 (0.70)	1 (1.07)
	width	0.33 (0.36)	0.11 (0.13)	0.15 (0.15)	0.11 (0.13)	0.33 (0.36)

Table 6 The parameters thet specifies the membership functions of the fuzzy system when the fuzzy partion is 5 in each subspace. The first values in each cell are the values obtained after step 2. Values within parentheses are the values obtained after step 3, i.e. the finial fine tuning.

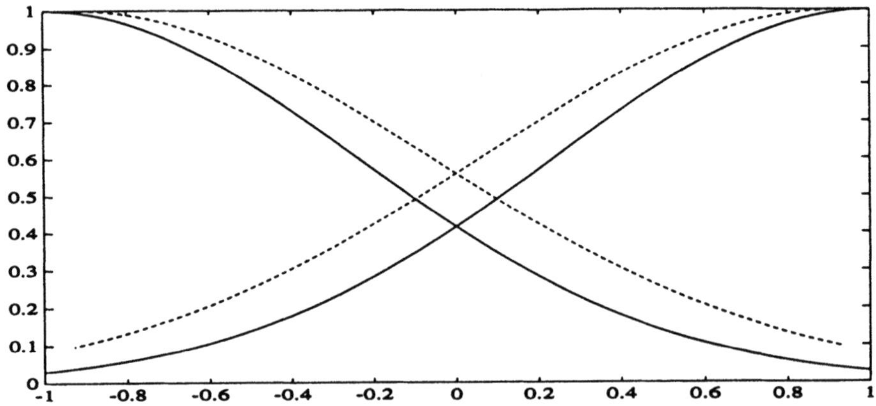

Figure 12 Membership functions in input space x and y when the fuzzy partition is 2. Dashed membership functions is the result after final tuning, i.e. step 3.

Different control strategies have been used: the state-variable model with linearization [35], neural networks [36, 37] as well as fuzzy logic controllers [29, 32, 38, 39]. Different versions of the inverted pendulum exists, but one of the most common is a pole hinged to a moving cart as shown in figure 19.

8.1 The Physical Model

The systems behavior is well known and can be mathematically described by the equations [40]

$$\ddot{\theta} = \frac{g\sin\theta + \cos\theta \cdot \frac{-F - m_p l\dot{\theta}^2 \sin\theta + \mu_c sgn\dot{x}}{M_c + m_p} - \frac{\mu_p \dot{\theta}}{m_p l}}{l\left(\frac{4}{3} - \frac{m_p \cos^2\theta}{M_c + m_p}\right)}, \tag{12.28}$$

$$\ddot{x} = \frac{F + m_p l\left(\dot{\theta}^2 \sin\theta - \ddot{\theta}\cos\theta\right) - \mu_c sgn\dot{x}}{M_c + m_p}. \tag{12.29}$$

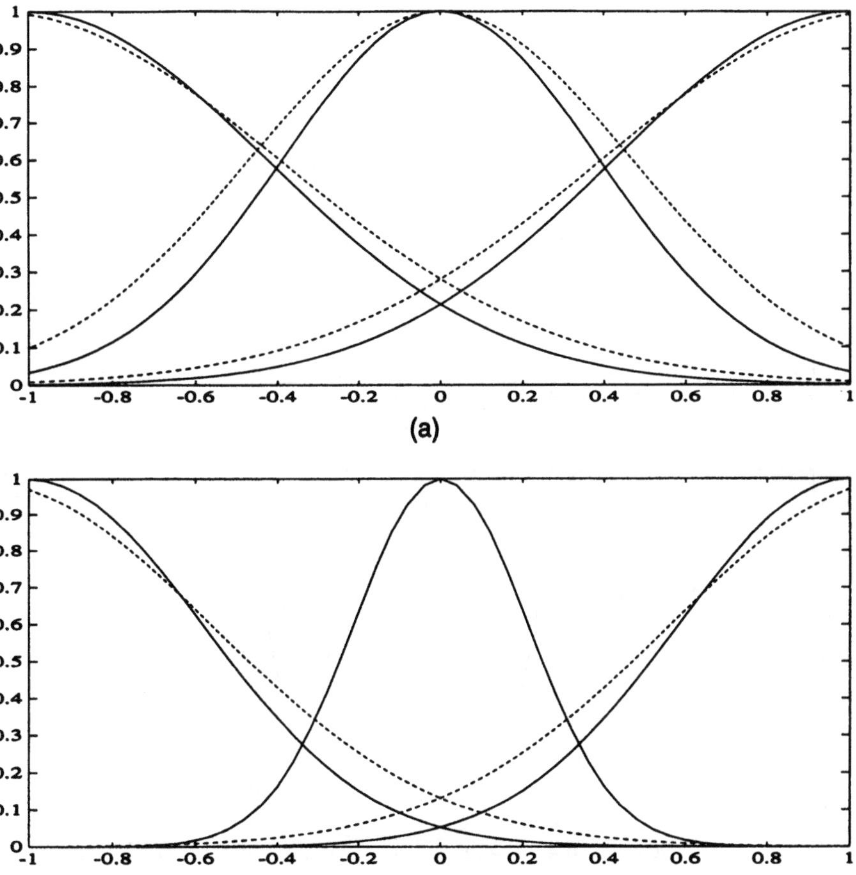

Figure 13 Membership functions when the fuzzy partition is 3. Dashed membership functions is the result after final tuning, i.e. step 3. (a) Input spaces x and y. (b) Output space z.

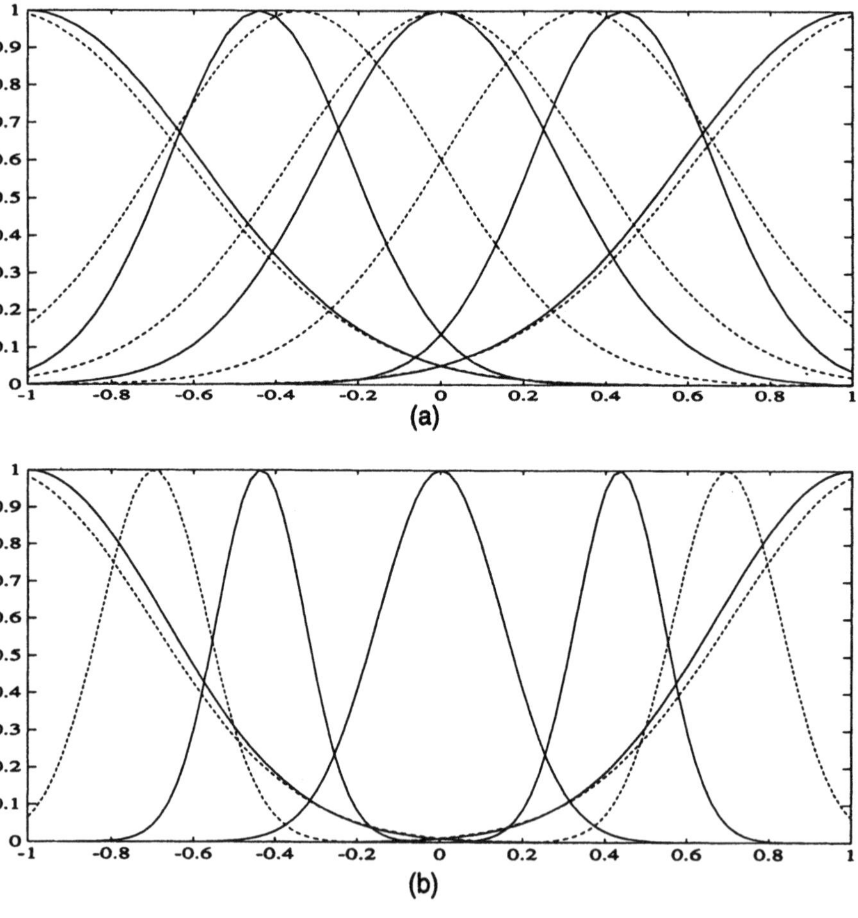

Figure 14 Membership functions when the fuzzy partition is 5. Dashed membership functions is the result after final tuning, i.e. step 3. (a) Input spaces x and y. (b) Output space z.

(a)

x \ y	-	+
-	+	-
+	-	+

(b)

x \ y	-	0	+
-	+	0	-
0	0	0	0
+	-	0	+

(c)

x \ y	-	-	0	+	+
-	⊕	+	0	-	Θ
-	+	+	0	-	-
0	0	0	0	0	0
+	-	-	0	+	+
+	Θ	-	0	+	⊕

Figure 15 The rule base that are produced by GESA. Θ and ⊕ represents the fuzzy sets very negative and very positive. (a) Fuzzy partition is 2. (b) Fuzzy partition is 3. (c) Fuzzy partition is 5.

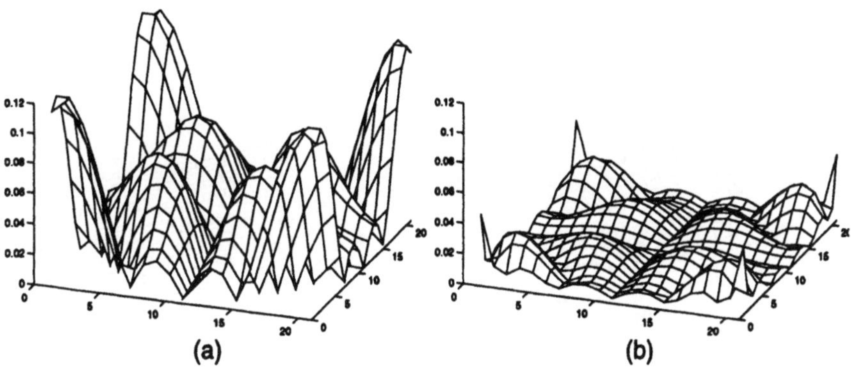

Figure 16 The absolute error compared to the real surface defined by (12.26). The parity-2 surface approximated by RWFS with fuzzy partition 2. (a) After step 2. (b) After step 3, i.e. final tuning.

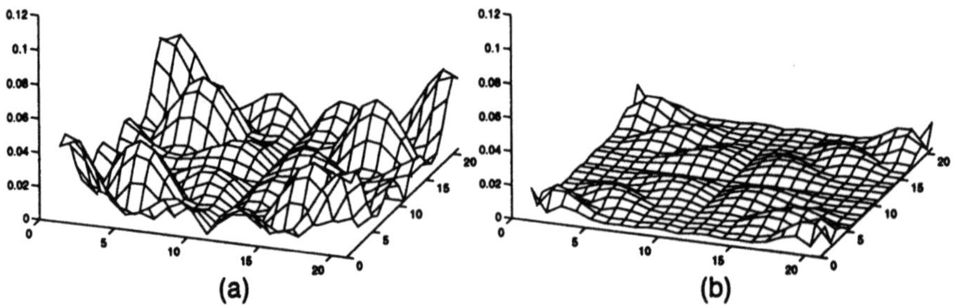

Figure 17 The absolute error compared to the real surface defined by (12.26). The parity-2 surface approximated by RWFS with fuzzy partition 3. (a) After step 2. (b) After step 3, i.e. final tuning.

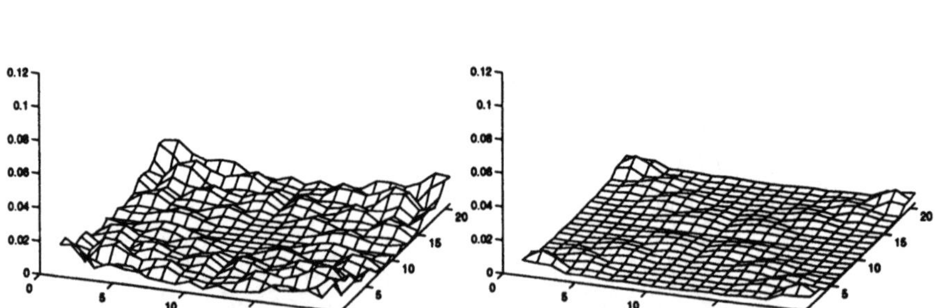

Figure 18 The absolute error compared to the real surface defined by (12.26). The parity-2 surface approximated by RWFS with fuzzy partition 5. (a) After step 2. (b) After step 3, i.e. final tuning.

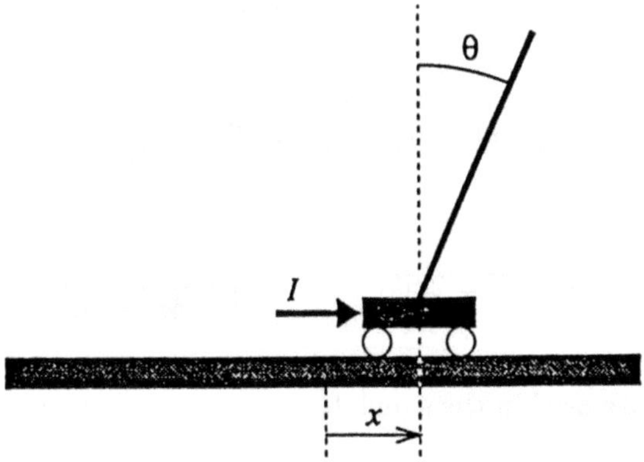

Figure 19 The inverted pendulum, hinged to a cart on a platform.

where θ is the pole angle, x is the position of the cart, g is the acceleration due to gravity, M_c is the mass of the cart, m_p is the mass of the pole, l is the length from hinge to the center of the mass of the pole, F is the applied force, μ_p is the hinge friction coefficient, and μ_c is the cart friction coefficient. This mathematical model could be used directly by applying an numerical integration method such as Runge-Kutta or Euler's [41]. However this would be relatively computer intensive so to decrease the computing involved, a simpler approximate representation is derived. First the friction is assumed to be neglectable. Second it is assumed that the mass m_p is small compared to M_c. This gives the expressions

$$\ddot{\theta} \approx \frac{g}{l} \sin\theta - \frac{F}{m_p} \cos\theta \tag{12.30}$$

$$\ddot{x} \approx \frac{F}{M_c + m_p}, \tag{12.31}$$

These continues second derivatives are approximated by the discrete analogy with an integrating stepsize of h seconds:

$$\ddot{x}(t) \approx \frac{x(t+h) - 2x(t) + x(t-h)}{h^2}. \tag{12.32}$$

Instead of a force F, an impulse $I = F \cdot h$ is used. Substituting (12.32) into (12.30) and (12.31), and solving for $\theta\,[t+h]$ and $x\,[t+h]$ results in

$$\theta\,[k+1] = 2\theta[k] - \theta[k-1] + \frac{h^2 g}{l}\sin\theta[k] - \frac{Ih}{M_c l}\cos\theta[k] \tag{12.33}$$

$$x[k+1] = 2x[k] - x[k-1] + \frac{Ih}{M_c + m_p} \tag{12.34}$$

These equation are used in the simulations and the parameters are set as follows:

$$\begin{aligned} g &= 9.81\frac{m}{s^2} \\ M_c &= 1.0kg \\ m_p &= 0.1kg \\ l &= 1.0m \\ h &= 0.05s \end{aligned}$$

The total length of the platform is chosen to $4m$, which means that the range of x is from -2 to 2.

8.2 The Control Model

Equations (12.33) and (12.34) provide the simulation model for the pendulum and the impulse is the control variable obtained from the fuzzy system used as a controller. A model of the controlled system is shown in figure 20. The inverted pendulum system has four state variables:

θ	angle of the pole
ω	angle velocity of the pole
x	position of the cart

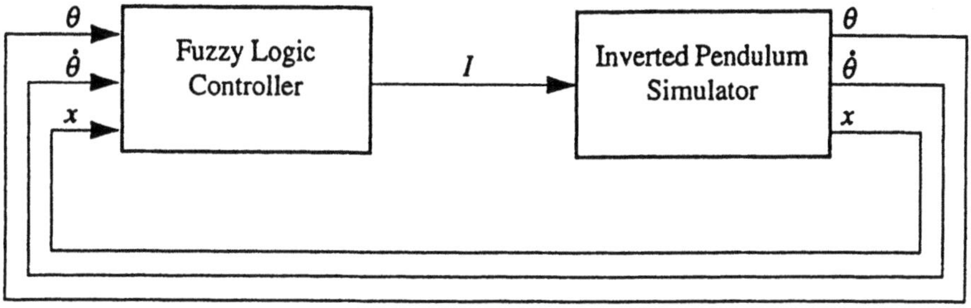

Figure 20 The system of the simulated inverted pendulum controlled by the fuzzy controller.

ν velocity of the cart

Of these, only θ, ω and x are chosen as inputs to the fuzzy system. The reason of not using ν is that it is desirable to keep the number of parameters down. The impulse I, which is applied every 0.1 second, is the output. The control goal is to keep the pendulum balanced and the cart on the platform.

8.3 Design of the Fuzzy System

The design strategy proposed in section 6.2 is followed:

1. predetermined parameters:

   (a) The number of input variables is 3, i.e. N = 3.

   (b) To avoid a too complicated optimization problem, the fuzzy partition of each subspace is chosen to be 3.

   (c) The maximum number of rules with all possible antecedents are chosen. That means K = 3 * 3 * 3 = 27.

   (d) The rule base is a 3-dimensional matrix. The 9 rules that contains the antecedent x = "zero", are easy to determine by intuition. The

3-dimensional rule matrix separated in three 2-dimensional matrices is shown in figure 21. The middle 2-dimensional matrix have the predetermined consequence and the other two matrices contain the consequence that need to be optimized. The positions of the membership functions in input space are predetermined. This can be done because it is relatively easy to estimate the range that need to be covered. The angle is for example always between $-45°$ and $45°$. This means that the center of the first and last membership function should be $-45°$ and $45°$ respectively. Similarly the size of the platform determines the centers of the membership function in x-position space. The centers of the membership functions in angle velocity space is not as intuitive but at least the magnitude can be estimated. When it comes to the width of the membership functions there is a rule of thumb [26] that says that the membership functions should intersect when the height is 0.25. This rule together with the estimated centers give all the widths of the membership functions. Both centers and widths are symmetric around the middle membership function, which means that the number of parameters are reduced to about the half. Finally the widths of the membership functions in output space, i.e. the impulse space, are assumed to be equal. All these predetermined parameters are shown in table 7, printed in bold. It could be argued that a more systematic, and scientific method should be used for predetermining these parameters. However experiment with fuzzy systems shows that they are very robust so even if a ad hoc approach is used, it is possible to obtain good results. This kind of reasoning is usually the standard when it comes to designing fuzzy systems in practice.

2. Before GESA is used to determine the rest of the rules and center of the impulse membership functions, some observations are made to simplify the optimization problem. There is a lot of symmetry in the inverted pendulum problem. The case when the cart is to the left is the same as the mirrored case when the cart is to the right. This knowledge is used when the 18 missing rules are optimized, and the search space is therefore decreased to 9 missing rules. Also the maximum impulse applied from the right and left should be equal. This means that the vector that is going to be optimized is:

$$\{c_{impulse}, C_1, C_2, \ldots, C_9\}$$

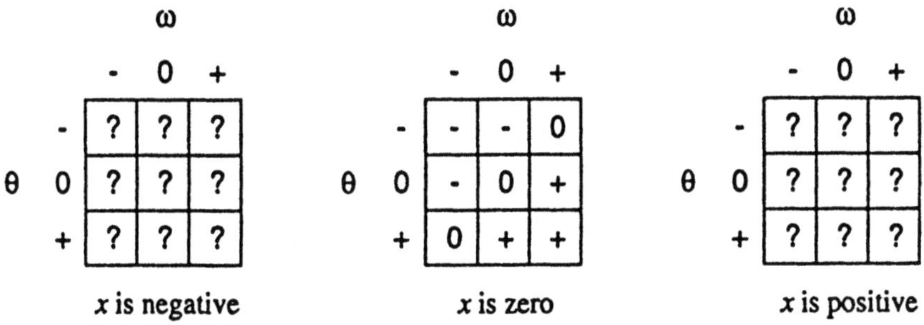

Figure 21 The 3-dimensional rule matrix represented as three 2-dimensional matrices. The consequence of the rules in the middle matrix are predetermined and the other consequence optimized by GESA.

subspace	param.	negative	zero	positive
angle	center	-45 (-54.1)	0 (0)	45 (54.1)
	width	22.5 (28.3)	11.2 (9.45)	22.5 (28.3)
angle velocity	center	-270 (-259)	0 (0)	270 (259)
	width	150 (153)	75 (77.4)	150 (153)
x position	center	-2 (-2.03)	0 (0)	2 (2.03)
	width	1 (1.01)	0.5 (0.43)	1 (1.01)
impulse	center	-3.31 (-3.55)	0 (0)	3.31 (3.55)
	width	-(1.03)	-(1.11)	-(1.03)

Table 7 The parameters of the fuzzy control system. Values within parentheses are the result after step 3 and the values outside the parentheses are predetermined.

GESA needs an objective function, whose value is going to be minimized. The loosely stated control goal is to keep the pendulum balanced and the cart on the platform. An objective function that reflect the capacity of satisfying the control goal can be constructed in many ways. The one that is used here is based on the formula

$$\begin{aligned} f(x) = c_\theta \left(\int_0^{t_{crash}} \theta^2(t)dt + (t_{max} - t_{crash})\, \theta^2_{max} \right) \\ + c_x \left(\int_0^{t_{crash}} x^2(t)dt + (t_{max} - t_{crash})\, x^2_{max} \right) \end{aligned} \tag{12.35}$$

where t_{crash} is the number of seconds that the control goal is satisfied, t_{max} is the maximum simulation time. The idea behind this function is that a controller that managed to keep the pendulum straight up in the center of the platform is a better controller than a controller that make the pendulum deviate from the angle zero or make the cart spend time close the ends of the platform. To assure a stable system, i.e. a system where the trajectory from all points in state space goes to a stable point, the initial values of the state variables should be all possible. This is of natural reasons not possible in reality where the number of initial points has to be finite. Because of computing time constraint, the number of initial points has to be low if the computer is not very fast. In this example, the number of initial points is 2. These two points are

$$\begin{array}{lll} \text{angle} = 30^\circ & \text{angle velocity} = 0^\circ/s & \text{x position} = 0\text{m} \\ \text{angle} = 0^\circ & \text{angle velocity} = 0^\circ/s & \text{x position} = 1.33\text{m} \end{array} \tag{12.36}$$

The total objective value is the sum of the individual objective values.

3. In this step, the whole system is fine tuned. This means that the optimized vector to be used by GESA is

$$\{c_{angle}, c_{angle-vel}, c_{x-pos}, c_{impulse}, \sigma_{1,angle}, \sigma_{2,angle}, \\ \ldots, \sigma_{1,impulse}, \sigma_{2,impulse}, C_1, C_2, \ldots, C_9\}$$

x is negative

θ \ ω	-	0	+
-	0	0	+
0	0	0	-
+	+	+	0

x is zero

θ \ ω	-	0	+
-	-	-	0
0	-	0	+
+	0	+	+

x is positive

θ \ ω	-	0	+
-	0	+	+
0	-	0	0
+	+	0	0

Figure 22 The rule base produced by GESA.

8.4 Results

The rule base produced by GESA is shown in figure 22. Step 3 doesn't change the rules from step 2. All the optimized parameters that determine the membership functions are shown in table 7 (the values belonging to step 3 are within parentheses). These membership functions are shown in figure 23, 23, 23 and 23. It can be seen that the values of the centers and the widths don't change much. This can be taken as an indication that the predetermined parameters were chosen well.

The simulations of the controlled pendulum from the 2 different initial positions (12.36) are shown in figure 27 for the controller produced after step 2. The same simulations but with the controller from step 3 are shown in figure 28. After step 3, all oscillation in the angle is reduced, but there is still a very low frequency oscillation in the x position. (The plotted curve looks like a linear decreasing function but if the behavior is studied over a longer period, it can be seen that this is really a low frequency oscillation.) The reason for this can be that the application of the proposed design strategy has found a local minimum instead of the global minimum. Another reason could be that to get a better control performance, more sensitivity has to be included in the system. That is the fuzzy partition of one or more subspace has to be increased.

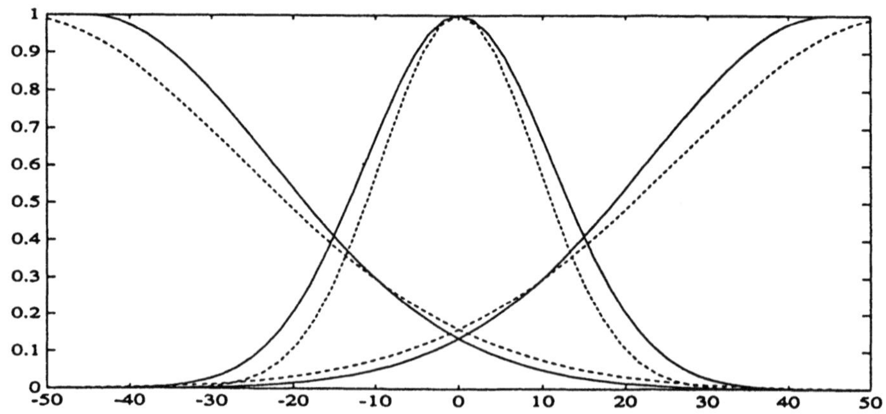

Figure 23 The membership functions of input space *angle*. Continues curves are after step 2 and dashed curves after step 3.

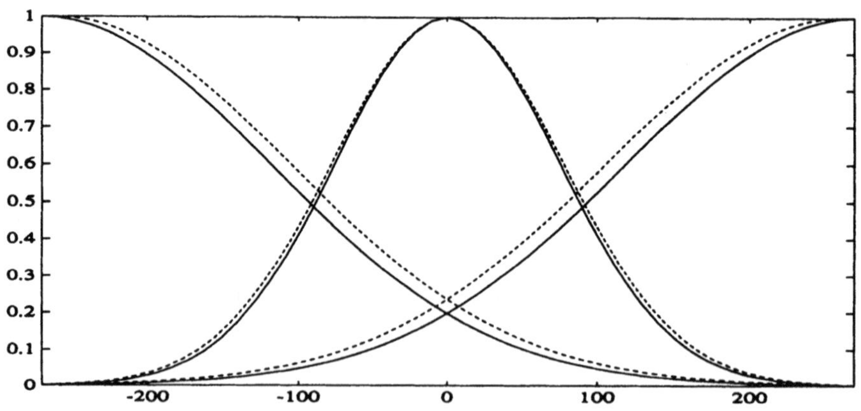

Figure 24 The membership functions of input space *angle velocity*. Continues curves are after step 2 and dashed curves after step 3.

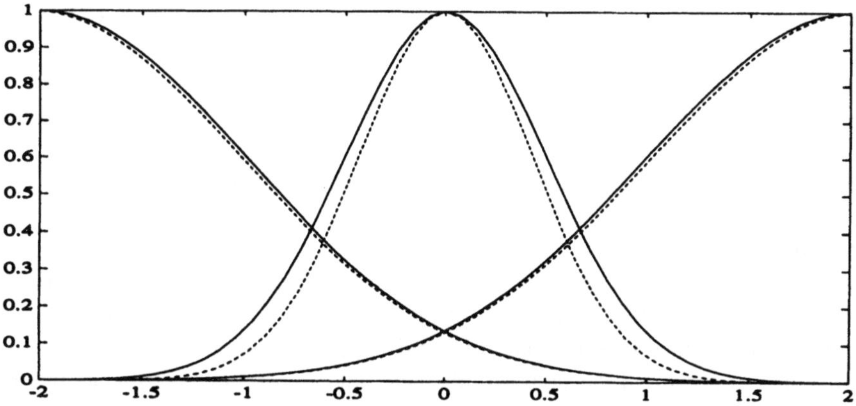

Figure 25 The membership functions of input space *x position*. Continues curves are after step 2 and dashed curves after step 3.

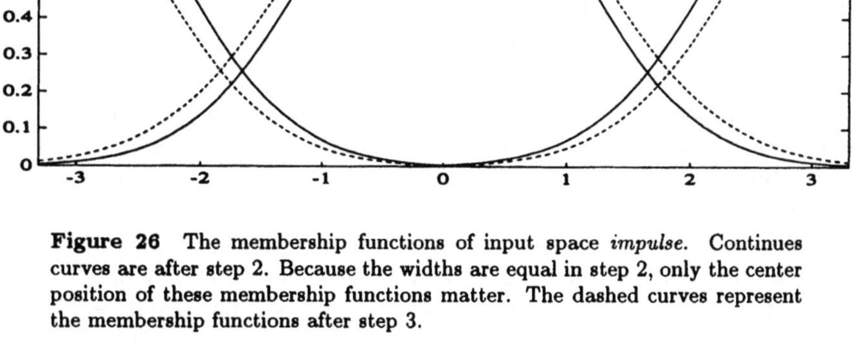

Figure 26 The membership functions of input space *impulse*. Continues curves are after step 2. Because the widths are equal in step 2, only the center position of these membership functions matter. The dashed curves represent the membership functions after step 3.

9 CONCLUSION AND FUTURE WORK

Due to the two applications investigated, it has been shown that the proposed design strategy works but has some limitations. Step 2 in section 6.2 is a usually a hard task because of the fuzzy system's nonlinear nature with many continuous and discrete variables. The experiments have also shown that the search space has many local minimas, often close to each other or to the global minima. Because the optimization problem is so hard, it is difficult for GESA to handle it. Step 3 is easier because in the neighborhood of the solution produced by step 2, there is usually only one minima.

Compared to the other methods for automatic fuzzy system design or tuning discussed in section 3, the present design strategy has a complete theoretical base which is not the case with the other methods. Also most other methods are more specialized to some part of the design. In terms of performance, as mentioned in section 3, these methods have not been reported to work better with full design than the one proposed in this chapter. Some other methods incorporates a mechanism for determining the number of rules and tries to minimize this number. This means that the search space becomes even more complex. If the task is full design with a global minimum goal, the experience from the work in this chapter give an indication that such a optimization is extremely difficult.

9.1 Optimizing with GESA

Even if GESA theoretically can handle the problem in step 2 and 3, it can require a huge amount of computing time. This makes it non practical for any bigger problem in which step 2 is responsible for most of the optimization. There are two reasons for long computing time. The first is a difficult optimization problem with complex search space, which is the case for all but the smaller fuzzy systems. To make GESA able to solve these optimization problems, a large amount of parents and children has to be used. In the two investigated examples, it was often necessary to have 30 families with 25 children each which makes a total of 750 children. The second reason is the fact the function that calculates the objective value is often very time consuming. In the two examples investigated, function approximation of parity-2 surface and control of inverted pendulum, each calculation of a objective value takes about 1 second and 10 seconds respectively on a SPARC-2 computer. 750 children times 10 seconds is 7500 seconds = 125 minutes for each generation! To get GESA to converge, it took often more

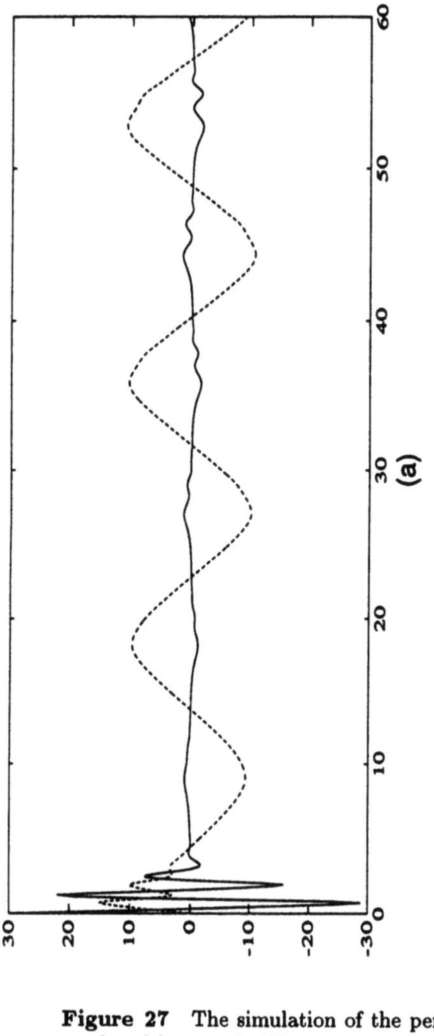

Figure 27 The simulation of the pendulum controlled by the fuzzy system produced by step 2. The continues curve is the angle θ (the axis shows values between -30 and 30) and the dashed curve is the x position (the axis shows values between -3.0m and 3.0m. (a) The first initial condition $\theta = 30°, x = 0m$. (b) The second initial condition $\theta = 0°, x = 1.33m$.

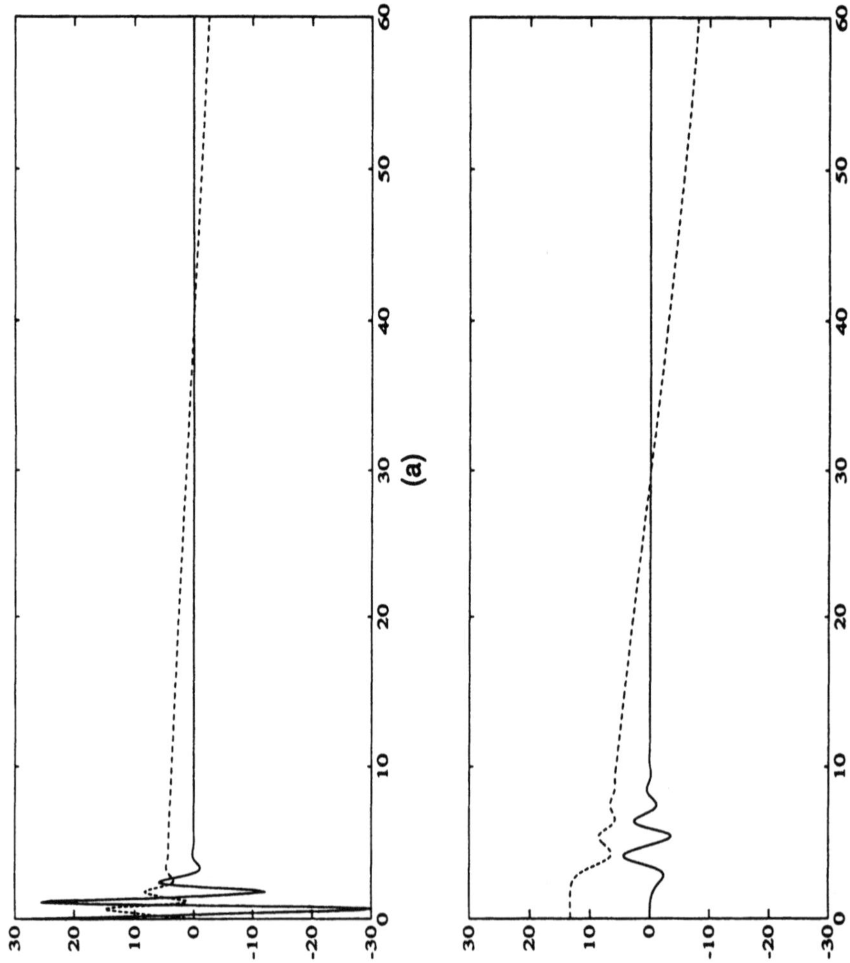

Figure 28 The simulation of the pendulum controlled by the fuzzy system produced by step 3. The continues curve is the angle θ(the axis shows values between -30 and 30) and the dashed curve is the x position (the axis shows values between -3.0m and 3.0m. (a) The first initial condition $\theta = 30°, x = 0m$. (b) The second initial condition $\theta = 0°, x = 1.33m$.

than 500 iterations, i.e. generations, which means 24 hours or more. However with the rapid evolution of computers, this problem should be smaller and smaller each year.

GESA is also a very complex method. Experiment from this and other project shows that it is very sensitive to the settings of its many parameters. Parameters in this context are for example the number of families, the number of children, temperature decrease factors, the implementation of the random change that generates new children. The only guidance in selecting these parameters are common sense and experience. This makes it a very unstable method for solving more difficult optimization problem. A slight change in any of the parameters can cause GESA to convert to a local minima instead of a global one. Another issue is the choice of objective function. It turns out that GESA can be very sensitive to this too. For example the use of square root of the objective value gives a qualitatively equal search space but it can mean a big difference in convergence rate or the ability to converge to the global minima.

These problems makes it difficult to perform a total fuzzy system design, i.e. optimizing all parameters at once, with the aim to find a global optimum. By relaxing these requirements and use the design strategy proposed in section 6.2, including step 1, GESA can be very helpful indeed. GESA should be viewed as not a tool by its own but a very useful technique together with regular methods, i.e. mostly human trial and error, engineering skills, and common sense.

9.2 Wang's fuzzy system

The WFS does not only have the advantage of being theoretically proved to be an universal approximator. The computational complexity of calculating the system output is very low. This is due to the COS defuzzification method instead of the COA method. Also because Gaussian membership functions are used, the functional (hyper-) surface of the system is very smooth. This is an advantage for example if the system is going to be used in a control application in which the smoothness is good for electro-mechanical equipment.

A disadvantage of the WFS is the fact that it can never give an output equal to the center of the first or last membership function in output space. This could be seen in the parity-2 problem and was described in section 7.3. However this is only an issue if a strict human interpretation of the system is required. For example in the parity-2 problem, by relaxing this requirement the system got much better performance but the intuitive interpretation of the fuzzy sets was lost.

9.3 Future Work

Because GESA is such a complex tool, more experience with the algorithm is required. A simplification of the algorithm is also desirable. Another area of future work is to show universal approximation capability for more fuzzy systems. If other fuzzy systems are found to have a theoretical base, these can be used with GESA in the same spirit as WFS. Another area of future work is to make a comparison study of GESA vs. other algorithms, e.g. GA. Also an more extensive evaluation of different types of fuzzy systems is needed.

REFERENCES

[1] L. A. Zadeh, "Fuzzy Sets," Information and Control, vol. 8, pp. 338-352, 1965.

[2] E. H. Mamdani and Assilian S. "An Experiment with Linguistic Synthesis with a Fuzzy Logic Controller," Int. Journal Man-Machine Studies, vol. 7, pp. 1-13, 1975.

[3] L. P. Holmblad and J. J. Ostergaard, "Control of a Cement Kiln by Fuzzy Logic," in Fuzzy Information and Decision Processes, Ed. M. M. Gupta and E. Sanchez, Amsterdam: North Holland 1982, pp. 389-399.

[4] M. Sugeno and K. Murakami, "An experimental study on fuzzy parking control using a model car," in Industrial Applications of Fuzzy Control, M. Sugeno, Ed. Amsterdam: North-Holland, 1985, pp. 125-138.

[5] S. Yasunobu, S. Miyamoto, "Automatic train operation by predictive fuzzy control," in Industrial Applications of Fuzzy Control, M. Sugeno, Ed. Amsterdam: North-Holland, 1985, pp. 1-18.

[6] C. C. Lee, "Fuzzy Logic in Control Systems: Fuzzy Logic Controller-Part I," IEEE Trans. Syst. Man Cybern., vol. SMC-20, no. 2, pp. 404-418, 1990.

[7] C. C. Lee, "Fuzzy Logic in Control Systems: Fuzzy Logic Controller-Part II," IEEE Trans. Syst. Man Cybern., vol. SMC-20, no. 2, pp. 419-435, 1990.

[8] M. Braae and D. Rutherford, "Fuzzy relations in control setting," Kybernetics, vol. 7, no. 3, pp. 185-188, 1978.

[9] M. S. Stachowicz and M. E. Kochanska, "Fuzzy modeling of the process," in Proc. 2nd IFSA Congress, Tokyo, Japan, July 1987, pp. 86-89.

[10] L. X. Wang, "Fuzzy Systems are Universal Approximators," Proc. IEEE Int. Conf. on Fuzzy Systems, San Diego, CA, 1992, pp. 1163-1169.

[11] J. Lee and S. Chae, "Completeness of Fuzzy Controller Carrying a Mapping $f : R1 \rightarrow R1$," 1993 IEEE Int. Conf. on Fuzzy Systems, San Fransisco, CA, March 1993, pp 231-235.

[12] B. Kosko, "Fuzzy Systems as Universal Approximators," Proc. IEEE Int. Conf. on Fuzzy Systems, San Diego, CA, 1992, pp. 1153-1162.

[13] H. Hellendoorn, "Design and Development of Fuzzy Systems at Siemens R&D," 1993 IEEE Int. Conf. on Fuzzy Systems, San Fransisco, CA, March 1993, pp. 1365-1370.

[14] D. Driankov, H. Hellendoorn and M. Reinfrank, An Introduction to Fuzzy Control, Berlin, Heidelberg: Springer-Verlag, 1993.

[15] T. Takagi and M. Sugeno, "Fuzzy identification of systems and its applications to modeling and control," IEEE Trans. Syst. Man Cybern., vol. SMC-15, no 1, pp. 116-132, 1985.

[16] J. Efstathiou, "Rule-based process control using fuzzy logic," in Approximate Reasoning in intelligent systems, decision and control, E. Sanchez and L. A. Zadeh, Ed. Oxford: Pergamon Press, 1987, pp. 145-158.

[17] Y. H. Pao, Adaptive Pattern Recognition and Neural Networks, Reading, MA: Addison- Wesley, 1989.

[18] D. Goldberg, Genetic algorithms in search, optimization, and machine learning, Addison-Wesley, Reading, MA: Addison-Wesley, 1989.

[19] L. J. Fogel, A. J. Owens and M. J. Walsh, Artificial intelligences through simulated evolution, New York: Wiley, 1966.

[20] N. Metropolis et al, "Equation of state calculations by fast computing machines," Journal of Chemical Physics, vol. 21, no. 6, pp. 1087-1092, 1953.

[21] S. Kirkpatrick, C. D. Gelatt Jr and M. P. Vecchi, "Optimization by simulated annealing," Science, vol. 220, pp. 671-680, 1983.

[22] Y. Takefuji, Neural network parallel computing, Boston: Kluwer Academic Publishers, 1992.

[23] P. P. C. Yip and Y. H. Pao, "A fast universal training algorithm for neural networks," Proc. World Congress on Neural Networks WCNN'93, Protland, Oregon, July 1993, pp. 614-621.

[24] P. P. C. Yip and Y. H. Pao, "Combinatorial optimization with use of guided evolutionary simulated annealing," IEEE Trans. Neural Network, accepted for publication, 1994.

[25] P. P. C. Yip and Y. H. Pao, "A guided evolutionary technique as function optimizer," Proc. the First IEEE Conf. on Evolutionary Computation, Orlando, Florida, June 1994, pp. 628-633.

[26] E. Aarts, Simulated annealing and Boltzmann machines: a stochastic approach to combinatorial optimization and neural computing, New York: Wiley, 1989.

[27] Feller, An Introduction to Probability Theory and its Applications, New York: Wiley, 1950.

[28] T. J. Procyk and E. H. Mamdani, "A linguistic self-organising process controller," Automatica, vol. 15, pp. 15-30, 1979.

[29] B. Kosko, Neural networks and fuzzy systems: a dynamical systems approach to machine intelligence, Englewood Cliffs, NJ: Prentice-Hall, 1991.

[30] D. Dubois and H. Prade, Fuzzy Sets and Systems: Theory and Applications, New York: Academic Press, 1980.

[31] A. Kandel, Fuzzy Mathematical Techniques with Applications, Reading, MA: Addison- Wesley, 1986.

[32] R. Jang, "Self-Learning Fuzzy Controllers Based on Temporal Back Propagation," IEEE Trans. Neural Networks, vol. 3, no. 5, 1992, pp. 714-723.

[33] M. Tsutomu et al, "Operator Tuning in Fuzzy Production Rules," 1993 IEEE Int. Conf. on Fuzzy Systems, San Fransisco, CA, March 1993, pp. 641-646.

[34] E. Khan and P. Venkatapuram, "Neufuz: Neural Network Based Fuzzy Logic Design Algorithms," 1993 IEEE Int. Conf. on Fuzzy Systems, San Fransisco, CA, March 1993, pp. 647-654.

[35] A. E. Bryson and D. G. Leuenberger, "The Synthesis of Regulator Logic Using State- Variable Concepts," Proceedings of the IEEE, vol. 58, no. 11, pp 1803-1811, 1970.

[36] B. Widrow, "The Original Adaptive Net Broom-Balancer," IEEE Int. Symposium on Circuits and Systems, May 1987, pp. 351-357.

[37] V. Williams and K. Matsuoka, "Learning to Balance the Inverted Pendulum using Neural Networks," 91 IEEE Int. Joint Conf. on Neural Networks (IJCNN'91), Singapore, Singapore, November 1991, pp 214-219.

[38] M. A. Lee and H. Takagi, "Integrating Design Stages of Fuzzy Systems using Genetic Algorithms," 1993 IEEE Int. Conf. on Fuzzy Systems, San Fransisco, CA, March 1993, pp 612-617.

[39] M. Jamshidi et al, "A comparison of an Expert and an Adaptive Fuzzy Control Approach," Proc. IEEE Conf. on Decision and Control, Brighton, England, December 1991, pp 1907-1908.

[40] A. G. Barto, R.S. Sutton, and C. W. Anderson, "Neuronlike Adaptive Elements that can solve Difficult Learning Control Problems," IEEE Trans. on Syst. Man Cybern., vol. SMC-13, no. 5, pp. 834-846, 1983.

[41] L. Råde and B. Westergren, BETA Mathematics Handbook, Lund, Sweden: Studentlitteratur, 1990.

[42] C. Karr and E. J. Gentry, "Fuzzy Control of pH using Genetic Algorithms," IEEE Trans. on Fuzzy Systems, vol. 1, no. 1, pp. 46-53, 1993.

[43] H. Ishibuchi, K. Nozaki and N. Yamamoto, "Selecting Fuzzy Rules by Genetic Algorithm for classification problems," 1993 IEEE Int. Conf. on Fuzzy Systems, San Fransisco, CA, March 1993, pp 1119-1124.

[44] P. Thrift , "Fuzzy Logic Synthesis with Genetic Algorithms," Proceedings of the Fourth Int. Conf. on Genetic Algorithms, San Mateo, CA, 1991, pp 502-513.

[45] A. M. Homaifar and E. McCormick, "Full Design of Fuzzy Controllers using Genetic Algorithms," Proceedings of SPIE,vol. 1766, San Diego, CA, July 1992, pp 393-404.

[46] B. Hu, "Cell State Algorithm and Neural Network Based Fuzzy Logic Controller Design," 1993 IEEE Int. Conf. on Fuzzy Systems, San Fransisco, CA, March 1993, pp 247-250.

[47] T. Whalen and B. Schott, "Lexicographic Tuning of a Fuzzy Controller using Box's 'Complex' Algorithm," 1993 IEEE Int. Conf. on Fuzzy Systems, San Fransisco, CA, March 1993, pp 285-290.

13

INTELLIGENT CONTROL USING DYNAMIC NEURAL NETWORKS WITH ROBOTIC APPLICATIONS

Liang Jin, Madan M. Gupta, Peter N. Nikiforuk

Intelligent Systems Research Laboratory
College of Engineering, University of Saskatchewan
Saskatoon, Saskatchewan, Canada S7N 0W0

ABSTRACT

The intelligent control of systems with complex, unknown, and high nonlinear dynamics, such as robotic systems, chemical engineering processes and space systems, has become a topic of considerable importance during recent years. The most concurrent advances in the area of artificial neural networks (ANNs) have provided the potential for dealing with such a challenging task. In this chapter, some new schemes of dynamic recurrent neural networks (DRNNs) are proposed to design robust learning control systems for a general class of multi-input and multi-output (MIMO) nonlinear systems with unknown dynamics. The detailed structure of the DRNNs and their learning capability are first discussed. The synthesis and design methods for the purposes of regulation, tracking and model reference control are then conducted. Based on the DRNNs approaches presented in this chapter, a new torque control scheme for robot manipulators is developed. The potentials of this scheme are demonstrated extensively by simulation studies.

1 INTRODUCTION

In recent years, there has been a dramatic increase in research on the intelligent systems which are required to have human-like capabilities of performing complex tasks such as processing sensory information and interacting with uncertain environments. In particular, the intelligent control of complex systems which contain unpredictable events, changing environments, and difficult to model internal dynamics, such as robotic systems, chemical engineering processes and space systems, is a challenging topic. In robotics, it is becoming obviously understood that the environment we want our robots to work in requires more intelligent control. Generally speaking, intelligent control systems are enhanced adaptive or self-organizing mechanisms that automatically adapt to plant and environmental changes without a priori knowledge of the changes or even of physical models of the process.

The difficulties that arise in the control of these complex systems can be classified under the three aspects: (i) system complexity: there exists a large number of degrees of freedom which have to be used to describe the system dynamic property; (ii) nonlinearity: nonlinear dynamics existing in a complex system is either modeled or unmodeled, however, there is a great deal of difficulties associated with designing nonlinear control systems using traditional concepts; (iii) uncertainty: only partial or even no a priori information concerning internal system structures can be used to mathematically model the system. Therefore, the system models are uncertainly in terms of the unknown parameters and dynamic structure. A control system which is capable of dealing with the above three categories of complexities is qualified as an intelligent control system. Moreover, the greater flexibility of the control schemes to deal with these factors implies more intelligence in the control systems.

Artificial neural networks, as models of specific biological structures, have the advantages of distributed information processing and the inherent potential for parallel computation. An artificial neural network consists of many interconnected identical simple processing units, called *neurons* or *nodes*, which form the layered configurations. A static feedforward neural network contains only feedforward synaptic connections, and its input-output relationship can be described by a parametrized nonlinear mapping from one finite dimensional input space to an output space. Such a class of neural networks are capable of approximating arbitrary continuous functions to a desired degree of accuracy on a compact set. Recent, extensive studies has demonstrated unusually effectiveness of the static feedforward neural networks for the identification and control of nonlinear dynamic systems.

On the other hand, there exists another class of neural networks, namely dynamic recurrent neural networks (DRNNs), which contain not only feedforward connections but also feedback or recurrent connections. The DRNNs with the adaptive learning capabilities are described by complex nonlinear systems with state feedback. From a computational point of view, a dynamic neural structure which contains a state feedback may provide more computational advantages than that of a purely feedforward neural structure.

Also, the nonlinear DRNNs have been found having potential for applications to problems such as identification, control, and pattern recognition. Recently, several studies have noted that an appropriate dynamic mapping may be realized by a dynamic recurrent neural network (DRNN) which is trained by means of a series-parallel or a parallel learning model similar to the case of the feedforward networks, so that a desired response is obtained (Narendra, and Parthasarathy, [24]). The DRNN consists of both feedforward and feedback connections between the layers and neurons forming complicated dynamics. This ability of such a recurrent neural network to approximate a continuous/discrete-time nonlinear dynamic system by the neural dynamics defined by a system of nonlinear differential/difference equations has the potential for application to adaptive control systems. When dynamic recurrent neural networks are used to approximate and control an unknown nonlinear system through on-line learning processes, they may be treated as subsystems of the adaptive control system. The weights of the networks need to be then updated using a learning algorithm during the control processes.

The DRNNs based intelligent control algorithms are presented in this chapter for a general class of discrete-time of multi-input and multi-output (MIMO) nonlinear systems which have either modeled or unmodeled dynamics. As a preliminary, some basic descriptions of nonlinear control systems are provided in Section 2, the control objectives such as state regulation and output tracking using dynamic feedback controllers are discussed in details. The mathematical models and capabilities of DRNNs for identification and control are explored in Section 3. Also, a new learning algorithm is presented in this section. Sections 4 and 5 deals with the problem of state regulation using DRNNs for both known and unknown nonlinear plants. Output tracking control for an unknown nonlinear system is addressed in Section 6 using only input-output data. Applications of the approaches presented to robotic control are discussed in Section 7. Some simulation studies are also given in this section to demonstrate the effectiveness of the proposed intelligent control schemes.

2 DESCRIPTION OF DISCRETE-TIME NONLINEAR SYSTEMS

2.1 Nonlinear MIMO State-Space Models

Consider a discrete-time MIMO nonlinear plant described by the following set of difference equations

$$P: \quad \begin{cases} \mathbf{x}(\mathbf{k}+1) &= \mathbf{f}(\mathbf{x}(\mathbf{k}), \mathbf{u}(\mathbf{k})) \\ \mathbf{y}(\mathbf{k}) &= \mathbf{h}(\mathbf{x}(\mathbf{k}), \mathbf{u}(\mathbf{k})) \end{cases} \tag{13.1}$$

where $\mathbf{x} \in \Re^{\mathbf{n}}$, $\mathbf{y} \in \Re^{\mathbf{l}}$, and $\mathbf{u} \in \Re^{\mathbf{m}}$ are state, output, and input vectors, respectively, and $\mathbf{f} : \Re^{\mathbf{n}} \times \Re^{\mathbf{m}} \longrightarrow \Re^{\mathbf{n}}$ and $\mathbf{h} : \Re^{\mathbf{n}} \times \Re^{\mathbf{m}} \longrightarrow \Re^{\mathbf{l}}$ are continuous and differentiable functions. The nonlinear plant (13.1) is assumed to be of the following characteristics:

1. $\mathbf{f}$ and $\mathbf{h}$ are known and the state vector $\mathbf{x}(\mathbf{k})$ is accessible at time k;
2. $\mathbf{f}$ and $\mathbf{h}$ are known and only the output vector $\mathbf{y}(\mathbf{k})$ is accessible at time k;
3. $\mathbf{f}$ and $\mathbf{h}$ are unknown and the state vector $\mathbf{x}(\mathbf{k})$ is accessible at time k;
4. $\mathbf{f}$ and $\mathbf{h}$ are unknown and only the output vector $\mathbf{y}(\mathbf{k})$ is accessible at time k.

2.2 Control Objectives

Like linear control systems, the control objectives for the above nonlinear systems can be summarized as follows:

(i) State Regulation Problem

For the problem of state regulation, also called stabilization, the task is to move, by using either state or output feedback control force, the state vector of the system from an arbitrary initial position to a specified final position which is the origin. In the other worlds, in the case of full state feedback, we have to design either a static state feedback controller

$$\mathbf{u(k)} = \mathbf{u(x(k))} \tag{13.2}$$

or a dynamic feedback control

$$\left\{ \begin{array}{lcl} \mathbf{z(k+1)} & = & \mathbf{f_c(z(k), x(k))} \\ \mathbf{u(k)} & = & h_c(\mathbf{z(k), x(k)}) \end{array} \right. \tag{13.3}$$

such that

$$\lim_{k \longrightarrow \infty} ||\mathbf{x(k)}|| = 0$$

(ii) Output Tracking Control

Such a task is to determine the control signal such that the output vector will track the given desired output vector $\mathbf{y_d(k)}$ with the error which goes zero as time becomes large; that is

$$\lim_{k \longrightarrow \infty} ||\mathbf{y(k)} - \mathbf{y_d(k)}|| = 0$$

The form of the static or dynamic controllers for such a task are given as follows

$$\mathbf{u(k)} = \mathbf{u(x(k), Y_d(k))} \tag{13.4}$$

or a dynamic feedback control

$$\begin{cases} \mathbf{z}(\mathbf{k}+1) = \mathbf{f_c}(\mathbf{z}(\mathbf{k}), \mathbf{x}(\mathbf{k}), \mathbf{Y_d}(\mathbf{k})) \\ \mathbf{u}(\mathbf{k}) = \mathbf{h_c}(\mathbf{z}(\mathbf{k}), \mathbf{x}(\mathbf{k}), \mathbf{Y_d}(\mathbf{k})) \end{cases} \tag{13.5}$$

where $\mathbf{Y_d}(\mathbf{k}) = [\mathbf{y_d^T}(\mathbf{k}), \mathbf{y_d^T}(\mathbf{k}+1), \ldots, \mathbf{y_d^T}(\mathbf{k}+\mathbf{r})]^{\mathrm{T}}$ and $r \geq 0$ is an integer.

(ii) **Model Reference Control:**

Given a stable reference model with form

$$\begin{cases} \mathbf{x_m}(\mathbf{k}+1) = \mathbf{f_m}(\mathbf{x_m}(\mathbf{k}), \mathbf{v}(\mathbf{k})) \\ \mathbf{y_m}(\mathbf{k}) = \mathbf{h_m}(\mathbf{x_m}(\mathbf{k}), \mathbf{v}(\mathbf{k})) \end{cases} \tag{13.6}$$

where $\mathbf{v}$ is a reference input. The model reference control demands on us to design a model based static or dynamic control law such that the system output $\mathbf{y}(\mathbf{k})$ will follow asymptotically the model output $\mathbf{y_m}(\mathbf{k})$ under such a control action; that is

$$\lim_{k \longrightarrow \infty} ||\mathbf{y}(\mathbf{k}) - \mathbf{y_m}(\mathbf{k})|| = \mathbf{0}$$

All of the three control strategies are applicable for both the known and unknown plants. For the control problems associated with know plants, the above approaches are well-known nonlinear control. On the other hand, control with the above objectives for uncertain or unknown plants are said to be adaptive control. Our main interest, however, will be concentrated on (i) and (ii) where the plants are either known or unknown. Recently, the structures of static controllers for either nonlinear or adaptive control have been studied extensively using static feedforward multilayered neural networks (Narendra and Parthasarathy, [24]; Jin, Nikiforuk and Gupta, [14], [18]) and Gaussian basis function networks (Hunt and Sbarbaro, [11]; Sanner and Slotine, [33]) since Narendra and Parthasarathy [24] first published their work. In this chapter, we will study only dynamic feedback control structures which are implemented using dynamic recurrent neural networks.

2.3 Existence of Dynamic Feedback Controller

It is not difficult to show that the static feedback controllers are the special cases of the dynamic feedback controllers. To demonstrate that a solution of such dynamic feedback controllers exists to the nonlinear control problems, we give the following basic explaination. In the case of the state regulation, one intends to design the following dynamic state feedback controller

$$\begin{cases} \mathbf{z}(\mathbf{k}+1) = \mathbf{f_c}(\mathbf{z}(\mathbf{k}), \mathbf{x}(\mathbf{k})) \\ \mathbf{u}(\mathbf{k}) = \mathbf{h_c}(\mathbf{z}(\mathbf{k}), \mathbf{x}(\mathbf{k})) \end{cases} \tag{13.7}$$

such that the closed loop system which consists of Eqs. (13.1) and (13.7) as follows

$$\begin{cases} \mathbf{z}(\mathbf{k}+1) = \mathbf{f_c}(\mathbf{z}(\mathbf{k}), \mathbf{x}(\mathbf{k})) \\ \mathbf{x}(\mathbf{k}+1) = \mathbf{f}(\mathbf{x}(\mathbf{k}), \mathbf{h_c}(\mathbf{z}(\mathbf{k}), \mathbf{x}(\mathbf{k}))) \end{cases} \tag{13.8}$$

is locally stable around the origin, where the internal state $\mathbf{z} \in \Re^{\mathbf{n_z}}$, and the nonlinear functions $\mathbf{f_c}$ and $\mathbf{h_c}$ are appropriately chosen.

The existence of such a dynamic feedback controller may be simply implied if the linearized system of the nonlinear system P around the origin $(\mathbf{x}, \mathbf{u}) = (\mathbf{0}, \mathbf{0})$ is controllable (Isidori, [13]; Nijmeijer and Van der Schaft, [28]). Since the linearized system may be represented as

$$\bar{\mathbf{x}}(k+1) = [\frac{\partial \mathbf{f}(\mathbf{0},\mathbf{0})}{\partial \mathbf{x}}]\bar{\mathbf{x}}(k) + [\frac{\partial \mathbf{f}(\mathbf{0},\mathbf{0})}{\partial \mathbf{u}}]\bar{\mathbf{u}}(k) \tag{13.9}$$

the controllability condition is easily given as

$$rank \left[(\frac{\partial \mathbf{f}}{\partial \mathbf{u}}), (\frac{\partial \mathbf{f}}{\partial \mathbf{x}})(\frac{\partial \mathbf{f}}{\partial \mathbf{u}}), \ldots, (\frac{\partial \mathbf{f}}{\partial \mathbf{x}})^{n-1}(\frac{\partial \mathbf{f}}{\partial \mathbf{u}}) \right] \Big|_{(\mathbf{0},\mathbf{0})} = n \tag{13.10}$$

In fact, controllability, as in the linear case, refers to the ability in terms of the control to transfer the system state from any initial state to any final state within a specified iterative steps. The above result shows that the nonlinear system P is controllable in some neighborhood of the equilibrium point in the state space if the associated linearized system is controllable. For the output tracking, if the system is locally controllable around the origin, and has stable zero-dynamics and well-defined relative degree around the origin, then such a dynamic feedback compensator exists. However, analytically finding such a dynamic controller is very complex task, we still expect that there is universal method to determine the nonlinear function $\mathbf{f_c}$ and $\mathbf{h_c}$. As a main idea throughout the chapter, we attempt to get help from dynamic neural networks to deal with this design and synthesis process since dynamic neural networks have powerful computational capabilities.

3 DYNAMIC RECURRENT NEURAL NETWORKS

A dynamic recurrent neural network (DRNN) contains both a huge number of feedforward and feedback synaptic connections (Hopfield, [6], [7]; Pineda, [29]-[31]). From the computational point of view, a dynamic neural structure which contains a state feedback may provide more computational advantages than that of a purely feedforward neural structure. For some problems, a small feedback system is equivalent to a large and possibly infinite feedforward system (Hush and Horne, [10]). A well-known example is that an infinite number of feedforward logic gates are required to emulate an arbitrary finite state machine, or that an infinite order FIR filter is required to emulate a single pole infinite impulse response (IIR). A nonlinear dynamic recurrent neural structure is particularly appropriate for identification, control, and filtering applications due to its ability of distributed and multiply-superposed information processing as the biological neural systems.

3.1 Structure and Model of DRNNs

A dynamic recurrent neural network (DRNN), as shown in Figures 1 and 2, is a complex nonlinear dynamic system described by a set of nonlinear differential or difference equations. In this chapter, we discuss only discrete-time version of analog DRNNs for control applications. A general expression of this type of DRNNs with n neural units is given by the following nonlinear system

$$DRNN: \quad \begin{cases} \mathbf{x(k+1)} = -\alpha\mathbf{x(k)} + \mathbf{f(A, x(k), B, u(k))} \\ \mathbf{y(k)} = \mathbf{Cx(k)} \end{cases} \tag{13.11}$$

where $\mathbf{x} \in \Re^{\mathbf{n}}$, $\mathbf{y} \in \Re^{\mathbf{l}}$, and $\mathbf{u} \in \Re^{\mathbf{m}}$ are neural state, output, and input vectors, respectively, $\mathbf{A} \in \Re^{\mathbf{n}\times\mathbf{n}}$ with , $\mathbf{B} \in \Re^{\mathbf{n}\times\mathbf{m}}$, and $\mathbf{C} \in \Re^{\mathbf{l}\times\mathbf{n}}$ are the connection weight matrices associated with the neural state, input, and output vectors, respectively, α is a constant for controlling state decaying and is chosen $0 \le \alpha < 1$, and $\mathbf{f} : \Re^{\mathbf{n}} \longrightarrow \Re^{\mathbf{n}}$ is appropriately chosen vector-valued nonlinear functions.

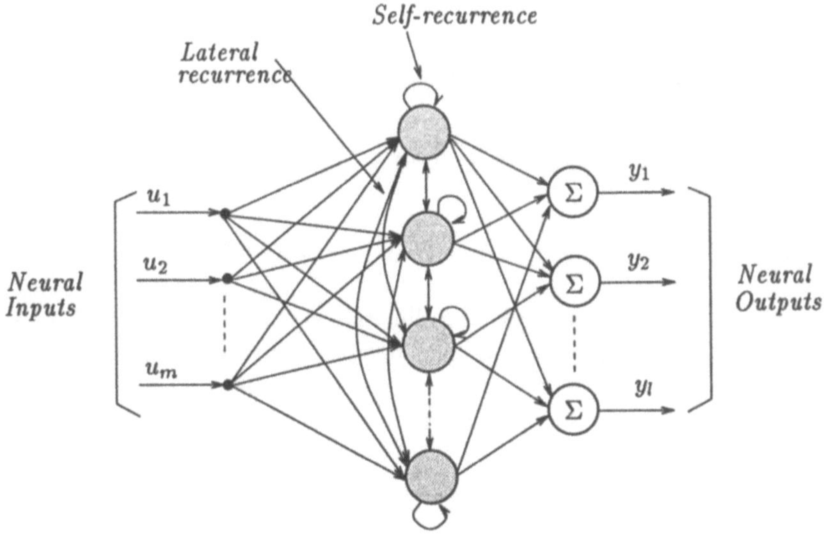

Figure 1 Schematic representation of dynamic recurrent neural network (DRNN).

Some interesting choices of the nonlinear function $\mathbf{f}$ are given as follows:

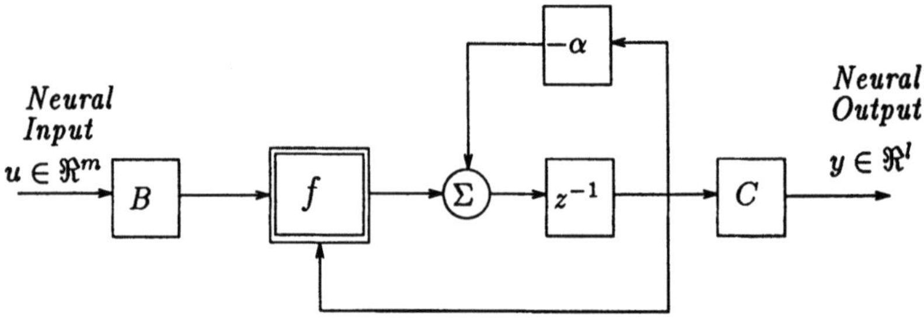

Figure 2 Block diagram of dynamic recurrent neural network (DRNN).

(i) Modified Hopfield type: $\mathbf{f}(\mathbf{A},\mathbf{x},\mathbf{B},\mathbf{u}) = \mathbf{A}\sigma(\mathbf{x}) + \mathbf{B}\mathbf{u}$

(ii) Modified Pineda type I: $\mathbf{f}(\mathbf{A},\mathbf{x},\mathbf{B},\mathbf{u}) = \sigma(\mathbf{A}\mathbf{x} + \mathbf{B}\mathbf{u})$

(iii) Modified Pineda type II: $\mathbf{f}(\mathbf{A},\mathbf{x},\mathbf{B},\mathbf{u}) = \sigma(\mathbf{A}\mathbf{x}) + \mathbf{B}\mathbf{u}$

where the nonlinear neural activation function $\sigma(.)$ may be chosen as a continuous and differentiable nonlinear sigmoidal function satisfying the following conditions: (i) $\sigma(x) \longrightarrow \pm 1$ as $x \longrightarrow \pm\infty$; (ii) $\sigma(x)$ is bounded with the upper bound 1 and the lower bound -1; (iii) $\sigma(x) = 0$ at a unique point $x = 0$; (iv) $\sigma'(x) > 0$ and $\sigma'(x) \longrightarrow 0$ as $x \longrightarrow \pm\infty$; (v) $\sigma'(x)$ has a global maximal value $\mu > 0$. Typical examples of such a function $\sigma(.)$ are

$$tanh(\mu x); \qquad \frac{1 - e^{-\mu x}}{1 + e^{-\mu x}}; \qquad \frac{2}{\pi} tan^{-1}(\frac{\pi\mu}{2}x).$$

where $\mu > 0$ is a constant which determines the slope of $\sigma(x)$ or the so-called *activation gain.*

For simplicity, we concentrate on the modified Hopfield type of DRNNs in the following studies. In this case, the network equation is given by

$$\begin{cases} \mathbf{x(k+1)} = -\alpha \mathbf{x(k)} + \mathbf{A}\sigma(\mathbf{x(k)}) + \mathbf{Bu(k)} \\ \mathbf{y(k)} = \mathbf{Cx(k)} \end{cases} \tag{13.12}$$

which is shown in Figure 3.

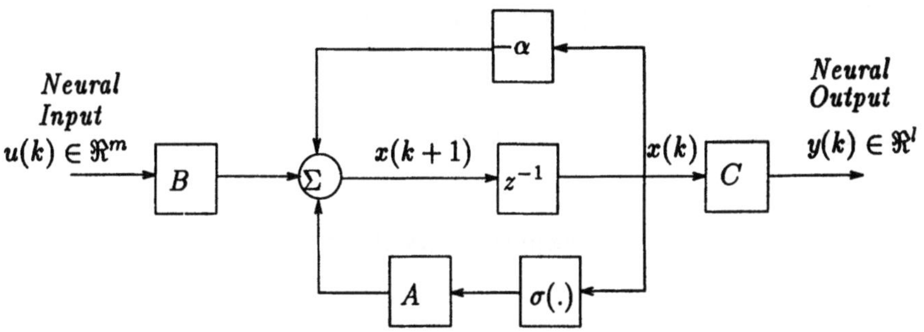

Figure 3 Block diagram of dynamic recurrent neural network (DRNN).

3.2 Approximation of Dynamic System Trajectories

Even if most theoretical studies on DRNNs have been mainly concentrated on the stability and convergence of the network trajectory to the equilibria. The nonlinear dynamic behavior of DRNNs is suitable of spatio-temporal information processing. Recently, Funahashi and Nakamura [4] studied the problem of approximation of dynamic systems by continuous-time DRNN while Li [20] conducted the problem in discrete-time domain. Several independent studies have found that DRNNs using the adaptive learning algorithms can approximate a wide range of input-output relationship of nonlinear systems to any desired degree of accuracy (Funahashi and

Nakamura, [4]; Li, [20]). Mathematically, this capability of DRNNs can be described in the following theorem:

Theorem 1 (Basic Approximation Theorem) *Let S be a compact subset of $\Re^l$, $\mathbf{g}:\mathbf{S}\longrightarrow\Re^{\mathbf{l}}$ be a continuous vector-valued function, and for arbitrary initial value $\mathbf{x}(0)\in\mathbf{S}$ the solution of a set of difference equations*

$$\zeta(k+1)=\mathbf{g}(\zeta(k)), \qquad \zeta\in S$$

be defined in S. Then, for an arbitrary $\epsilon>0$, there exist an integer n and a DRNN

$$\begin{cases} \mathbf{x}(k+1)=-\alpha\mathbf{x}(k)+\mathbf{A}\sigma(\mathbf{x}(\mathbf{k})) \\ \mathbf{y}(k)=\mathbf{C}\mathbf{x}(k) \end{cases}$$

with an appropriate initial state such that

$$\max_k ||\mathbf{y}(k)-\zeta(k)||<\epsilon$$

The above theorem gives no information regarding the number of dynamic neural units that we have to use for the purpose of an approximation, however, the existence of such a nonlinear scheme is ensured. The above results obviously imply that the DRNN is not only capable of approximating arbitrary nonlinear dynamic systems but also continuous functions defined on compact sets through appropriate choice of the number of the units and the connection weights. In practical applications, most commonly-used scheme for approximating an input-output dynamic system using a DRNN with an adaptive weight learning scheme is shown in Figure 4. As a matter of fact, this is a standard parameter identification process using input-output data. Some criteria for selecting error function, input signal and other factors of this identification process have been well reported in the literature (Liung and Soderstrom, [21]).

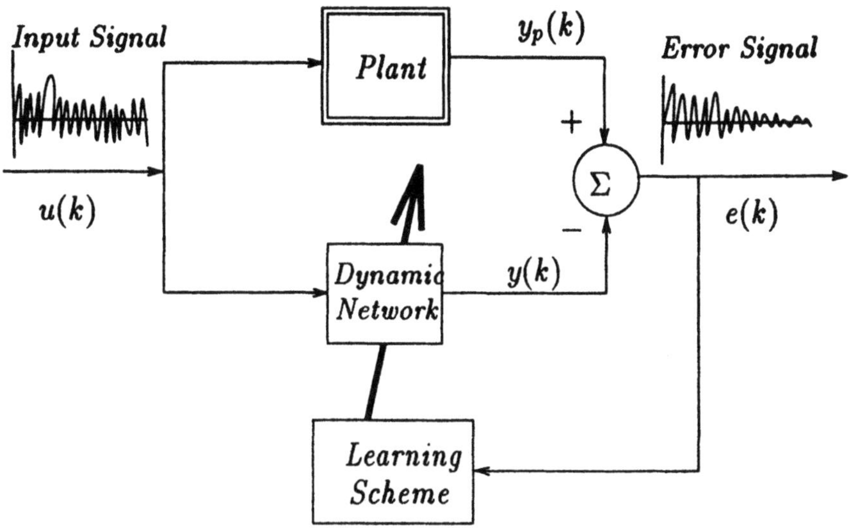

Figure 4 Identification process using DRNN.

3.3 Supervised Weight Learning

In the supervised learning of the DRNNs, the purpose of the weight learning of the DRNN is to estimate the weights such that the output error defined as $||\mathbf{y}(k) - \mathbf{y}_d(k)||$ converges to zero as $k \longrightarrow \infty$. Hence, if the weights of a DRNN are taken into account as the unknown parameters of a nonlinear input-output system, the weight learning problem of the DRNN can be phrased as a parameter identification problem for the dynamic nonlinear system. As a matter of fact, a simple and natural extension of the back-propagation (BP) algorithm for the multilayered feedforward neural networks (MFNNs) is the dynamic back-propagation (DBP) algorithm for the DRNN, and this learning approach was first studied by Williams and Zipser [35], and Narendra and Parthasarathy [25]. Recently, this algorithm has been applied to study an adaptive control problem by Jin, Nikiforuk and Gupta [15]. However, with some simplification, a simple static learning algorithm may also be applicable for weight learning of such a dynamic neural network. This algorithm is outlined as follows.

For convenience, one re-lists the equations of a DRNN as follows

$$\begin{cases} \mathbf{z}(k+1) = -\alpha\mathbf{z}(k) + \mathbf{A}\sigma(\mathbf{x}(k)) + \mathbf{B}\mathbf{u}(k) \\ \mathbf{y} = \mathbf{C}\mathbf{x}(k) \end{cases}$$

where the matrices **A**, **B** and **C** are partitioned as follows

$$\mathbf{A} = \begin{bmatrix} \mathbf{a}_1^T \\ \mathbf{a}_2^T \\ \vdots \\ \mathbf{a}_{n_z}^T \end{bmatrix}, \quad \textit{with} \quad \mathbf{a}_i = \begin{bmatrix} a_{i1} \\ a_{i2} \\ \vdots \\ a_{in_z} \end{bmatrix}$$

$$\mathbf{B} = \begin{bmatrix} \mathbf{b}_1^T \\ \mathbf{b}_2^T \\ \vdots \\ \mathbf{b}_{n_z}^T \end{bmatrix}, \quad \textit{with} \quad \mathbf{b}_i = \begin{bmatrix} b_{i1} \\ b_{i2} \\ \vdots \\ b_{in} \end{bmatrix}$$

$$\mathbf{C} = \begin{bmatrix} \mathbf{c}_1^T \\ \mathbf{c}_2^T \\ \vdots \\ \mathbf{c}_m^T \end{bmatrix}, \quad \textit{with} \quad \mathbf{c}_j = \begin{bmatrix} c_{j1} \\ c_{j2} \\ \vdots \\ c_{jn_z} \end{bmatrix}$$

Given a desired output $\mathbf{y_d}(\mathbf{k})$, like conventional back-propagation for multilayered neural networks, one may define a least-square error function as

$$E(k) = \frac{1}{2}||\mathbf{y_d}(\mathbf{k}) - \mathbf{y}(\mathbf{k})||^2 \tag{13.13}$$

using the gradient descent technique, the weight updating formulations for a purpose of minimizing the error function can be obtained as

$$\mathbf{a_i}(\mathbf{k}+1) = \mathbf{a_i}(\mathbf{k}) - \eta \left.\frac{\partial \mathbf{E}(\mathbf{k})}{\partial \mathbf{a_i}}\right|_{\mathbf{k}}$$

$$\mathbf{b_i}(\mathbf{k}+\mathbf{1}) = \mathbf{b_i}(\mathbf{k}) - \eta \left.\frac{\partial \mathbf{E}(\mathbf{k})}{\partial \mathbf{b_i}}\right|_{\mathbf{k}}$$

$$\mathbf{c_j}(\mathbf{k}+\mathbf{1}) = \mathbf{c_j}(\mathbf{k}) - \eta \left.\frac{\partial \mathbf{E}(\mathbf{k})}{\partial \mathbf{c_j}}\right|_{\mathbf{k}}$$

where $\eta > 0$ is a learning rate, and the partial derivatives on the right-hand side of the above equations are computed at instant k. Using the assumption that $\mathbf{a_i}(\mathbf{k}) \approx \mathbf{a_i}(\mathbf{k}-\mathbf{1})$ and $\mathbf{b_i}(\mathbf{k}) \approx \mathbf{b_i}(\mathbf{k}-\mathbf{1})$ in deriving the partial derivatives on the right-hand side of the above equations yields

$$\mathbf{a_i}(\mathbf{k}+\mathbf{1}) = \mathbf{a_i}(\mathbf{k}) + \eta \mathbf{y}^{\mathbf{T}}(\mathbf{k}) \left.\frac{\partial \mathbf{y}(\mathbf{k})}{\partial \mathbf{a_i}}\right|_{\mathbf{k}}$$

$$\mathbf{b_i}(\mathbf{k}+\mathbf{1}) = \mathbf{b_i}(\mathbf{k}) + \eta \mathbf{y}^{\mathbf{T}}(\mathbf{k}) \left.\frac{\partial \mathbf{y}(\mathbf{k})}{\partial \mathbf{b_i}}\right|_{\mathbf{k}}$$

$$\mathbf{c_j}(\mathbf{k}+\mathbf{1}) = \mathbf{c_j}(\mathbf{k}) + \eta \mathbf{y}^{\mathbf{T}}(\mathbf{k}) \left.\frac{\partial \mathbf{y}(\mathbf{k})}{\partial \mathbf{c_j}}\right|_{\mathbf{k}}$$

with

$$\frac{\partial \mathbf{y}(\mathbf{k})}{\partial \mathbf{a_i}} = \begin{bmatrix} 0 \\ \vdots \\ 0 \\ \sigma^{\mathbf{T}}(\mathbf{z}(\mathbf{k}-\mathbf{1})) \\ 0 \\ \vdots \\ 0 \end{bmatrix} \quad \longleftarrow ith$$

$$\frac{\partial \mathbf{y}(\mathbf{k})}{\partial \mathbf{b}_i} = \begin{bmatrix} 0 \\ \vdots \\ 0 \\ \mathbf{u}^T(\mathbf{k}-1) \\ 0 \\ \vdots \\ 0 \end{bmatrix} \leftarrow ith$$

$$\frac{\partial \mathbf{y}(\mathbf{k})}{\partial \mathbf{c}_j} = \begin{bmatrix} 0 \\ \vdots \\ 0 \\ \mathbf{x}^T(\mathbf{k}) \\ 0 \\ \vdots \\ 0 \end{bmatrix} \leftarrow jth$$

4 REGULATION OF KNOWN NONLINEAR SYSTEMS USING DRNNS

So far there is no effective and universal approach for designing a nonlinear regulator for an arbitrary nonlinear plant due to the complexities of the nonlinear dynamics, even when the model of such a nonlinear plant is known. However, an attempt will be made in this section towards developing a general design approach of nonlinear regulators for a general class of known nonlinear systems, this method involves an adaptive weight learning process.

4.1 State Regulation using Dynamic State Feedback

Let the origin be an equilibrium point of the nonlinear plant P with zero input; that is $\mathbf{f}(\mathbf{0},\mathbf{0}) = \mathbf{0}$. To regulate the nonlinear system given by Eq. (13.1) such that the state vector $\mathbf{x}(\mathbf{k})$ converges to the origin for an arbitrary initial value as k becomes large, one may use the following DRNN based *dynamic state feedback controller*

$$\begin{cases} \mathbf{z(k+1) = -\alpha z(k) + A\sigma(z(k)) + Bx(k)} \\ \mathbf{u(k) = Cz(k)} \end{cases} \tag{13.14}$$

which is also called a parametrized nonlinear state compensator. If the system (13.1) is known the objective of designing the dynamic controller can be carried out by appropriately selecting the weight matrices **A**, **B** and **C** such that the following closed loop dynamic system is asymptotical stable in the sense of the Lyapunov

$$\begin{cases} \mathbf{z(k+1) = -\alpha z(k) + A\sigma(z(k)) + Bx(k)} \\ \mathbf{x(k+1) = f(x(k), Cz(k))} \end{cases} \tag{13.15}$$

However, presently, there is no analytical method to do so. The classical linearization method may provide a local solution. In fact, local stability around $\mathbf{x} = \mathbf{0}$ and $\mathbf{z} = \mathbf{0}$ is guaranteed if the all eigenvalues of the Jacobian $\mathbf{J(0,0)}$ are located inside in the unit circle in the complex plane

$$-1 < |\lambda_i(\mathbf{J(0,0)})| < \mathbf{1} \tag{13.16}$$

with

$$\mathbf{J(0,0)} = \begin{bmatrix} \mathbf{-\alpha I + A} & \mathbf{B} \\ \mathbf{Cf_u(0,0)} & \mathbf{f_x(0,0)} \end{bmatrix}$$

where $\mathbf{f_x} = [\frac{\partial \mathbf{f}}{\partial \mathbf{x}}]$ and $[\frac{\partial \mathbf{f_u}}{\partial \mathbf{u}}]$. The detailed algorithms for determining the matrices **A**, **B** and **C** have been studied extensively in linear control systems theory. However, these algorithms give only some local solutions which are based on linearization around the origin $\mathbf{x} = \mathbf{0}$ and $\mathbf{z} = \mathbf{0}$. This is not a point of interest in this chapter. Next, we present an on-line weight learning algorithm for dealing with the above state regulation problem, this learning algorithm incorporates both the plant model and input-output data which is supposed to be available at every instant. Let us assume that an error function is formulated as

$$E(k) = \frac{1}{2}||\mathbf{x}(\mathbf{k})||^2 = \frac{1}{2}\sum_{i=1}^{n}\mathbf{x}_i^2(\mathbf{k})$$

Minimization of the error function by a standard steepest-descent technique yields the following set of learning equations

$$\mathbf{a}_i(\mathbf{k}+1) = \mathbf{a}_i(\mathbf{k}) + \eta \left.\frac{\partial \mathbf{E}}{\partial \mathbf{a}_i}\right|_{\mathbf{k}}$$

$$\mathbf{b}_i(\mathbf{k}+1) = \mathbf{b}_i(\mathbf{k}) + \eta \left.\frac{\partial \mathbf{E}}{\partial \mathbf{b}_i}\right|_{\mathbf{k}}$$

$$\mathbf{c}_i(\mathbf{k}+1) = \mathbf{c}_i(\mathbf{k}) + \eta \left.\frac{\partial \mathbf{E}}{\partial \mathbf{c}_i}\right|_{\mathbf{k}}$$

where $\eta > 0$ is a learning rate, and the partial derivations on the right-hand side are computed at the instant k. Using Eqs. (13.1) and (13.2), we may conduct the partial derivatives of the error function with respect to the network parameters as follows:

$$\frac{\partial E(k)}{\partial \mathbf{a}_i} = \mathbf{x}^T(\mathbf{k})\frac{\partial \mathbf{x}(\mathbf{k})}{\partial \mathbf{a}_i} = \mathbf{x}^T(\mathbf{k})\frac{\partial \mathbf{x}(\mathbf{k})}{\partial \mathbf{z}(\mathbf{k}-1)}\frac{\partial \mathbf{z}(\mathbf{k}-1)}{\partial \mathbf{a}_i}$$

$$= \mathbf{x}^T(\mathbf{k})\frac{\partial \mathbf{x}(\mathbf{k})}{\partial \mathbf{u}(\mathbf{k}-1)}\frac{\mathbf{u}(\mathbf{k}-1)}{\partial \mathbf{z}(\mathbf{k}-1)}\frac{\partial \mathbf{z}(\mathbf{k}-1)}{\partial \mathbf{a}_i}$$

$$\frac{\partial E(k)}{\partial \mathbf{b}_i} = \mathbf{x}^T(\mathbf{k})\frac{\partial \mathbf{x}(\mathbf{k})}{\partial \mathbf{b}_i} = \mathbf{x}^T(\mathbf{k})\frac{\partial \mathbf{x}(\mathbf{k})}{\partial \mathbf{z}(\mathbf{k}-1)}\frac{\partial \mathbf{z}(\mathbf{k}-1)}{\partial \mathbf{b}_i}$$

$$\frac{\partial E(k)}{\partial \mathbf{c}_j} = \mathbf{x}^T(\mathbf{k})\frac{\partial \mathbf{x}(\mathbf{k})}{\partial \mathbf{u}(\mathbf{k}-1)}\frac{\partial \mathbf{u}(\mathbf{k}-1)}{\partial \mathbf{c}_i}$$

where

$$\frac{\partial \mathbf{x}(\mathbf{k})}{\partial \mathbf{u}(\mathbf{k}-1)} = \mathbf{f}_{\mathbf{u}}(\mathbf{x}(\mathbf{k}-1), \mathbf{C}\mathbf{z}(\mathbf{k}-1))$$

$$\frac{\partial \mathbf{u}(\mathbf{k}-1)}{\partial \mathbf{z}(\mathbf{k}-1)} = \mathbf{C}$$

$$\frac{\partial \mathbf{z}(\mathbf{k}-1)}{\partial \mathbf{a}_{\mathbf{i}}} = \begin{bmatrix} 0 \\ \vdots \\ 0 \\ \sigma^{\mathbf{T}}(\mathbf{z}(\mathbf{k}-2)) \\ 0 \\ \vdots \\ 0 \end{bmatrix} \begin{matrix} \\ \\ \\ \longleftarrow ith \\ \\ \\ \\ \end{matrix}$$

$$\frac{\partial \mathbf{z}(\mathbf{k}-1)}{\partial \mathbf{b}_{\mathbf{i}}} = \begin{bmatrix} 0 \\ \vdots \\ 0 \\ \mathbf{x}^{\mathbf{T}}(\mathbf{k}-2) \\ 0 \\ \vdots \\ 0 \end{bmatrix} \begin{matrix} \\ \\ \\ \longleftarrow ith \\ \\ \\ \\ \end{matrix}$$

$$\frac{\partial \mathbf{u}(\mathbf{k}-1)}{\partial \mathbf{c}_{\mathbf{j}}} = \begin{bmatrix} 0 \\ \vdots \\ 0 \\ \sigma^{\mathbf{T}}(\mathbf{z}(\mathbf{k}-2)) \\ 0 \\ \vdots \\ 0 \end{bmatrix} \begin{matrix} \\ \\ \\ \longleftarrow jth \\ \\ \\ \\ \end{matrix}$$

The learning controller introduced above is an on-line scheme, in the other words, the DRNN is on-line trained so that the states of overall system converge to the origin as the time becomes large. A block diagram of the system is shown in Figure 5.

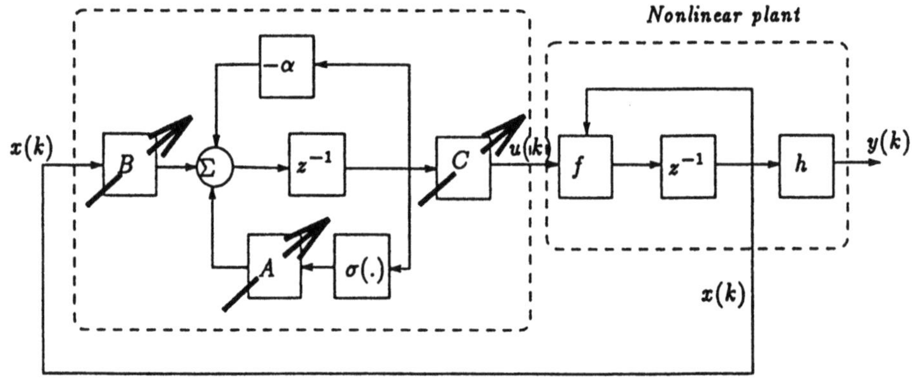

Figure 5 Block diagram of dynamic state feedback controller using a DRNN.

4.2 Output Regulation using Dynamic Output Feedback

The concept of output feedback is commonly used in control systems . A DRNN based dynamic output feedback compensator for the nonlinear system (13.1) is given as

$$\left\{ \begin{array}{c} \mathbf{z}(\mathbf{k}+\mathbf{1}) = -\alpha\mathbf{z}(\mathbf{k}) + \mathbf{A}\sigma(\mathbf{z}(\mathbf{k})) + \mathbf{B}\mathbf{y}(\mathbf{k}) \\ \mathbf{u}(\mathbf{k}) = \mathbf{C}\mathbf{z}(\mathbf{k}) \end{array} \right. \tag{13.17}$$

In this case, the closed loop equations of the entire system are

$$\left\{ \begin{array}{c} \mathbf{z}(\mathbf{k}+\mathbf{1}) = -\alpha\mathbf{z}(\mathbf{k}) + \mathbf{A}\sigma(\mathbf{z}(\mathbf{k})) + \mathbf{B}\mathbf{h}(\mathbf{x}(\mathbf{k}), \mathbf{C}\mathbf{z}(\mathbf{k})) \\ \mathbf{x}(\mathbf{k}+\mathbf{1}) = \mathbf{f}(\mathbf{x}(\mathbf{k}), \mathbf{C}\mathbf{z}(\mathbf{k})) \end{array} \right. \tag{13.18}$$

To analytically choose the weight matrices involved in above dynamic output feedback controller, the Jacobian around the origin may be given as follows

$$J(0,0) = \begin{bmatrix} -\alpha\mathbf{I} + \mathbf{A} + \mathbf{BCh_u}(0,0) & \mathbf{Bh_x}(0,0) \\ \mathbf{Cf_u}(0,0) & \mathbf{f_x}(0,0) \end{bmatrix}$$

In the context of learning control scheme, an error function used for training DRNN is defined as

$$E(k) = \frac{1}{2}||\mathbf{y}(\mathbf{k})||^2 = \frac{1}{2}\sum_{\mathbf{i}=1}^{l}\mathbf{y}_{\mathbf{i}}^2(\mathbf{k})$$

The gradient descent technique based learning formulations may also be implied using the same method. The resulting formulations are omitted here. However, the block diagram of the above control structure is given in Figure 6.

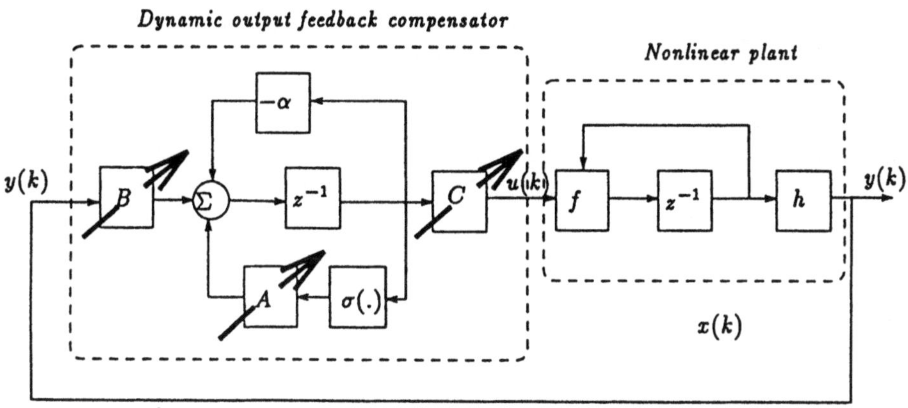

Figure 6 Block diagram of dynamic output feedback controller using a DRNN.

5 STATE REGULATION OF UNKNOWN PLANTS USING DRNNS

When the plants are unknown the control systems design in terms of state regulation becomes more complex . As known in adaptive control system theory, to deal with a system which contains some uncertain parameters or is totally unknown (black box), some adaptive algorithms towards parameters estimation or system modeling are necessary based on input-output data. Note that the DRNNs based state regulation schemes proposed in the last section, explicit information of the model used in the control algorithm is only the partial differential matrix $\mathbf{f_u}$ which describes the change of the system right-hand side vector-valued function $\mathbf{f}$ with respect to the change of the control vector input $\mathbf{u}$. Hence, if the elements of this partial differential matrix are known or ever their signs are known, the above schemes are still valid. However, this assumption does not always make sense in general, consequently, some universal methods for dealing with such unknown plants have to be exploited.

5.1 Dynamic State Feedback Structure

The basic idea here is introduced one more DRNN to on-line model the plant through accessible information such as measurable state vector or output vector provided by the system. In this case, one assumes that the plant (13.1) is totally unknown except for the state information $\mathbf{x(k)}$. Then, the following two DRNNs may be employed to implement a dynamic state feedback controller C and approximate model M for the plant P in terms of the input-output relationship

$$C: \left\{ \begin{array}{c} \mathbf{z_c(k+1) = -\alpha_c z_c(k) + A_c \sigma(z_c(k)) + B_c x(k)} \\ \mathbf{u(k) = C_c z_c(k)} \end{array} \right. \tag{13.19}$$

and

$$M: \left\{ \begin{array}{c} \mathbf{z_m(k+1) = -\alpha_m z_m(k) + A_m \sigma(z_m(k)) + B_m u(k)} \\ \mathbf{x_m(k) = C_m z_m(k)} \end{array} \right. \tag{13.20}$$

As shown in Figure 7, the output $\mathbf{u(k)}$ of the DRNN-C is the dynamic state feedback input to the unknown plant while the output $\mathbf{x_m(k)}$ of the BRNN M is used to approach the real state vector $\mathbf{x(k)}$ of the plant which is assumed to be available at the instant k. If the unknown plant P is perfectly identified by the DRNN-M after dynamic learning process which will discussed in detailed later, one has the following relationship

$$\lim_{k \longrightarrow \infty} ||\mathbf{x(k)} - \mathbf{x_m(k)}|| = 0 \tag{13.21}$$

Next, one may treat the DRNN M as a known model of the plant. Using the approach discussed in the last section, one is able to design a DRNN based controller such that the output of the DRNN M converges to zero as the time becomes large; that is

$$\lim_{k \longrightarrow \infty} ||\mathbf{x_m(k)}|| = 0 \tag{13.22}$$

In this case, one may verify that the state vector of the plant will converge to the origin since

$$\begin{aligned}\lim_{k \longrightarrow \infty} ||\mathbf{x(k)}|| &= \lim_{\mathbf{k} \longrightarrow \infty} ||\mathbf{x(k)} - \mathbf{x_m(k)} + \mathbf{x_m(k)}|| \\ &= \lim_{k \longrightarrow \infty} \Big(||\mathbf{x(k)} - \mathbf{x_m(k)}|| + ||\mathbf{x_m(k)}|| \Big) \\ &= 0\end{aligned} \tag{13.23}$$

Define the error functions for training both the DRNNs as follows

$$E_c(k) = \frac{1}{2}||\mathbf{x_m(k)}||^2 = \frac{\mathbf{1}}{\mathbf{2}}\sum_{\mathbf{i=1}}^{\mathbf{n}} \mathbf{x^2_{mi}(k)} \tag{13.24}$$

and

$$E_m(k) = \frac{1}{2}||\mathbf{x(k)} - \mathbf{x_m(k)}|| \tag{13.25}$$

where $E_c(k)$ is used to train the DRNN-C used as a dynamic feedback controller while the $E_m(k)$ is chosen to train the DRNN M used as a model of the unknown plant. Moreover, the updating formulations for the weight parameters in both the DRNNs C and M may be given as follows

$$\mathbf{a_{mi}(k+1) = a_{mi}(k)} + \eta \left.\frac{\partial \mathbf{E_m(k)}}{\partial \mathbf{a_{mi}}}\right|_{\mathbf{k}}$$

$$\mathbf{b_{mi}(k+1) = b_{mi}(k)} + \eta \left.\frac{\partial \mathbf{E_m(k)}}{\partial \mathbf{b_{mi}}}\right|_{\mathbf{k}}$$

$$\mathbf{c_{mj}(k+1) = c_{mj}(k)} + \eta \left.\frac{\partial \mathbf{E_m(k)}}{\partial \mathbf{c_{mj}}}\right|_{\mathbf{k}}$$

and

$$\mathbf{a_{ci}(k+1) = a_{ci}(k)} + \eta \left.\frac{\partial \mathbf{E_c(k)}}{\partial \mathbf{a_{ci}}}\right|_{\mathbf{k}}$$

$$\mathbf{b_{ci}(k+1) = b_{ci}(k)} + \eta \left.\frac{\partial \mathbf{E_c(k)}}{\partial \mathbf{b_{ci}}}\right|_{\mathbf{k}}$$

$$\mathbf{c_{cj}(k+1) = c_{cj}(k)} + \eta \left.\frac{\partial \mathbf{E_c(k)}}{\partial \mathbf{c_{cj}}}\right|_{\mathbf{k}}$$

where the partial derivatives may be implied using Eqs. (13.19) and (13.20). The overall performance of this control structure depends upon both the DRNNs. In particular, the convergence of the DRNN C is closely related with the representation of the plant by DRNN M. It is worth to mention that the identification process of the unknown plant using the DRNN M may be carried out in either a on-line or off-line way. When the on-line identification is adapted, the convergence period of the overall control system may take quite long time. Hence, if possible from the practical point of view, a off-line identification procedure is recommended to improve the convergence speed of the learning control process.

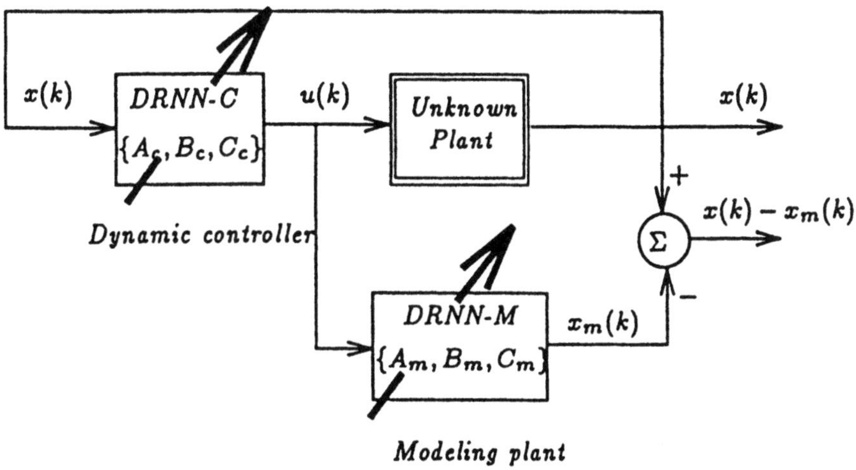

Figure 7 Block diagram of dynamic state feedback controller using a DRNN for an unknown plant.

5.2 Dynamic Output Feedback Scheme

Let nonlinear system P be a square system; that is, the number of the inputs is equal to the number of the outputs. If only the output vector $\mathbf{y(k)}$ of the plant is available, and the control objective becomes as an output regulation, we have to re-consider the above control scheme using a dynamic output feedback structure. As a matter of fact, one may easily modify the equations of the DRNNs given in Eqs. (13.19) and (13.20) as follows

$$C: \begin{cases} \mathbf{z_c(k+1)} = -\alpha_c \mathbf{z_c(k)} + \mathbf{A_c}\sigma(\mathbf{z_c(k)}) + \mathbf{B_c y(k)} \\ \mathbf{u(k)} = C_c \mathbf{z_c(k)} \end{cases} \tag{13.26}$$

and

$$M: \left\{ \begin{array}{rcl} \mathbf{z_m(k+1)} & = & -\alpha_m \mathbf{z_m(k)} + \mathbf{A_m}\sigma(\mathbf{z_m(k)}) + \mathbf{B_m u(k)} \\ \mathbf{y_m(k)} & = & C_m \mathbf{z_m(k)} \end{array} \right. \tag{13.27}$$

where the output $\mathbf{y(k)}$ of the plant becomes the input of the DRNN-C which is used to produce a dynamic feedback control law. To build the weight updating equations, the error functions used for training the above two networks are defined as follows

$$E_c(k) = \frac{1}{2}||\mathbf{y_p(k)}||^2 \tag{13.28}$$

and

$$E_m(k) = \frac{1}{2}||\mathbf{y(k)} - \mathbf{y_m(k)}||^2 \tag{13.29}$$

The system structure is shown in Figure 8.

The above control configuration consists of primarily of three main components: a DRNN-C, as a dynamic output feedback controller, to produce the suitable control action, another DRNN-M to emulate the system input-output behavior, and the associated on-line learning algorithms to adjust the connection weights of the DRNNs according to the available system information.

6 OUTPUT TRACKING OF UNKNOWN PLANTS USING DRNNS

As known in the previous sections, output tracking control in nonlinear systems is an ubiquitous control objective. In the field of the nonlinear systems, many sufficient conditions for solving various forms of the tracking

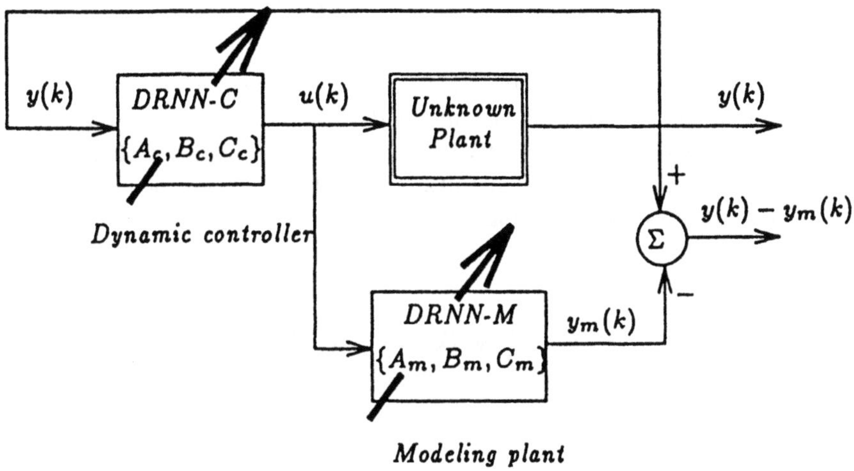

Figure 8 Block diagram of dynamic output feedback controller using a DRNN for an unknown plant.

problem have been obtained. Also, neural networks based static controllers as approximate solutions have been discussed extensively, this type of the tracking control performs an asymptotic tracking. Let $\mathbf{y_d(k)}$ be a given desired output data. Our objective is to control the output of the unknown plant such that $\mathbf{y(k)} \longrightarrow \mathbf{y_d(k)}$ when k becomes large. Using the DRNNs, the dynamic state feedback controller C for such a objective is given

$$C: \quad \left\{ \begin{array}{lcl} \mathbf{z_c(k+1)} & = & -\alpha_c \mathbf{z_c(k)} + \mathbf{A_c}\sigma(\mathbf{z_c(k)}) + \mathbf{B_c^1 x(k)} + \mathbf{B_c^2 Y_d(k)} \\ \mathbf{u(k)} & = & \mathbf{C_c z_c(k)} \end{array} \right. \tag{13.30}$$

where

$$\mathbf{Y_d(k)} = [\mathbf{y_d^T(k)} \ \mathbf{y_d^T(k+1)} \ \ldots \ \mathbf{y_d^T(k+r)}]^\mathbf{T}$$

with an integer $r \geq 0$. Another DRNN-M used to model the unknown plant is given as

$$M: \quad \begin{cases} \mathbf{z_m}(\mathbf{k}+1) & = & -\alpha_m \mathbf{z_m}(\mathbf{k}) + \mathbf{A_m}\sigma(\mathbf{z_m}(\mathbf{k})) + \mathbf{B_m}\mathbf{u}(\mathbf{k}) \\ \mathbf{y_m}(\mathbf{k}) & = & \mathbf{C_m}\mathbf{z_m}(\mathbf{k}) \end{cases} \tag{13.31}$$

As shown in Figure 9, the DRNN based controller C receives the state information of the unknown plant and the desired output as its inputs at instant k, and produces an output which is sent to the unknown plant and the model DRNN P as input signal simultaneously. In this case, the errors used to train the DRNNs are defined as

$$E_c(k) = \frac{1}{2}||\mathbf{y_d}(\mathbf{k}) - \mathbf{y_m}(\mathbf{k})||^2 \tag{13.32}$$

and

$$E_m(k) = \frac{1}{2}||\mathbf{y}(\mathbf{k}) - \mathbf{y_m}(\mathbf{k})||^2 \tag{13.33}$$

If the learning process is successful; that is

$$\lim_{k \longrightarrow \infty} E_c(k) = 0$$

and

$$\lim_{k \longrightarrow \infty} E_m(k) = 0$$

we have

$$\lim_{k \longrightarrow \infty} ||\mathbf{y_d}(\mathbf{k}) - \mathbf{y}(\mathbf{k})|| = 0$$

which carries out an asymptotic output tracking. If only the output signal of the unknown plant is available, we use an output feedback in the DRNN C, as shown in Figure 10 resulting the following system

$$C: \quad \begin{cases} \mathbf{z_c(k+1)} &= -\alpha_c \mathbf{z_c(k)} + \mathbf{A_c}\sigma(\mathbf{z_c(k)}) + \mathbf{B_c^1 y(k)} + \mathbf{B_c^2 Y_d(k)} \\ \mathbf{u(k)} &= \mathbf{C_c z_c(k)} \end{cases} \quad (13.34)$$

The above dynamic controller uses both the desired output given by the design and the plant output measured as input signals, and produces a control signal to both the unknown plant and the model DRNN-M.

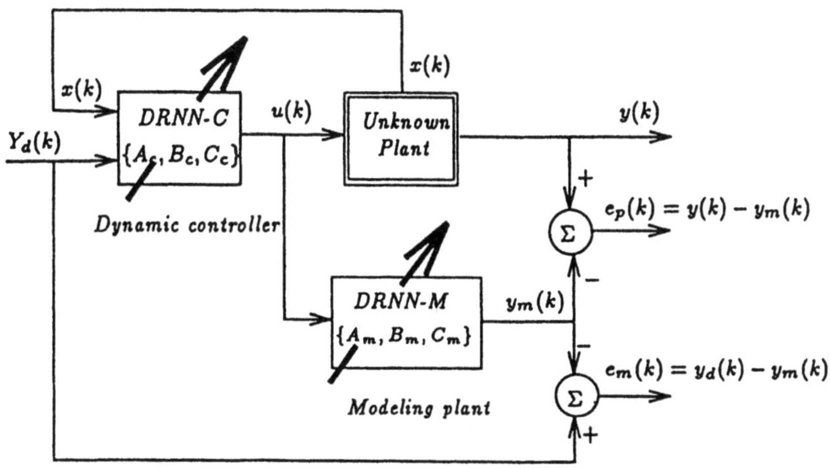

Figure 9 Block diagram of output tracking control using DRNNs with state feedback for an unknown plant.

7 APPLICATIONS TO ROBOTIC CONTROL

Recently, there has been considerable attention in developing efficient adaptive control algorithms for robot manipulators under a uncertain environment. The complexity of the control problem for manipulators arises mainly from their nonlinear dynamics and uncertain characristics. The dynamics of articulated mechanisms in general, and of robot manipulators in particular, involves strong nonlinear coupling effects between joints as well as centrifugal

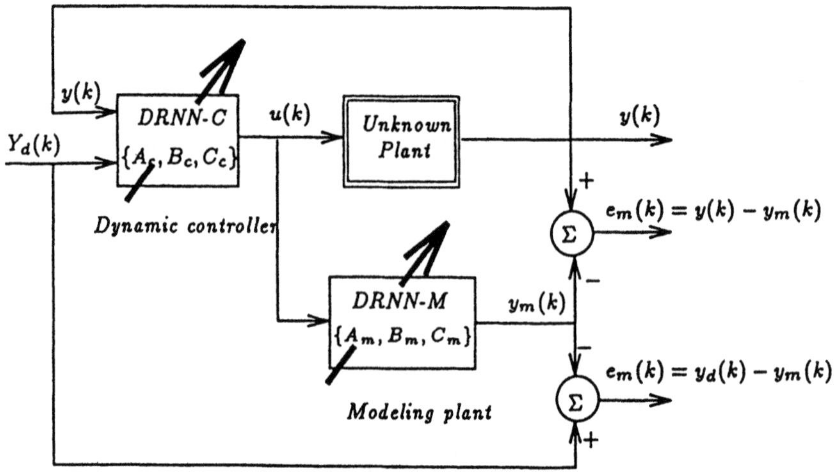

Figure 10 Block diagram of output tracking control using DRNNs with output feedback for an unknown plant.

and Coriolis forces. In this section, the nonlinear control schemes presented previously will be applied to deal with control tasks of robot manipulators. It is well-known that the equations of motion for a multi-joint manipulator may be represented in the following compact form (Spong and Vidyasagar, [34])

$$\mathbf{M}(\theta)\frac{\mathbf{d}^2\theta}{\mathbf{dt}^2} + \mathbf{N}(\theta, \frac{\mathbf{d}\theta}{\mathbf{dt}})\frac{\mathbf{d}\theta}{\mathbf{dt}} + \mathbf{g}(\theta) = \tau \tag{13.35}$$

where θ is a vector of the angles of the joints, $\mathbf{M}(\theta) \in \Re^{\mathbf{n}}$ is said to be the manipulator inertia-mass matrix which is symmetric and invertible, $\mathbf{N}(\theta, \dot{\theta}) \in \Re^{\mathbf{n}\times\mathbf{n}}$ represents forces arising from the Coriolis force and centrifugal force, $\mathbf{g}(\theta) \in \Re^{\mathbf{n}}$ is associated with gravitational potential energy, and $\tau \in \Re^{\mathbf{n}}$ is a vector of the joint torques supplied by the actuators. Introducing the state vectors $\mathbf{x_1} = \theta$ and $\mathbf{x_2} = \dot{\theta}$ yields

$$\frac{d\mathbf{x_1}}{\mathbf{dt}} = \mathbf{x_2}$$

$$\frac{d\mathbf{x_2}}{\mathbf{dt}} = -\mathbf{M}^{-1}(\mathbf{x_1})\mathbf{N}(\mathbf{x_1}, \mathbf{x_2})\mathbf{x_2} + \mathbf{M}^{-1}(\mathbf{x_1})\tau \tag{13.36}$$

This is a state space expression of the motion equations of the manipulator.

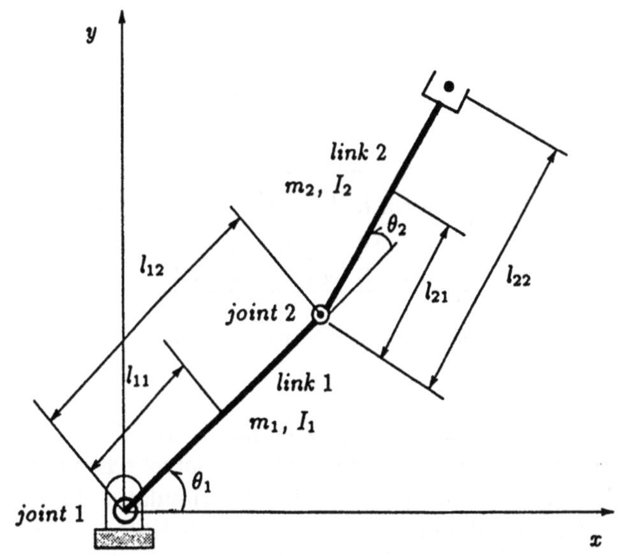

Figure 11 Two-link planar manipulator.

In this chapter for simulation studies we consider only a two-link planar manipulator as shown in Figure 11. Then, the elements of the matrices $\mathbf{M}(\theta)$ and $\mathbf{N}(\theta, \dot{\theta})$ may given as (Spong and Vidyasagar, [34])

$$\begin{cases} m_{11} = I_1 + I_2 + m_1 l_{11} + m_2[l_{12}^2 + l_{21}^2 + 2l_{12}l_{21}cos(\theta_2)] \\ m_{12} = m_{21} = I_2 + m_2[l_{21}^2 + l_{12}l_{21}cos(\theta_2)] \\ m_{22} = I_2 + m_2 l_{21}^2 \end{cases}$$

$$\begin{cases} n_{11} = 0 \\ n_{12} = -m_2 l_{12} l_{21} sin(\theta_2)(2\dot{\theta}_1 + \dot{\theta}_2) \\ n_{21} = m_2 l_{12} l_{21} sin(\theta_2)\dot{\theta}_1 \\ n_{22} = 0 \end{cases}$$

$$\begin{cases} g_1 = (m_1 l_{11} + m_2 l_{12}) g \, cos(\theta_1) + m_2 l_{21} g \, cos(\theta_1 + \theta_2) \\ g_2 = m_2 l_{21} cos(\theta_1 + \theta_2) \end{cases}$$

where I_i is the moment of inertia for the link i, m_1 is the mass of the link i, l_{i2} is the length of the link i, l_{i1} is the distance from joint i to the center of gravity of the link i, and g is a gravitional constant. The actual values for the parameters of the two-link planar manipulator shown in Figure 11 are given in Table 1.

Two-Link Robot		Links	
		Link 1	Link 2
Parameters	Mass m_i [kg]	5.5	5.9
	Inertia I_i [Kg.m^2]	0.15	0.14
	Length l_{i1} [m]	0.3	0.3
	Length l_{i2} [m]	0.6	0.6

Table 1 Actual parameter values of the manipulator.

Let the joint angles θ_1, θ_2 and their differentials $\dot{\theta}_1$ and $\dot{\theta}$ are measurable at every instant so that we can use state feedback control schemes in the following studies. The sampling rate used in the simulation for discrete-time controllers is 0.001 second and the 4th-order Runge-Kutta numerical method is employed to integrate system differential equations with the integration step 0.001 second. Both the state regulation and output tracking problems are studied in the simulation. First, state regulation problem was conducted for the case of the known plant using a DRNN as a dynamic state feedback controller as shown in Figure 12. The number of the dynamic neurons involved in the DRNN was selected as $n_z = 8$ and the initial condition of the DRNN is zero; that is $z(0) = 0$. The simulation results for regulating an initial state of $\theta_1(0) = \theta_2 = 0.47$ $(rad.)$ and $\dot{\theta}_1 = 0$ and $\dot{\theta}_2 = 0$ (rad/sec) for the manipulator to the origin are given in Figure 14. In this case, the states of the system converge quickly to the origin with such a dynamic feedback controller. Moreover, let the dynamics model of the manipulator be unknown. The state regulation problem was studied using the two DRNNs as shown in Figure 13, where one of the DRNNs with 8 dynamic neurons is acted as dynamic controller while another with 10 neurons is used to model the manipulator dynamics on-line. The satisfactory control results were obtained and shown in Figure 15. The observation on these simulation results shows that the process of the on-line modeling usually reduce the convergence speed of the control system. However, if an appropriate choice of the structure of the DRNNs, such as the number of the neurons and learning rates, is obtained the system with the control action may still have a good convergence property.

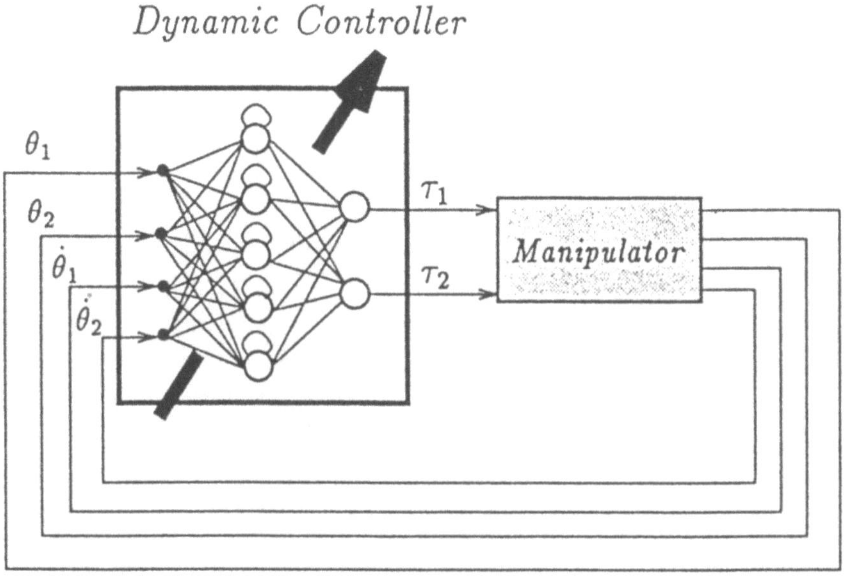

Figure 12 Schematic representation of state regulation for a two-link planar manipulator with known dynamics using a DRNN.

The second task here is to implement an output tracking control for the case of unknown plant. Here the two outputs are chosen as the two joint angles θ_1 and θ_2 of the manipulator. Thus the output tracking here means that θ_1 and θ_2 are controlled to follow the desired trajectories θ_{d1} and θ_{d2}. The control structure for this control objective is given in Figure 16 where the two DRNNs are used to produce the control signal and model the plant on-line, respectively. Let the desired output signals $\mathbf{y_d(k)}$ be slow change so that one may assume $\mathbf{Y_d(k)} \approx \mathbf{y_d(k)}$. The initial conditions of the system is $\theta_1 = \theta_2 = 0$ and $\dot{\theta}_1 = \dot{\theta}_2 = 0$. The number of the control DRNN is 8 while that of the model DRNN is 10. However, increasing number of the neurons does not significantly improve the control performance. If the desired outputs signals are designed as the square waves with amplitude 0.5 and the cycle 200 the simulation results are shown in Figure 17. For the sine wave output signals given by

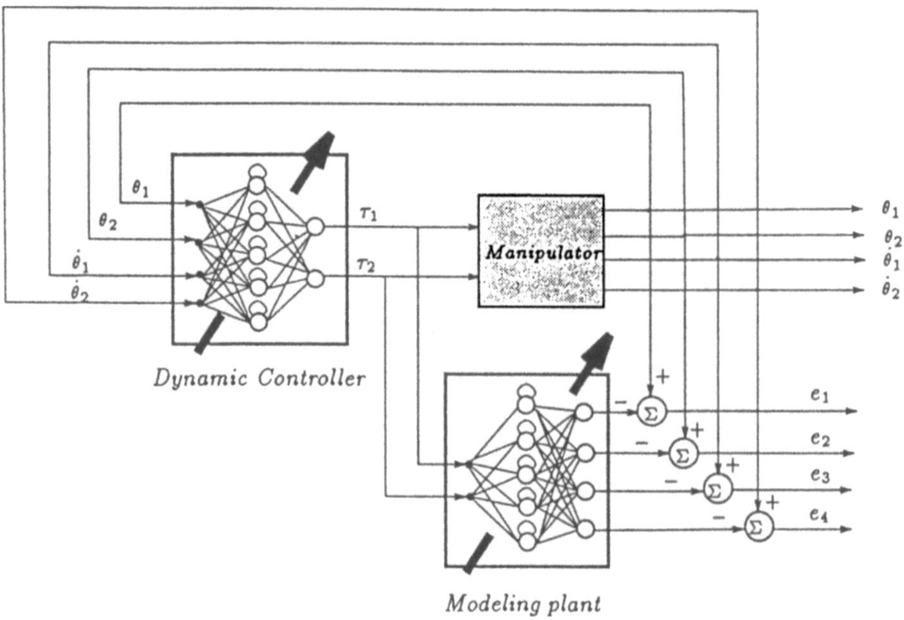

Figure 13 Schematic representation of state regulation for a two-link planar manipulator with unknown dynamics using two DRNNs.

$$\theta_{d1}(k) = sin(2\pi k/100)$$

$$\theta_{d2}(k) = -sin(2\pi k/100)$$

the simulation results are given in Figure 18. These results indicate that the outputs of the controlled plant can track perfectly the desired outputs by the DRNNs based dynamic state feedback system. On the other hand, since the modeling and control processes are started simultaneously, the reference input signal which is the persistent excitation is needed to assure the convergence of the neural networks to the true models and also improve the convergence speed of the control processes.

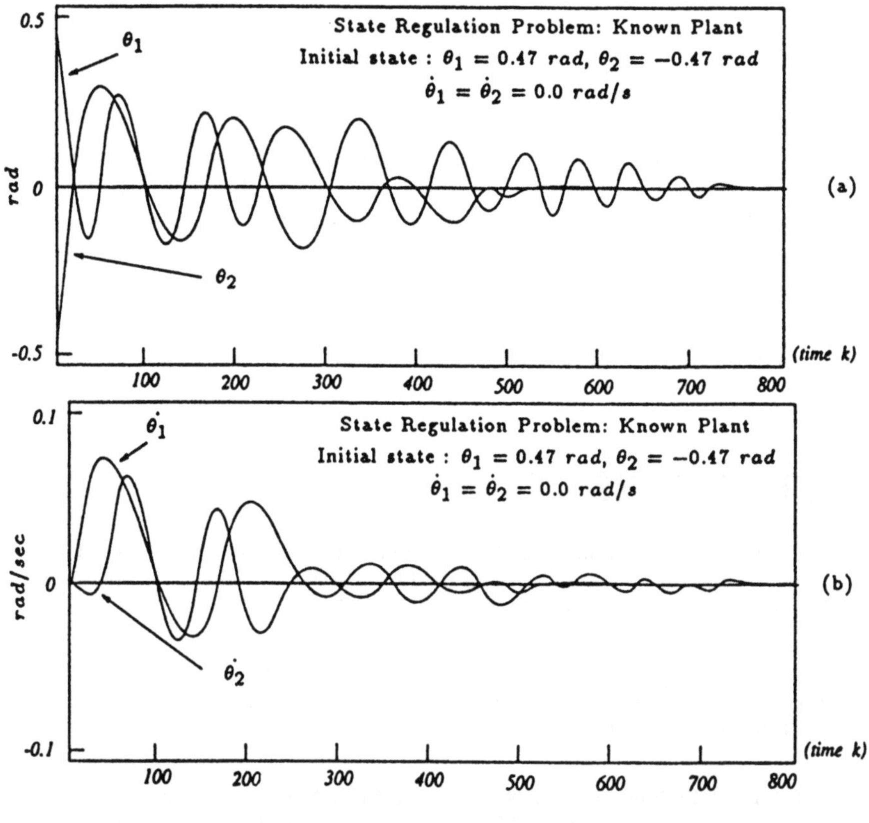

Figure 14 Simulation results of state regulation for a known plant using a DRNN.

8 CONCLUSIONS

It is beneficial to consider the need for new techniques of control in terms of complex tasks similar to that discussed in this chapter. These new developments of control should build on the powers and tools that the field of adaptive control systems has shown already, but it must extend its capabilities beyond the classical adaptive control schemes, so that it can work in a large range of domains, and under more dynamic and more realistic conditions. This initial feasibility study reported in this chapter demonstrates that it is possible to implement an intelligent control for a general class of nonlinear systems using dynamic neural network structures. In particular, it was shown how state regulation and output tracking can be realized by dynamic feedback controllers constructed using DRNNs. The use of dynamic recurrent neural systems, rather than conventional feedforward neural networks, may provide a great deal of advantages for dynamic feedback control using only input-output information of a unknown plant.

Comparison of the static neural networks based adaptive control schemes and the DRNNs based dynamic feedback controllers shows that the latter requires less knowledge of the plant. From nonlinear control theory point of view, the existence of such dynamic feedback control structures is easily ensured in practical systems. Further experimentation and development are required to both theoretical and application topics such as stability issue of overall system consisting of plant and neural controller.

REFERENCES

[1] L. B. Almeida, "A Learning Rule for Asynchronous Perceptions with Feedback in a Combinatorial Environment," Proc. of the First *IEEE* International conference on Neural Networks (ICNN), Vol. 1, pp. 609-618, 1987.

[2] S.A. Billings, H.B. Jamaluddin and S.Chen, "Properties of Neural Networks with Applications to Modeling Non-linear Dynamical Systems ", Int. J. Control, Vol. 55, No. 1, pp. 193-224, 1992.

[3] F.-C Chen, " Back-propagation Neural Networks for Nonlinear Self-Tuning Adaptive Control", *IEEE* Control System Magazine, April 1990,pp. 44-48.

[4] K. Funahashi and Y. Nakamura, "Approximation of Dynamical Systems by Continuous Time Recurrent Neural Networks," *Neural Networks*, Vol. 6, pp. 801-806, 1993.

[5] E. Hernandez and Y. Arkun, "Study of the Control-Relevant Properties of Backpropagation Neural Models of Nonlinear Dynamical Systems," *Computer Chem. Engng*, Vol. 6, No. 4, pp.227-240, 1992.

[6] J. Hopfield, "Neural networks and physical systems with emergent collective computational abilities," Proc. Nat. Acad. Sci. USA, Vol. 79, pp. 2554-2558, 1982.

[7] J. Hopfield, "Neurons with graded response have collective computational properties like those of two state neurons," Proc. Nat. Acad. Sci. USA, Vol. 81, pp. 3088-3092, 1984.

[8] J. Hopfield and D. W. Tank, "Neural computational of decisions in optimization problems," Biolog. Cybernetics, Vol. 52, pp. 141-152, 1985.

[9] J. Hopfield and D. W. Tank, " Computing with neural circuits : a model," *Science*, Vol. 233, pp. 625-633, 1986.

[10] D.R. Hush and B.G. Horne, " Progress in Supervised Neural Networks," *IEEE* Signal Processing Magazine, No. 1, pp. 8-39, 1993.

[11] K.J.Hunt and D.Sbarbaro, "Neural Networks for Nonlinear Internal Model Control," *IEE* Proc.-D, Vol. 138, No. 5, pp. 431-438, Sept., 1991.

[12] K.J.Hunt, D.Sbarbaro, R. Zbikowski, and P.J. Gawthrop, "Neural Networks for Control Systems-A Survey," *Automatica*, Vol. 28, No. 6, pp. 1083-1112, 1992.

[13] A. Isidori, Nonlinear Control System, Springer Verlag, New York, 1989.

[14] L. Jin, P.N. Nikiforuk and M.M. Gupta, "Adaptive Tracking of SISO Nonlinear Systems Using Multilayered Neural Networks", Proc. *1992 American Control Conference (ACC)*, pp. 56-60, Chicago, June 24-26, 1992.

[15] L. Jin, P.N. Nikiforuk and M.M. Gupta, "Direct Adaptive Output Tracking Using Multilayered Neural Networks," *IEE Proceedings-D: Control Theory and Applications*, Vol. 140, No. 6, pp. 393-398.

[16] L. Jin, P.N.Nikiforuk and M.M.Gupta, "Adaptive Control of Discrete-Time Nonlinear Systems Using Recurrent Neural Networks", *IEE Proceedings-D: Control Theory and Applications*; Vol. 141, No. 3, pp. 169-176, 1994.

[17] L. Jin, P.N. Nikiforuk and M.M. Gupta, " Multilayered Recurrent Networks for Learning and Control of Unknown nonlinear Systems", *ASME Journal of Dynamic Systems, Measurement, and Control* (In Press).

[18] L. Jin, M.M. Gupta and P.N. Nikiforuk, "Computational Neural Architectures for Control Applications," in *Soft Computing: Fuzzy Logic, Neural Networks, and Distributed Artificial Intelligence*, edited by F. Aminzadeh and M. Jamshida, Chapter 6, pp. 121-152, Prentice Hall, Englewood cliffs, New Jersey, 1993.

[19] L. Jin, M.M.Gupta and P. N. Nikiforuk, "Neural Networks and Fuzzy Basis Functions for Functional Approximation," in this book.

[20] K. Li, " Approximation Theory and Recurrent Networks," Proc. of 1992 IJCNN, Vol. II, 266-271, 1992.

[21] L.Liung and T.Soderstrom, Theory and Practice of Recursive Identification , Cambridge, MA:MIT Press, 1983.

[22] W.T. Miller, III, R.S. Sutton, and P.J. Werbos, Ed., "Neural Networks for Control," The MIT Press, London, England, 1990.

[23] S. Mukhopadhyay and K.S. Narendra, "Disturbance Rejection in Nonlinear Systems Using Neural Networks," *IEEE* Trans. on Neural Networks, Vol. 4, No.1, pp. 63-72, 1993.

[24] K.S. Narendra and K.Parthasarathy, "Identification and Control of Dynamical Systems using Neural Networks," *IEEE* Trans. Neural Networks, Vol. NN-1, No. 1, pp. 4-27, March, 1990.

[25] K.S. Narendra and K.Parthasarthy, "Gradient Methods for the Optimization of Dynamical Systems Containing Neural Networks", *IEEE* Trans. Neural Networks, Vol. 2, No. 2, pp. 4-27, 1991.

[26] K.S. Narendra and A.M. Annaswamy, "Robust adaptive control in the presence of bounded disturbances," *IEEE* Trans. on Automatic Control, AC-31, pp. 306-315, April 1986.

[27] K.S. Narendra and S. Mukhopadhyay, "Adaptive Control of Nonlinear Multivariable Systems Using Neural Networks," *Neural Networks*, Vol. 7, No. 5, pp. 737-752, 1994.

[28] H.Nijmeijer and A.J.Van der Schaft, Nonlinear Dynamical Control Systems, Springer-Verlag, New York Inc. 1990.

[29] F.J. Pineda, "Generalization of back-propagation to recurrent neural networks," *Physical Review Letters,* Vol. 59, No. 19, pp. 2229-2232, 1987.

[30] F. J. Pineda, "Dynamics and architecture for neural computation. J. Complexity, Vol. 4, pp. 216-245, 1988.

[31] F.J. Pineda, "Recurrent backpropagation and the dynamic approach to adaptive neural computation," *Neural Computation*, Vol. 1, pp. 161-172, 1989.

[32] D.E. Rumelhart and J.L. McCelland, " Learning Internal Representations by Error Propagation ", Parallel Distributed Processing : Explorations in the Microstructure of Cognition, Vol. 1: Foundations, MIT Press, 1986.

[33] R.M. Sanner and J.J.E.Slotine, "Gaussian Networks for Direct Adaptive Control," *IEEE* Trans. Neural Networks, Vol. 3, No. 6, pp. 837-863.

[34] M. Spong and M. Vidyasagar, *Robot Dynamics and Control*, John Wiley & Sons, 1989.

[35] R. Williams and D. Zipser, "A Learning Algorithm for Continually Running Fully Recurrent Neural Networks", *Neural Computation*, 1, 270-280, 1989.

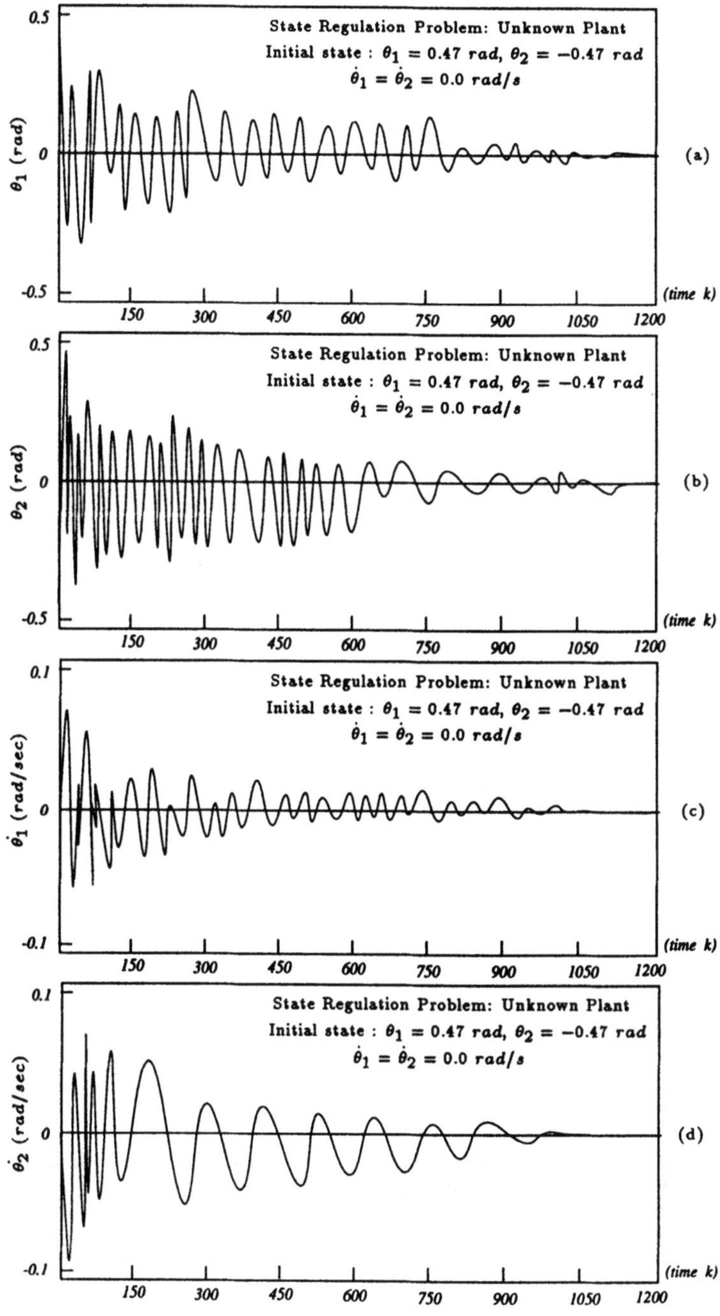

Figure 15 Simulation results of state regulation for an unknown plant using two DRNNs.

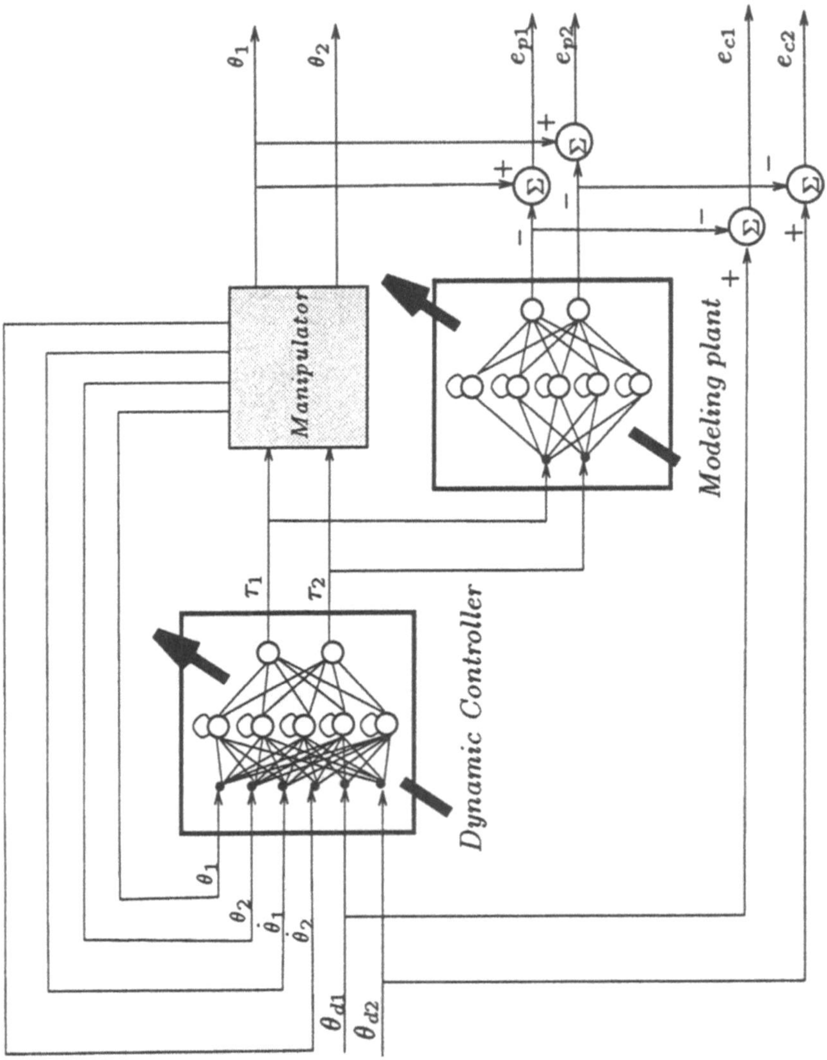

Figure 16 Schematic representation of output tracking for a two-link planar manipulator with unknown dynamics using two DRNNs.

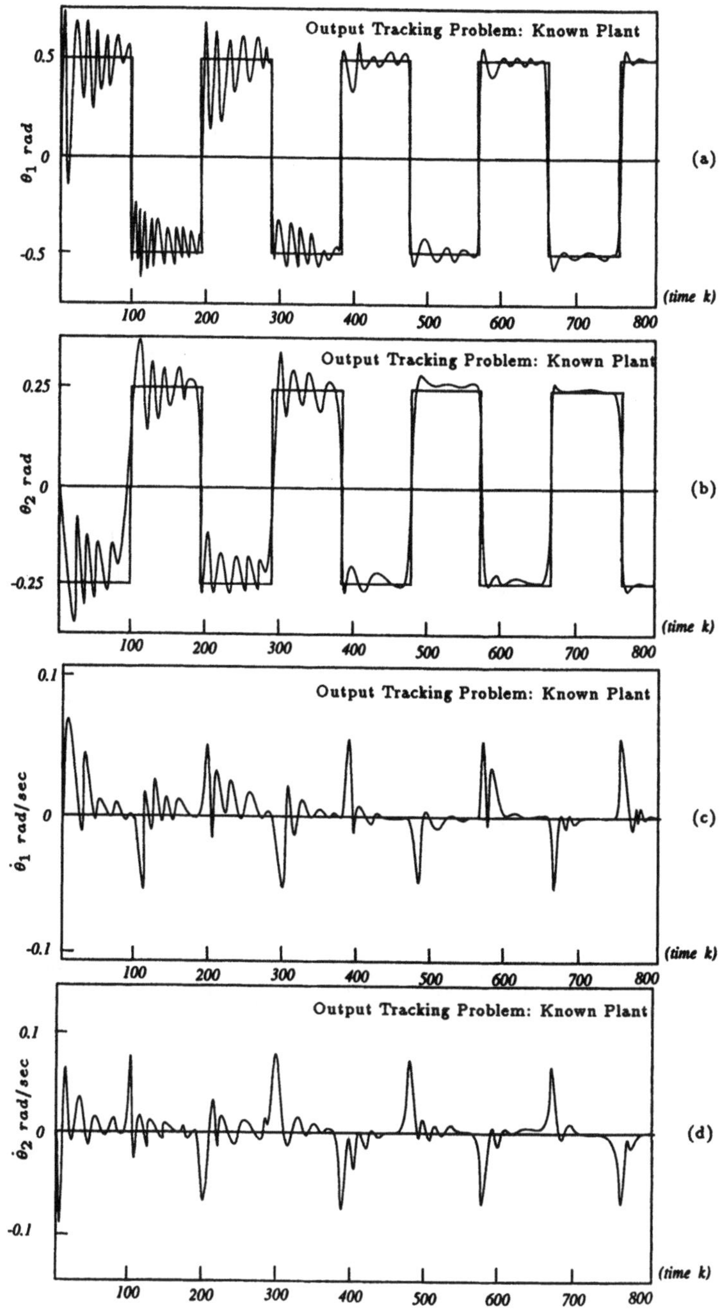

Figure 17 Simulation results of output tracking a square wave for an unknown plant using two DRNNs.

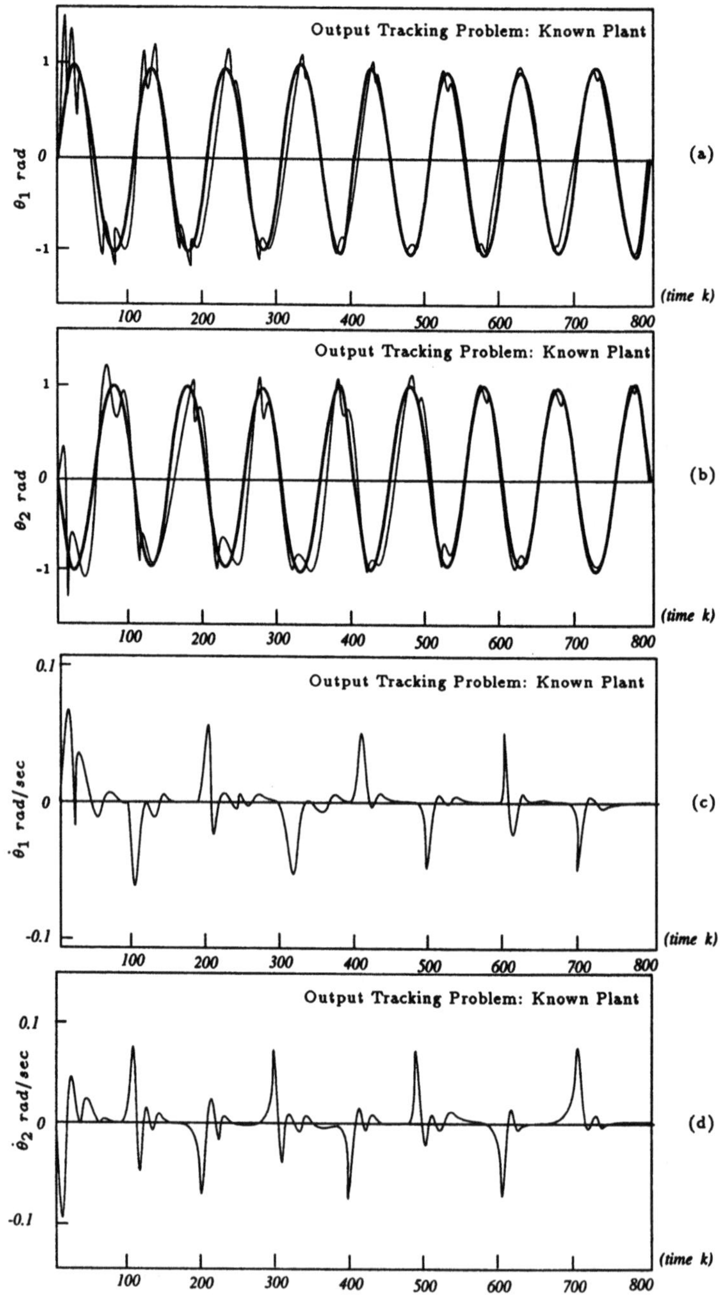

Figure 18 Simulation results of output tracking a sine wave for an unknown plant using two DRNNs.

14

CAMCORDER OPERATION JUDGEMENT USING A NEURAL COMPUTING APPROACH

Kitahiro Kaneda and Paul Wang*

Video Products Development Center and
Fuzzy Logic Research Laboratory
CANON INC., Tokyo, Japan
**Department of Electrical Engineering*
Duke University
Durham, NC 27708, USA

ABSTRACT

In this study, the neural computing approach is applied to the judgment of the camcorder operations which include fix-shot, panning and tilting. The data used in this study is the moving vector obtained from the real camcorder image. Also, seven features are carefully selected as possible inputs to this neural network. As a result of this approach, 96.88% accuracy of the operation judgment is obtained by using the length, angle of the moving vector and the time variation of the moving vector length as a set of three features. This result indicates that the approach proposed here can indeed classify three classes in the camcorder operations and it is also very easy to be utilized to control the automatic functions in the camcorder as well.

1 INTRODUCTION

In the past few years, a camcorder has become very popular for most people. Usually the camcorder users are using several camcorder operations in order to take high quality videotapes. Strictly speaking, there are seven basic camcorder operations such as fix-shot, panning, tracking, tilting, booming, zooming, dollying. Table 1 explains these operations briefly.

Among those operations, most users use only four operations, fix-shot, panning, tilting, zooming, because other operations are very difficult to be handled for amateur camera operators. Figure 1 describes those four operations graphically.

On the other hand, camcorder operations are important to camcorder designers in a different sense. That is to say, most automatic functions in the camcorder are needed to change their control parameters according to the camcorder's operation state. For instance, in the auto-image-stabilizer, the controller has to stop the operation of the image-stabilizing during panning or tilting for the purpose of avoiding excess control. For this reason, the

Name	Operation	
Fix-shot	Fixed	
Panning	Right and	Change of the camcorder direction
Tracking	Left	Change of the camcorder position
Tilting	Up and	Change of the camcorder direction
Booming	Down	Change of the camcorder position
Zooming	Back and	Change of the image angle
Dollying	Forth	Change of the camcorder position

Table 1 Definitions of the basic camcorder operations

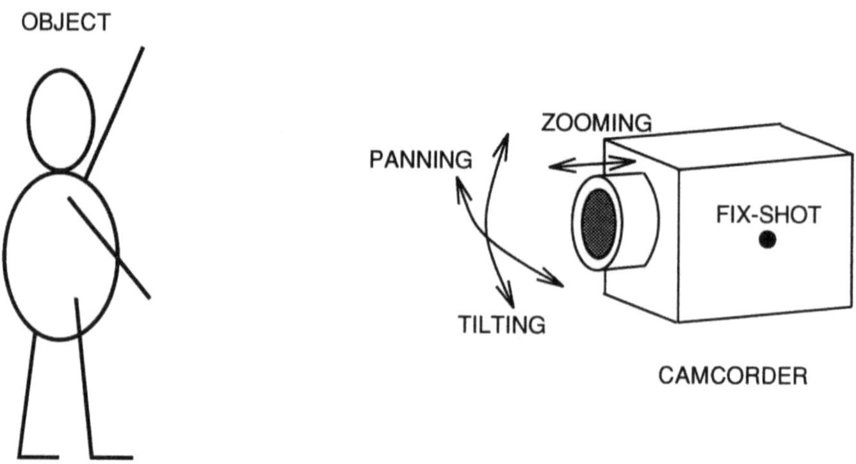

Figure 1 Four common camcorder operations.

camcorder designers have been eager to establish the automatic camcorder operation judgment, however it is not an easy task. Among the four popular camcorder operations mentioned above, zooming is easy to be identified, because normally this is controlled by the zooming switch. Other operations, fix-shot, panning and tiling, are not easy to be managed, since they are controlled by the only user's motion.

One attempt to judge those camcorder operations has been suggested by Kaneda [1]. The objective of this work was to find the simple and effective way to identify the three main camcorder operations, fix-shot, panning, tilting, so that it can be utilized to control the automatic functions in the camcorder with low cost. In his work, this problem was treated as a pattern recognition issue in the moving vector distribution. And a perceptron algorithm and a neural computing approach were applied. In consequence, several experiences for solving the camcorder operations judgment such as the nonlinear characteristics of the camcorder operations, the comvergence of the neural computing approach were obtained.

The main objective of this study is basically the same as our previous work. In addition to this, we would like to particularly focus on determining how to apply the neural computing approach as a viable method to the camcorder operation judgment.

The results of this study should be useful and significant for camcorder designers in order to develop a high performance robust camcorder.

2 PROBLEM FORMULATION AND METHODOLOGY

2.1 Input Data Selection

The input data selection is one of the important parts in this study. We choose to calculate a moving vector of the successive images obtained from the real videotape [1]. Because of its plentiful information about the image, this data turned out to be very meaningful for our study.

The moving vector set used here consists of 50 moving vectors per each image. Each moving vector stands for the relative motion of two sequential images in the small segment. Each image comes in every 1/60 second[2].

Figure 2 shows the definition of a moving vector. The data set for this study consists of 800 moving vector sets. This data set is further divided into 4 groups labeled as scene 1, 2, 3 and 4. Scene 1 and 2 contains two subgroups which represent static operation and panning operation. Also Scene 3 and 4 contains two subgroups which represent static operation and tilting operation. Each subgroup has 100 moving vector sets. This means the sampling period of each subgroup is 100/60 second.

Since these moving vector sets are obtained from the real videotapes, they are already containing several kinds of noises such as human uncertainty, environmental noises (dark, bright), data processing noises, and so on.

---

[1] K.Kaneda, "Camcorder Operation Judgment Using Several Pattern Recognition Methods," term paper, Duke University, Durham, NC, 1993

[2] This period is often called 1 field

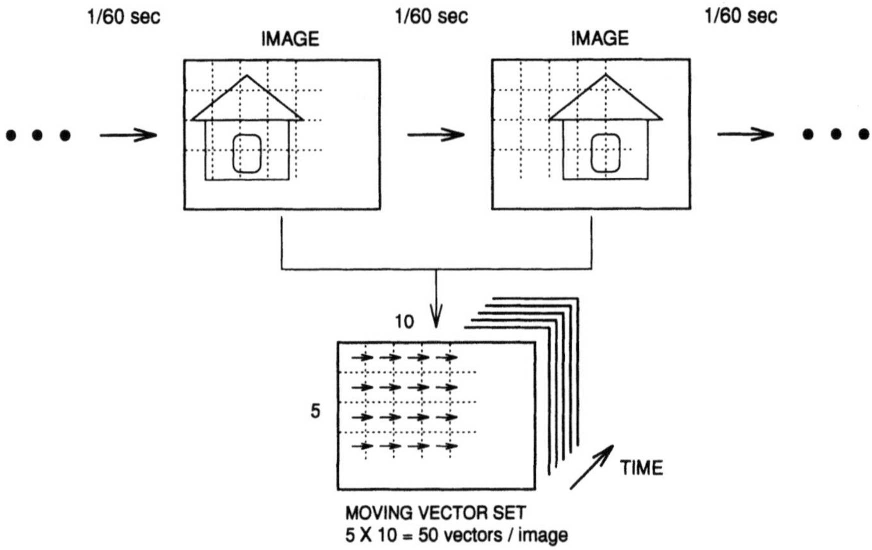

Figure 2 The definition of a moving vector.

2.2 Feature Extraction

Fix-shot, panning and tilting operations can be visualized by utilizing the information related to the moving vector. Figure 3 shows a typical example of the moving vector distributions of fix-shot, panning and tilting operations.

In Figure 3, there seems to be some correlations between the camcoder operations and the moving vector distributions. This means this problem can be interpreted to as a three-class neural network categorizing problem which uses the moving vector measurements as an input.

The highest priority work for achieving the optimal result in the neural network is to find the good input data in this case. The structure of the network, undoubtly, will have strong impact on the performance of correct classification. In this study, however, we shall concentrate first on the feature extraction issue.

Based upon the previous design experiences, we heuristically selected seven features which seemed to be relevant to our design objective. The representative moving vector appearing in the following explaination represents the average moving vector of each 50 moving vector contained in one image. These features are briefly described in the following.

1. The length of the representative moving vector,

$$\mathbf{x}_{1,i} = \sqrt{avx_i{}^2 + avy_i{}^2},$$

where i = field number, avx_i = x component of the ith field's average moving vector, avy_i = y component of the ith field's average moving vector.

2. The angle of the representative moving vector,

$$\mathbf{x}_{2,i} = tan^{-1}\frac{avy_i}{avx_i}.$$

3. The variation of the length in the space,

$$\mathbf{x}_{3,i} = \sqrt{\frac{\sum_{j=1}^{50}(length_{i,j} - avelength_i)^2}{n-1}},$$

where j = vector number in the same image, $length_{i,j}$ = length of the ith field's jth moving vector , $avelength_i$ = length of the ith field's average moving vector.

4. The variation of the angle in the space,

$$\mathbf{x}_{4,i} = \sqrt{\frac{\sum_{j=1}^{50}(angle_{i,j} - aveangle_i)^2}{n-1}},$$

where $angle_{i,j}$ = angle of the ith field's jth moving vector, $aveangle_i$ = angle of the ith field's average moving vector.

5. The variation of the length in time,

$$\mathbf{x}_{5,i} = \frac{\sum_{k=i-4}^{i}(avelength_k)}{5} - avelength_i,$$

6. The variation of the angle in time,

$$\mathbf{x}_{6,i} = \frac{\sum_{k=i-4}^{i}(aveangle_k)}{5} - aveangle_i.$$

Group	Operation	Field No.		Desired result
		training	testing	
scene 1	fix-shot	33-40	41-48	00
scene 1	panning	33-40	41-48	01
scene 2	fix-shot	33-40	41-48	00
scene 2	panning	33-40	41-48	01
scene 3	fix-shot	33-40	41-48	00
scene 3	tilting	33-40	41-48	10
scene 4	fix-shot	33-40	41-48	00
scene 4	tilting	33-40	41-48	10

Table 2 Training and testing patterns

7. The Fourier transform[3] of the length in the space,

$$\mathbf{x}_{7,i} = FFT(length_{i,j}).$$

8. The Fourier transform of the angle in the space,

$$\mathbf{x}_{8,i} = \mathbf{FFT}(\mathbf{angle}_{i,j}).$$

The ordering of the features and desirable combination of features which offer high performance of judging the camcorder operations shall be discussed later.

2.3 Neural Computing Approach

Neural network is a very powerful tool to solve the pattern recognition problems of this type . Since there are so many good references for neural network, for example [3], [4], we will not try to summarize the whole theory here. Rather, we just provide the brief explanaton of our network structure and learning.

The neural network presented here uses fully-connected three layer (input, output and hidden layers) feed-forward structure. The back-propagation error learning algorithm (5) was used in order to judge the camcorder operations. The sigmoid activation function was applied in this network. Figure 4 shows the structure used in this study.

Back-propagation error learning algorithm is a supervised learning algorithm. We need to provide a set of training patterns, testing patterns and desired results. The data selected for the training and testing are shown in Table 2.

After extensive searches, the software package of "O'inca" has been chosen for building, training and simulating this neural network. "O'inca" is a software tool for the design and

[3]The "MATLAB" FFT function is used for calculating this feature [2].

Group	Operation	Field No.	Desired result
scene 1	fix-shot	49-56	00
scene 1	panning	49-56	01
scene 2	fix-shot	49-56	00
scene 2	panning	49-56	01
scene 3	fix-shot	49-56	00
scene 3	tilting	49-56	10
scene 4	fix-shot	49-56	00
scene 4	tilting	49-56	10

Table 3 Evaluation patterns

implementation of intelligent systems which allows the integration of fuzzy logic, neural networks and user-defined modules in one framework [6].

2.4 Evaluation Of The Proposed Technique

Some feature combinations including up to 4 features among the seven features mentioned above are carefully selected for this study. Due to the computational complexity in practical use, the cases of more than 5 features were not considered. The learning results were evaluated by the moving vector groups tabulated in Table 3.

In order to clarify the propriety of each simulation results, the following evaluation functions were created in this study.

1. The average error between the the simulation result and the desired result (AE):

$$AE = \frac{\sum_{i=1}^{64}(desired\ result_i - simulation\ result_i)}{total\ number\ of\ the\ simulation\ data(=64)},$$

where i = Evaluation pattern number.

2. The satisfying ration(SR):

$$SR = \frac{total\ number\ of\ the\ correct\ results}{total\ number\ of\ the\ simulation\ data(=64)}.$$

Combination features	Average error	Satisfying ratio %
$\mathbf{x}_1, \mathbf{x}_2$	0.09666	90.63
$\mathbf{x}_2, \mathbf{x}_7$	0.09217	95.31

Table 4 Converged 2-feature case

Combination features	Average error	Satisfying ratio %
$\mathbf{x}_1, \mathbf{x}_2, \mathbf{x}_7$	0.1293	93.75
$\mathbf{x}_1, \mathbf{x}_2, \mathbf{x}_3$	0.1085	90.63
$\mathbf{x}_1, \mathbf{x}_2, \mathbf{x}_4$	0.1237	90.63
$\mathbf{x}_1, \mathbf{x}_2, \mathbf{x}_5$	0.08924	96.88
$\mathbf{x}_1, \mathbf{x}_2, \mathbf{x}_6$	0.1116	90.63
$\mathbf{x}_1, \mathbf{x}_2, \mathbf{x}_8$	0.1347	92.19
$\mathbf{x}_2, \mathbf{x}_7, \mathbf{x}_3$	0.1585	95.31
$\mathbf{x}_2, \mathbf{x}_7, \mathbf{x}_4$	0.1407	95.31
$\mathbf{x}_2, \mathbf{x}_7, \mathbf{x}_5$	0.1732	93.75
$\mathbf{x}_2, \mathbf{x}_7, \mathbf{x}_6$	0.1412	95.31
$\mathbf{x}_2, \mathbf{x}_7, \mathbf{x}_8$	0.1325	95.31

Table 5 Converged 3-feature case

3 EXPERIMENTAL RESULTS

The expression of the feature symbol in the following sections will be cut its field number part. For instance, $\mathbf{x}_{1,i}$ is going to be abbreviated to $\mathbf{x}_1$.

1. With 1 feature. The neural network with one input feature about all seven features were trained, however none of them converged in this case.

2. With 2 features. The neural network with all 2-feature combinations (28 patterns) were trained. Only 2 combinations were able to converge. The rest of them did not converge. The results of the 2 converged case are listed in Table 4.

3. With 3 features. In the 3-feature neural network, we tried to eliminate the combinations which have no possibility of convergence based upon the 2-feature results. From this point of view, 11 combinations were picked up based on the combinations of $\mathbf{x}_1, \mathbf{x}_2$ and $\mathbf{x}_2, \mathbf{x}_7$ shown in Table 5. All 11 combinations have converged in this experiment.

Combination features	Average error	Satisfying ratio %
$\mathbf{x}_1, \mathbf{x}_2, \mathbf{x}_7, \mathbf{x}_3$	0.1753	90.63
$\mathbf{x}_1, \mathbf{x}_2, \mathbf{x}_7, \mathbf{x}_4$	0.1663	89.06
$\mathbf{x}_1, \mathbf{x}_2, \mathbf{x}_7, \mathbf{x}_5$	0.1844	89.06
$\mathbf{x}_1, \mathbf{x}_2, \mathbf{x}_7, \mathbf{x}_6$	0.1701	89.06
$\mathbf{x}_1, \mathbf{x}_2, \mathbf{x}_7, \mathbf{x}_8$	0.1707	87.50

Table 6 Converged 4-feature case

4. With 4 features. In the 4-feaure combination, only 5 combinations which are most likely to converge in the same sence as that of the 3-feature case were selected. In the result, all 5 combinations have converged. These results are listed in Table 6 for comparison.

Figure 5 indicates how the results of all converged case look like in terms of the AE and SR.

3.1 Discussion

According to our results presented in the previous section, the best feature combination for judging the camcorder operations turns out to be the length, angle of the average moving vector and the time variation of the moving vector length. This best results marked 96.88% in the satisfying ratio which would be sufficient in practical use.

This result is quite interesting since the best result is not the 4-feature case. It means that the number of the features is not as important as the combination of the features so far as the identifying the camcorder operations is concerned.

Figure 5 reveals this fact even clearer and in a more subtle way. We can see that there is no relation between the number of the features used and the accuracy of the operation judgment. This should be very big advantage for a practical use, because if the number of the features is proportion to the accuracy, the computational comlexicity will increase in a more dramatical manner.

From the feature's point of view, in addition to the best possible combination, feature $\mathbf{x}_2$ (angle of the average moving vector) and $\mathbf{x}_7$ (the Fourier transform of the length in the space) combination also marked high accuracy in Figure 5. The common fact between these observations is that angle of the moving vector indeed plays very important role for this problem. On the other hand, this result should not be surprising, because the biggest difference between fix-shot, panning and tilting, indeed, is the direction of the moving vector.

4 CONCLUSION

In this study, the neural computing approach is applied to the judgment of the camcorder operations which include fix-shot, panning and tilting. The data used in this study is the moving vectors obtained from the real camcorder image. Also seven features are carefully selected as possible inputs to this neural network.

As a result of our investigation, 96.88% accuracy of the operation judgment is achieved by using the length, angle of the average moving vector and the time variation of the moving vector length as a set of three features. These results indicate that the approach proposed here can indeed classify the three classes in relatively high performance for the camcorder operations. Since it requires only three features in each image to achieve this accuracy, it should be easy to justify its utilization in controlling the automatic functions in the camcorder.

REFERENCES

[1] . Sekine, T. Kondou, and H. Hrose, "Motion vector detecting system for video image stabilizers," *Proceedings of IEEE 1994 International Conference on Consumer Electronics (Digest of Technical Papers)*, pp. 268-269, 1994.

[2] atlab User Guide, the MATH WORKS inc., Natick, Massachusetts, 1992.

[3] . Minsky and S. Papert, *Perceptrons: An Introduction to Computational Geometry*, MIT Press, Cambridge, Mass., 1969.

[4] .J. Hopfield, "Neural networks and physical systems with emergent collective computational abilities," *Proceedings of Natl. Sci. USA, Vol. 323, pp. 533-536, 1986.*

[5] .E. Rumelhart, G.E. Hinton, and R.J. Williams, "Learning representations by back-propagating errors," Nature, Vol. 323, pp. 533-536, 1986.

[6] 'inca User Manual, Intelligent Machines, Inc., Sunnyvale, California, 1994.

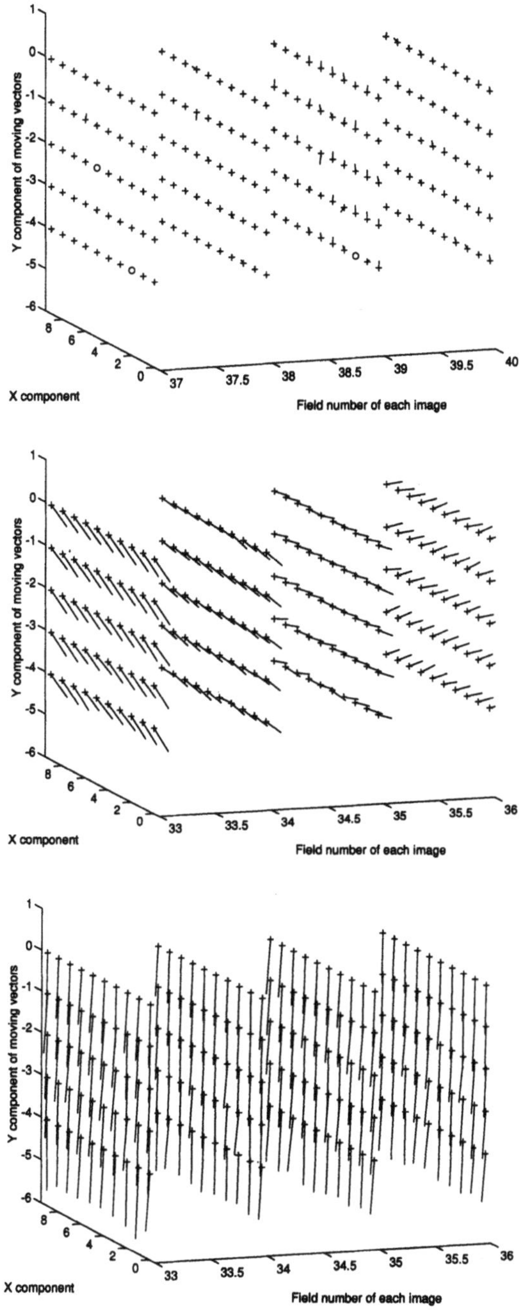

Figure 3 The moving vector distributions (a) Fix-shot; (b) Panning; (c) Tilting.

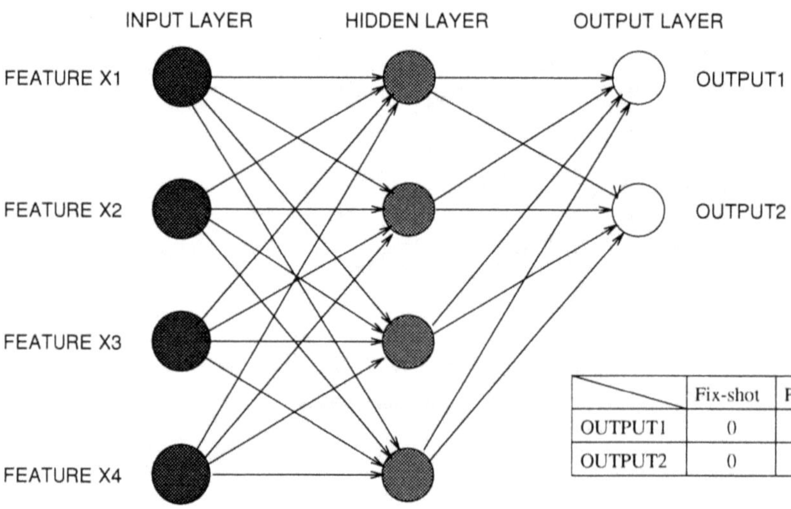

	Fix-shot	Panning	Tilting
OUTPUT1	0	0	1
OUTPUT2	0	1	0

Figure 4 The neural network structure for 4-feature case.

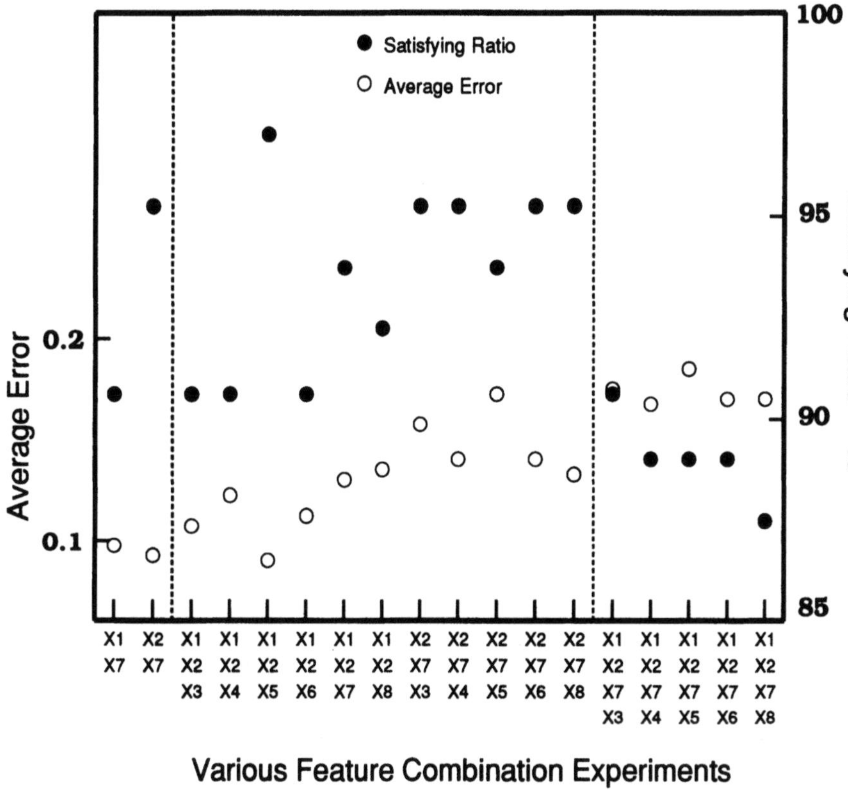

Figure 5 Integration of the results

15

MODEL REDUCTION AND CONTROL OF MULTISTAGE FLASH (MSF) DESALINIZATION PLANTS

Srinivas Ramamurthy, Jayanta Pal, Ak Sinha, Darwish Al Gobaisi †, Ganti Rao

Department of Electrical Engineering
India Institute of Technology
Kharagpur 721302, India
† Water and Electricity Department
Government of Abu Dhabi, UAE

ABSTRACT

Multistage flash (MSF) desalinization plants are a major means of desalting seawater for human use in several arid regions of the world in the present times. The MSF plants are physically large and their control usually involves more than twenty control loops. According to the present practice, the controllers are of the PI or PID type and their tuning is largely based on experience rather than on systematic modeling of the plant. Plant modeling based on physical principles gives rise to a large and complex set of coupled nonlinear differential equations which has to be linearized about a chosen set of operating conditions. As the operating point changes, the resulting linearized model also changes. This requires retuning of the controllers in certain loops depending on the changing linear plant model in the related loops. The linearized model happens to be enormously large in size requiring reduction for controller design and practical implementation. There exist several model reduction methods and they have to be chosen to meet the objectives of adequate modeling. In a nonlinear plant, the linearized model parameters vary with the operating conditions. To make a controller to simultaneously meet the demands of model reduction and variable operating conditions, the conventional approaches of control are either inadequate or too involved. In this chapter, a technique based on artificial neural networks (ANN) for model reduction under plant parameter perturbations is proposed. The complexity of analysis, reduction, computation and controller design in the variable conditions of operation is avoided by simply training an ANN to give the parameters of a reduced model for use in controller design. An automated decision support may be provided to choose the best ANN configuration for reduced order modeling of a large, complex and variable plant which provides the basis for a robust design of a simple controller. Certain well established model reduction methods are employed in the mainstream of the procedure and the related results are impressive. Based on these results, a scheme based on an added ANN, for direct controller implementation under the discussed conditions is proposed. The results of this modest attempt point out to the strong possibility of more intelligent control of large complex plants under uncertainty and/or variable plant dynamics. The present discussion is centred on SISO loop designs, which does not rule out the possibility of simple extensions to MIMO designs.

1 INTRODUCTION

Multistage flash desalinization plants are used to produce potable water from the sea in many parts of the world today. Some of these plants have a capacity of production of several millions of gallons of water per day. The world's largest unit of 10-12 millions of gallons per day (MGD) is coming up in the Abu Dhabi, UAE area. These plants are energy expensive. One source of energy is steam from nearby electrical power plants and electricity from the power plants is also drawn to drive the large number of pumps and control systems. Fig. 1 illustrates the principle MSF desalinization. The brine is heated by LP steam up to a temperature, called the top brine temperature (TBT), in a brine heater unit and led into a cascade of large chambers where the brine flashes to produce water vapour along with certain noncondensables. The vapour rises to the top of the flashing chambers through demisters, giving away heat to the incoming brine. The vapour condenses on the tube bundles carrying the brine towards the brine heater. The distillate is collected in trays below the tube bundles and is sent to the storage tanks from where it is distributed to the consumers. In order to operate the MSF plant stably, efficiently and according to the distillate demand, several variables of the MSF process have to be controlled. Fig. 1 does not show any of these control systems but Fig. 2 does. One finds as many as 20 control loops in a typical MSF plant whose schematic is shown in Fig. 2.

2 MSF CONTROL SYSTEMS

The following is a typical list of controlled variables in a modern MSF plant.

Brine heater section

1. Top brine temperature (TBT)
2. Temperature of low pressure (LP) steam
3. Pressure of LP steam
4. Level of brine heater condensate
5. Conductivity of brine heater condensate

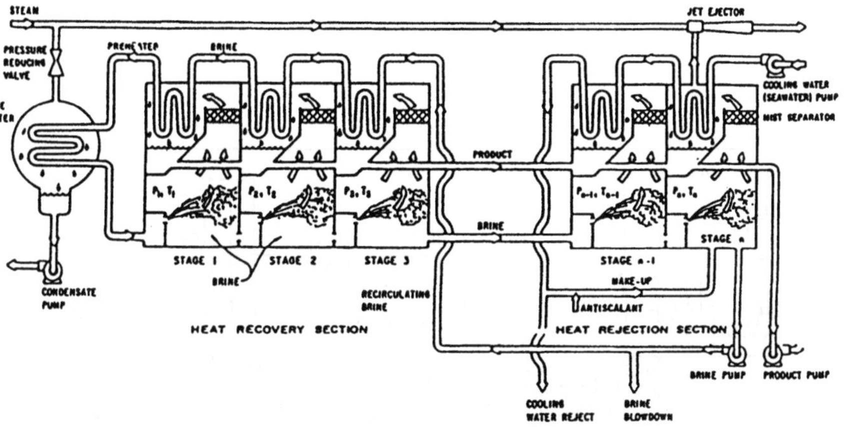

Figure 1 A typical flow schematic of brine recirculation MSF plant (courtesy of Sasakura).

Condenser section (recirculation and make up flow)

6. Flow rate of brine recirculation

7. Make-up flow rate

8. Antiscale dosing (or antiscale/make-up ratio)

9. Sodium sulphite injection into brine recirculation stream

Evaporation section

10. Brine level in the last stage

11. Distillate level in the last stage

12. Flow rate of blowdown

13. Conductivity of distillate

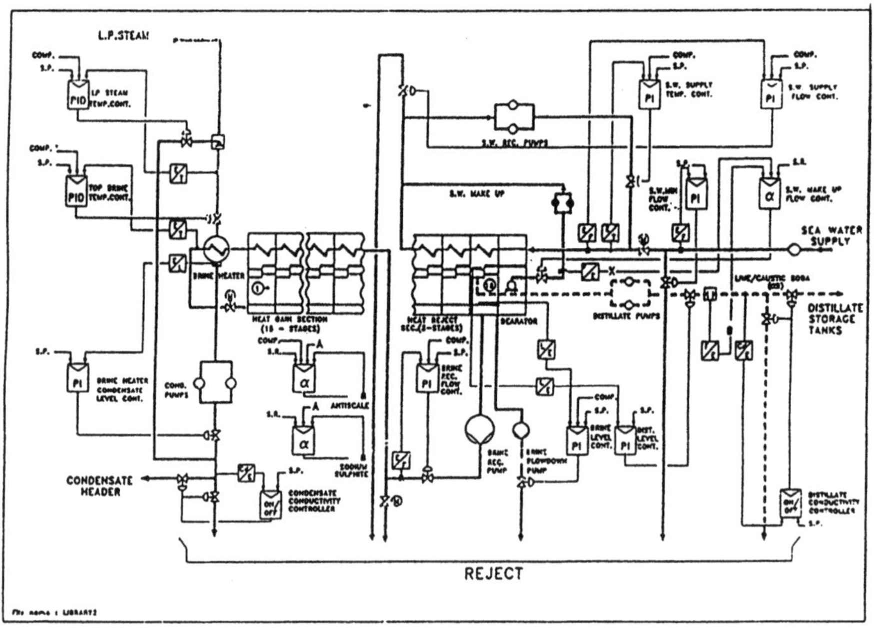

Figure 2 Different control loops in MSF desalinization plant.

14. Chloride injection into distillate

15. pH-value of output distillate (lime/caustic soda injection into distillate)

Cooling section

16. Flow of seawater to heat reject section

17. Inlet temperature of cooling water

18. Minimum flow of seawater

Ejector and venting section

19. Level of condensate in ejector condenser

20. Conductivity of ejector condensate.

5, 13 and 20 are on-off type controllers. Al-Gobaisi et al [1] discussed the action of these control systems and suggested improvements to the conventional schemes.

The present day practice is to operate these control systems with PI or PID type controllers which are designed by thumb rules and the actual tuning is often manual and based on operator experience. There has been little research and development effort towards improving the control systems in MSF plants by transfer of technology from the enormously advanced field of automatic control. This may be attributed mainly to the fact that the product of these plants is not a commercial commodity aimed at profit making. Water production is the responsibility and commitment of a municipality or a government. Today the scenario is changing. R&D efforts in desalinization are finding growing support. The International Foundation For Water Science and Technology (IFFWSAT) at Abu Dhabi is supporting R&D efforts in educational institutions abroad, in addition to its noble pursuit at home where extensive facilities for practical tests on MSF plants are also possible.

3 MSF PLANT MODELLING

Starting from physical principles, the various elements and subsystems in an MSF process can be model. When the connections among these are provided, the resulting model represents the overall system. Such models involve energy and mass balance conditions. In a commercially available flowsheet simulator with a user friendly interface, the various units can be described by appropriate equations, correlations and interconnective relations. The overall system of equations is usually based on lumped parameter descriptions although the process has distributed parameter type of phenomena. However, in places where better approximations are desired, as in the brine heater section, the unit can be segmented and each segment can then be described by a lumped model. The resulting, usually nonlinear, system of ordinary differential equations is solved in the simulators by well established numerical routines. At first steady-state simulations provide detailed sets of operating conditions. At any of such operating conditions, the nonlinear dynamic model can be linearized to get the standard linear state space form of dynamic model. Commercial flowsheet simulators have the feature of such a description. In addition to choosing a set of operating conditions, it will be necessary to provide a chosen set of manipulated and controlled variables (inputs and outputs). Then the state space description in the linearized form becomes available for further analysis and control design. Under certain conditions, the related steady-state transfer function matrix can also be obtained from the commercial simulators. To give an idea of the authors' experience with a simulation of a plant actually operating in Abu Dhabi, the following broad picture would suffice:

1. Brine heater is segmented into 10 sections.
2. There are 18 flash stages. Out of these, the first 15 are heat recovery stages and the last 3 are heat rejection stages.
3. Number of state variables = 155.
4. Number of inputs = Number of outputs = 8 (chosen for study).

Each element of the transfer function matrix happens to be a ratio of two very high order polynomials with some possible pole-zero cancellations. However, the problem of high dimension still persists and for simple controller design in the PI or PID form, these transfer functions are unwieldy. Experimental black box modeling in the standard forms suitable for PID tuning is an altogether different approach with its attendant complexity.

It behaves us at this stage to consider model reduction as an important aspect of study. A special feature of the dynamic model obtained from flowsheet simulations deserves special mention here. The controller variable TBT (Top brine temperature) has a nonlinear relationship with the manipulated variables. This was observed in the model simulated at 60%, 70%, 80%, 90% and 100% of load which is the distillate output flowrate. The dynamic model significantly varies in this range of operating conditions. If model reduction has to be applied in the conventional way, the exercise has to be repeated at each operating condition. This is rather practically unattractive proposal. In view of this complexity and large size of this problem we discuss here an Artificial Neural Network (ANN) based method of reduced order modeling for possible use in such situations in general.

4 REDUCED ORDER MODELLING

Mathematical models of physical dynamic processes are often formed either by using physical laws and cause-effect procedures, or by identification algorithms in state-space or curve-fitting the experimentally obtained frequency/time responses. In practical MSF processes, the resultant dynamic models consist of a very large number of nonlinear differential and algebraic equations. After linearization, often the orders of such models are so large as to be inconvenient or impractical for many purposes, including simulation, control system design and implementation. Consequently, a real need exists for systematic procedures for deriving reduced order models (ROM) from high order systems.

Many methods for model order reduction have been proposed during the last few decades. Representative methods in this area are eigenvalue preservation [2], error minimization in the time domain [3,4] or frequency domain [5], singular perturbation [6], balanced realizations [7] Padé approximations [8,9], H^∞ method [10], continued fractions [11,12], time-moments matching [13,14], Routh approximation [15,16], root-clustering technique [17] etc. These methods differ in mathematical complexity, computational requirement, programming effort, quality of model and general applicability. In practical situations, however, a nominal high order system may undergo changes in its parameters depending on the operating point or because of aging effects and other environmental factors. In such a case a reduced order modeling technique has to be repeatedly applied for arriving at the parameters of the family of reduced order models to account for all such variations in the original high order system. This may not be practically feasible proposition for complex systems with many inputs, outputs and interactions. It is in this context that an artificial neural network (ANN) based model reduction/identification technique is proposed that will lead to a reduced order model which is expected to account for the perturbations of the nominal high order system within the range presented in the training data. In the sequel, we suggest an extension of this ANN based reduced order modeling approach for combined modeling and robust control of high order complex plants under parameter variations.

5 NEURAL NETWORK MODEL

ANNs consist of highly interconnected simple processing elements called neurons. A block diagram of a typical neuron is given in Fig. 3.

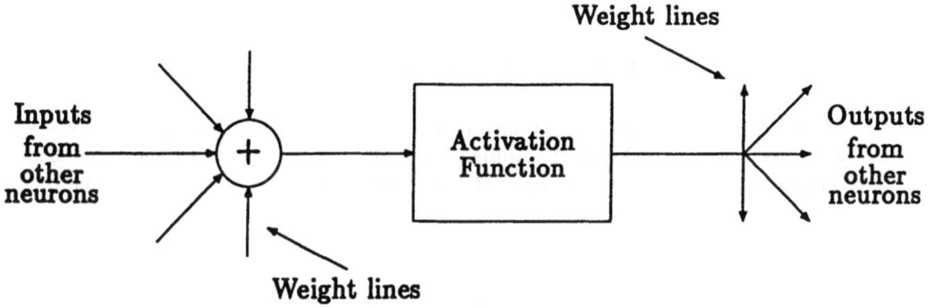

Figure 3 Block diagram of a neuron.

Each neuron consists of a summing junction, which adds together the weighted inputs from the other neurons, and an activation function, which generates the neuron output from the summing junction output. The output from the neuron is directed as inputs to other neurons. neurons transmit signals to each other via weighted links, which attenuate or amplify the transmitted signal depending on the value of the weight.

ANNs can be grouped into various classes depending on their feedback link connection structure or various architectures. In this work we use feedforward neural network (FNN) with backpropagation for training. The FNN consists of layers of neurons with weighted links connecting the outputs of neurons in one layer to the inputs of neurons in the next layer. One or more layers exist between the input layer and the output layer. These layers are called the hidden layers. A typical feedforward network is shown in Fig. 4. A three-layered network is shown, but in principle there could be more than one hidden layer .

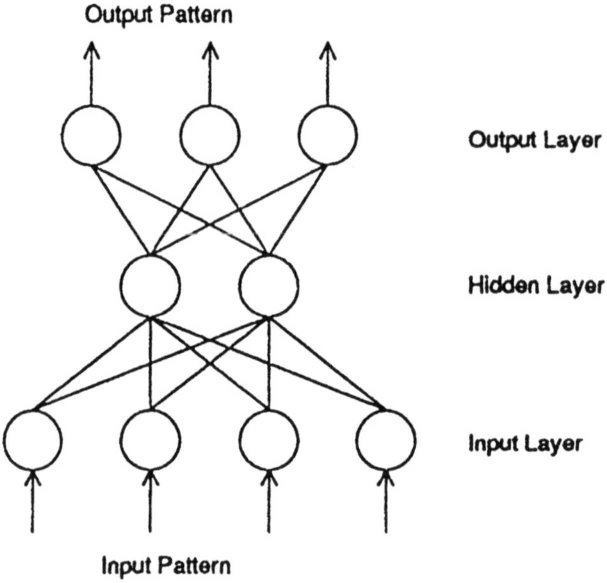

Figure 4 Multilayer feedforward neural network.

The output of units in layer 1 are multiplied by appropriate weights w_{ij} and these are fed as inputs to the next layer, the hidden layer. If i is the input layer then the output of a unit O_i will be equal to the input of that particular unit i.e., $O_1 = X_1$. The total input to a unit in layer j is

$$Net_j = \sum_i w_{ij} O_j, \tag{15.1}$$

and the output of a unit in layer j is

$$O_i = f(Net_j), \tag{15.2}$$

where f is an activation function. A convenient logistic activation function is given by the relation

$$V_j = f(Net_j) =$$

The function f yields an output that varies continuously from 0 to 1. The quantity Φ_j serves as a "threshold" and positions the transition region of the f function. The quantity Φ_0 denotes the abruptness of this transition. An example of such a function is shown in Fig. 5.

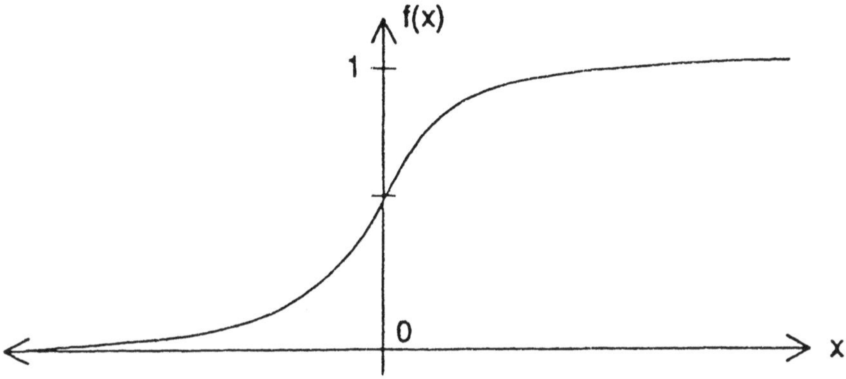

Figure 5 Sample sigmoidal function.

In learning the hidden representation (i.e. the weights and the threshold values), the network functions solely in a feedforward manner as shown by eqns (15.1) and (15.2). In the learning process, the network is fed with two sets of patterns, an input pattern and a corresponding output pattern. Using the (initialized) weights and threshold values, the network produces its own output pattern that is then compared with the desired (target) output pattern. The error at any output unit in layer k is

$$e_k = t_k - O_k, \tag{15.3}$$

where t_k is the desired output for that unit in layer k and O_k is the actual output. The total error function may be written as

$$E = \frac{1}{2}\sum (t_k - O_k)^2. \tag{15.4}$$

Learning comprises changing the weights and thresholds so as to minimize the error function in a gradient descent manner. The analytical continuous nature of the activation function allows errors to be traced backwards. For the activation function of eqn (15.3), the 'delta rule' of iterative convergence towards improved values for the weights and the threshold may be stated as

$$\Delta w_{kj} = \eta d_k O_j, \tag{15.5}$$

where the error signal d_k at an output unit k is given by

$$d_k = (t_k - O_k)O_k(1 - O_k), \tag{15.6}$$

and the error signal d_j for an arbitrarily hidden u_j is given by

$$d_j = O_j(1 - O_j)\sum_k d_k w_{kj}. \tag{15.7}$$

In (15.6) η is called the learning rate parameter. In practice, it has been found that one way to increase the learning rate without causing oscillations is to modify (15.6) to include a momentum term α, that is

$$\Delta w_{kj}(n+1) = \eta d_k O_j + \alpha \Delta w_{kj}(n), \tag{15.8}$$

where n is the number of times for which a set of input patterns have been presented to the network, α is a constant which relates how the past weights change to the present ones.

6 MODEL REDUCTION

In this section, we describe an ANN based method for directly obtaining the parameters of a ROM from the input-output information of a high-order SISO system. The dynamics of the high order system are assumed to undergo changes due to variations in the operating conditions. The method uses ANN with back-propagation learning. The main advantage of ANN is that it does not require any explicit mathematical description relating the input and output quantities. Various ANN based approaches have been proposed in the area of identification and control [18,19,20,21]. A state-space model- based neural-net identifier has been described by Pham and Liu [18]. A Hopfield network has been used by Chu et al [19] to identify the parameters of the mathematical model of a linear time-varying or time-invariant system. In reference [20] a series-parallel model is used for an ANN to identify system dynamics.

In the proposed approach, we start with the identified high order models of a plant at several key operating points. It is assumed that these high order models are available and have been obtained by using existing identification algorithms or experimental procedures. These nominal high order models are used for training an ANN called the ANN identifier. It is desired that the ANN identifier directly yields the parameters of the ROM when the outputs from the plant under varying operating conditions are used as its inputs. A two-step procedure is adopted for training the ANN identifier.

In the first step a high order SISO plant transfer function model S is identified at one operating point. All the coefficients of the transfer function S are randomly perturbed within ±20% of the nominal values to obtain the family of perturbed systems $S_1, S_2, \ldots, S_n$. As shown in Fig. 6, a step input is applied to the plant S_i and its output response is obtained. Discrete Fourier Transform is applied to the output response to obtain the Fourier coefficients. The significant Fourier components DFC_i for the system S_i are used as inputs for training the ANN.

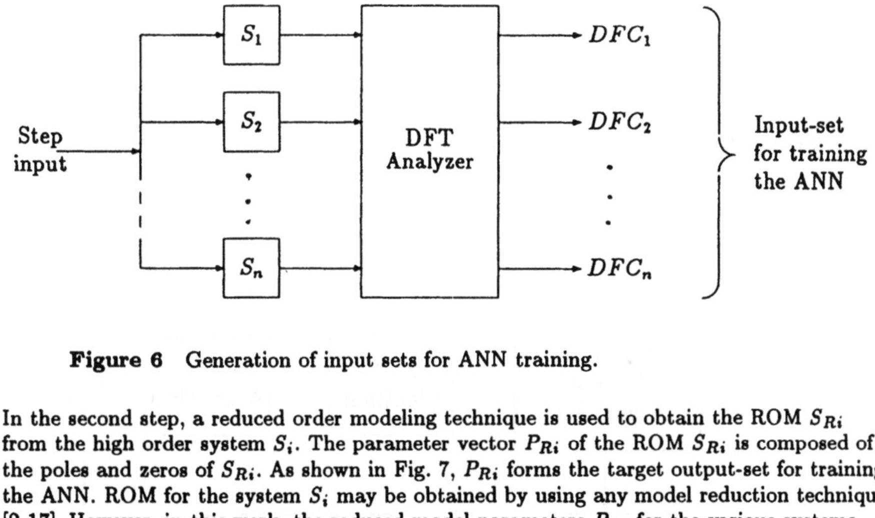

Figure 6 Generation of input sets for ANN training.

In the second step, a reduced order modeling technique is used to obtain the ROM S_{Ri} from the high order system S_i. The parameter vector P_{Ri} of the ROM S_{Ri} is composed of the poles and zeros of S_{Ri}. As shown in Fig. 7, P_{Ri} forms the target output-set for training the ANN. ROM for the system S_i may be obtained by using any model reduction technique [2-17]. However, in this work, the reduced model parameters P_{Ri} for the various systems were obtained by using the clustering technique [17]. The basic modeling scheme is illustrated by the block diagram of Fig. 8.

It may be noted that the ROM parameters P_{Ri} are obtained from S_i by using standard order reduction methods. The ANN is however trained with the significant DFT coefficients DFC_i as the input-set and the corresponding ROM parameters P_{Ri} as the target output-set. Thus, in effect, after proper training, the net learns and stores the mappings between the DFC_i and the P_{Ri}. Since the net is trained for the family of plant transfer functions within the range of (say) ±20% perturbations, the ANN identifier will act as an effective plant emulator that takes such plant perturbations into consideration.

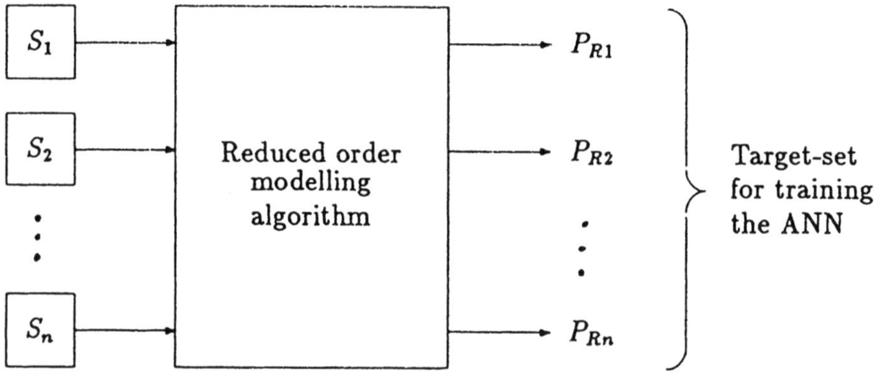

Figure 7 Generation of output sets for ANN training.

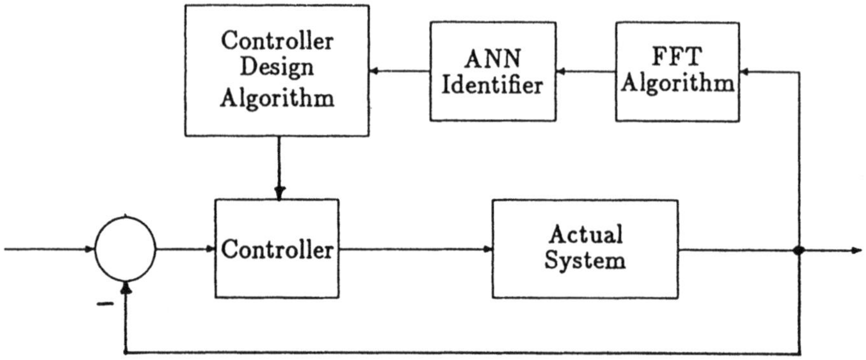

Figure 8 Proposed ROM identification scheme.

6.1 Generation of Input and Output Data-sets for ANN Training

ANN learns the complex functional relationships between DFC_i and P_{Ri} from the input-output data-sets presented during the learning phase. Since the mapping achieved by neural networks are approximations, errors are to be expected. Once the ANN has been trained, it can be used for generalization, i.e. obtaining outputs given the inputs for which the ANN was not trained. However, a good performance cannot be expected of the ANN for DFC_i values outside the range presented in the training data- sets. For the ANN to yield good interpolations, a sufficiently large number of input-output pairs must be available during the training phase.

We have chosen three different transfer functions for the purpose of training the ANN. These systems have different numbers of finite poles and zeros and show different (fast, non-minimum phase, oscillatory) responses when excited by a step input. These are typical responses that are commonly found in the process industries and have been chosen to test and illustrate the viability of the reduced order modeling technique.

Details of the System Studied

System number 1: We consider the SISO system $G(s) = \frac{N(s)}{D(s)}$ where

$$\begin{aligned} N(s) &= 37500(s + 0.08333), \\ D(s) &= s^7 + 83.64s^6 + 4097s^5 + 70342s^4 + 853703s^3 \\ &\quad + 2814271s^2 + 3310875s + 281250 \\ &= \left(s + 9.193238 \times 10^{-2}\right)(s + 2.024395 \pm j0.964649) \\ &\quad (s + 7.674372 \pm j13.44615)(s + 32.07526 \pm j38.84927). \end{aligned}$$

This is a seventh order system with a pole-zero excess of six. It shows a very fast response to a unit step input. A ROM of order three having one finite zero is found to be adequate.

System number 2: We consider the system $G(s) = \frac{N(s)}{D(s)}$ where

$$\begin{aligned} N(s) &= (s-1)(s+4)(s+7.5)(s+12)(s+15)(s+25), \\ D(s) &= (s+1+j2)(s+1-j2)(s+2)(s+7)(s+10)(s+10)(s+12.5) \\ &\quad (s+18)(s+20)(s+30). \end{aligned}$$

This is a non-minimum phase system having a pair of repeated poles. A ROM having three zeros and four poles is chosen in this case.

System number 3: We consider the system $G(s) = \frac{N(s)}{D(s)}$, where

$$\begin{aligned} N(s) &= 19.82s^7 + 429.26156s^6 + 4843.8098s^5 + 45575.892s^4 \\ &\quad + 241544.75s^3 + 905812.05s^2 + 1890443.1s + 842597.95 \\ &= 19.82(s + 0.5882)\left(s^2 + 1.968s + 74.756\right)\left(s^2 + 3.459s + 22.29\right) \\ &\quad (s + 3.587)(s + 12.055), \\ D(s) &= s^8 + 30.41s^7 + 358.4295s^6 + 2913.8638s^5 + 18110.567s^4 \\ &\quad + 67556.983s^3 + 173383.58s^2 + 149172.19s + 37752.826 \end{aligned}$$

$$= \left(s^2 + 0.7s + 46.3625\right)(s + 0.46)(s + 0.75)\left(s^2 + 4.4s + 17.8\right)$$
$$(s + 8.5)(s + 15.6).$$

This eighth order system shows a typical "ringing" type oscillation in the step response. A fourth order ROM is found to reproduce this oscillatory behavior of the original system.

Generation of Input Data-sets

The coefficients of the s terms in the numerator and denominator of the above three nominal transfer functions determine the position of its poles and zeros. To create a family of transfer functions from the nominal one, each numerator and the denominator coefficient is randomly perturbed by $\pm 20\%$ of its nominal value. These transfer functions are found to show similar attributes in the step or impulse response behaviors. These three families of transfer functions (called perturbed system number 1 through 3, henceforth) are used to train the ANN for the particular nominal system under consideration.

Given the step response of a SISO stable system, the proposed method identifies a reduced order transfer function model by using the back propagation ANN. The step response of the original high order system is found and sampled in the interval 0 to 10s with a sampling interval of 0.02s. The time horizon of 10s is chosen as all the systems considered settled down to the final steady-state value within that period. This gives 500 samples for each system. For obtaining the discrete Fourier transform coefficients, twelve additional samples are generated at the end using the final steady-state value at $t = 10s$. The discrete Fourier transforms of theses 512 samples are found using the FFT algorithm as:

$$X(k) = \sum_{0}^{511} x(n)\exp(-j(2\pi/N)k_n), \quad k = 0, 1, \ldots, 511,$$

where $X(k)$ is the kth Fourier component and $x(n)$ is the sampled value of the step response at $t = (n \times 0.02)s$.

Three different networks ANN1, ANN2 and ANN3 were trained for identifying the three different SISO systems described in section 6.1. The networks were trained with the significant (complex) Fourier coefficients $X(k)$ obtained from the sampled time response of the original high order system as inputs. For testing purpose, both impulse and step responses were used. Twenty Fourier components were used in the case of impulse input while seven Fourier components were used for the step input. After exhaustive experimentations, it was found that the networks performed better with step inputs.

Generation of output data sets

For the perturbed system S_i the corresponding ROM S_{Ri} of specified order r may be obtained by any standard reduced order modeling technique (see section 6.4). In this work, the ROMs were obtained by using the clustering technique [17]. The poles and zeros of the ROM S_{Ri} formed the parameter vector P_{Ri} that was used as the target output set for training the ANN.

6.2 Practical Considerations

Three different SISO systems as detailed in section 6.1 were used for training three different networks ANN1, ANN2 and ANN3. The desired outputs of the ANN are the s-domain poles and zeros of the ROM, while the inputs are the DFT coefficients. For each system, 25 perturbed systems and their corresponding ROMs and DFCs were found; 15 pairs of input-output data were used for training while the rest 10 were used for testing.

Normalization of Data

It was generally found that the ANN learns faster if all the inputs and outputs are normalized so that their values lie within a range of 0 to 1 and are as close as possible to 0.5. For this purpose, it was necessary to normalize all the input-output data so that they lie within the 0-1 range.

All the complex-valued Fourier coefficients $X(k)$ are normalized before being fed to the ANN as inputs. The normalized $X(k)$ is given by

$$[X(k)]_{NORM} = \frac{Re\,[x(k)] + |x(0)|}{2|x(0)|} + j\frac{Im\,[x(k)] + |x(0)|}{2|x(0)|}$$

In the case of step response, the first seven Fourier coefficients were found to be significant and were used as inputs to the network.

Normalization of the output poles and zeros of the ROM was done over an interval 50-100 using the relation:

$$P_{NORM} = \frac{P + 50}{100}$$

where P =pole or zero, and P_{NORM} =normalized pole or zero.

Data Management

The normalized complex Fourier components that are used as inputs are arranged so that the real part of each component is followed by the modulus of its imaginary part. Seven complex Fourier components were fed into the 14 input nodes of the three nets.

The normalized poles and zeros of the reduced models were used as the desired outputs for the ANN. The poles and zeros were fed in condensed from, i.e. only the real part of a real pole (zero) was fed and a pair of complex poles (zeros) were represented by two outputs viz. the real part and the absolute value of the imaginary part. The outputs were arranged in the following order: real zeros, real part of complex zeros, imaginary part of complex zeros, real poles, real part of complex poles and imaginary part of complex poles.

ANN Training and Testing

As mentioned in section 6.1, an adequate ROM for system number 1 has one real pole, one real zero and a pair of complex poles. Thus ANN1 has 4 outputs and 14 inputs (for the seven complex valued Fourier components). In the case of system number 2, the fourth order ROM has three real zeros, two real poles and a pair of complex poles. So the network ANN2 is designed for 7 outputs and 14 inputs. The ROM for system number 3 is of order 4 and has one real zero, a pair of complex zeros, two real poles and a pair of complex poles. The network ANN3 thus has 7 outputs and 14 points.

After testing various alternatives, we finally used one topology of multilayer perceptions for storing the maps. All the networks considered had one input layer, a hidden layer and one output layer. The number of neurons in the hidden layer for each network was initially chosen by the thumb rule

$$Number\ of\ neurons\ in\ the\ hidden\ layer = \sqrt{no.\ of\ inputs \times no.\ of\ outputs}$$

The nearest integer may be slightly adjusted for best results. In this study the number of neurons in the hidden layer was chosen as 6 as this gave the best results.

The error between the desired and actual output was squared and summed up for all the outputs over all the learning samples. This was then averaged to give the rms error for training. The same procedure was repeated for the testing samples. The iterations or complete cycles of pattern presentation for training each network was carried out till the error E_{rms} became less than the specified tolerance. E_{rms} is given by

$$E_{rms} = \left[\frac{1}{N_s O_n} \sum_{j=1}^{N_s} \sum_{k=1}^{O_n} (t_k - O_k)^2 \right]^{\frac{1}{2}},$$

where

$$O_n = \text{number of neurons in the output layer,}$$
$$N_s = \text{number of training samples.}$$

The following rule was used to get a dynamic variation of η (learning gain) and a α (momentum gain) parameters: if the rms error, E_{rms} failed to decrease after an iteration, then η and α values were divided by $\sqrt{2}$; provided η and α were greater than some minimum value, say 0.05 or 0.1. The initial values of η and α were chosen to get a rapid decrease in the rms error.

6.3 Simulation Results

Tables 1 to 3 and Figs. 9-11 present the results for the system number 1 to number 3 that correspond to the networks ANN1 to ANN3 respectively. From Figs. 9-11 it is found that

Inputs	Desired outputs	Calculated outputs	Percentage error
56.59352	-0.02983	-0.0298	-0.08716
0.00000	-0.03395	-0.03327	-2.02627
-5.16161	-1.63609	-1.67504	2.38065
1.17932	0.99624	0.97026	-2.60787

Table 1

Inputs	Desired outputs	Calculated outputs	Percentage error
411.69809	0.64971	0.73549	13.20258
0.00000	-3.9723	-3.96864	-0.09209
-93.17631	-11.79273	-11.41063	-3.24014
38.53078	-3.84597	-3.58595	-6.76076
-67.46623	-15.43156	-14.20983	-7.91709
75.97988	-0.76864	-0.7534	-1.98228
-12.25065	1.51729	1.56315	3.02184

Table 2

the step responses of the identified reduced models in each case come close to the desired ones.

For system number 1, training was done using step response for 20% perturbed samples. The best result in training gives: Average error is 1.7754 and Rms error is 2.0360.

For system number 2, the training was done using step response for 20% perturbed samples. The best result in training gives average error equal to 5.1738, and Rms error equal to 6.6145.

For system number 3, the training was done using step response for 20% perturbed samples. The best result in training gives average error equal to 6.1327 and Rms error equal to 9.3026.

7 ROBUST CONTROLLER DESIGN

In section 6, we have illustrated a scheme for identifying the ROM parameters of a high order system by using the significant discrete Fourier components of the plant output signal. In this section, we discuss the potential applicability of ANNs to robust controller design. Using the transfer-function parameters P_{Ri} of the family of ROMs, S_{Ri}, standard robust

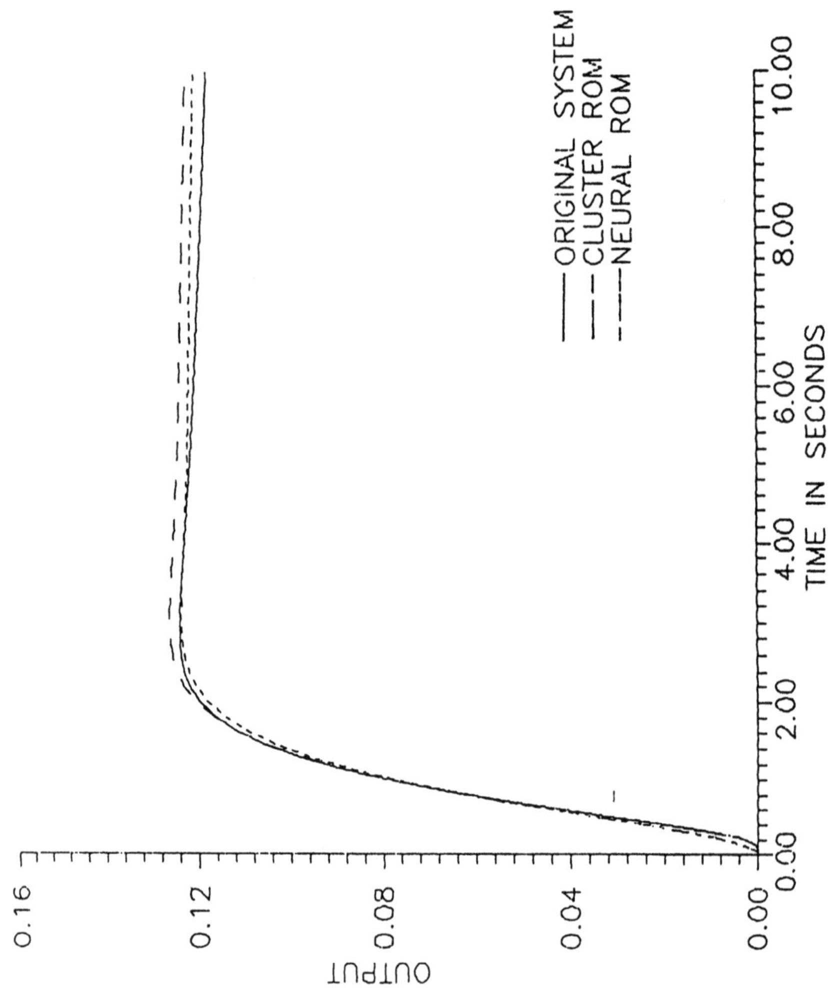

Figure 9 Step response comparisons for system number 1.

controller design procedures like LQG/LTR [22] or H_∞ [23] may be used to find a family of robust controllers with parameters P_{ci}. These parameters P_{ci} form the target output-sets while the corresponding parameters P_{Ri} may be used as input data-sets for training the neural controller. Since the parameters P_{ci} of a large number of robust controllers for the perturbed models S_{Ri} are used for training the network, the neural controller is also expected to be robust. Since, the mapping achieved by the neural controller will be approximations, enough stability margins must be ensured at the offline design stage.

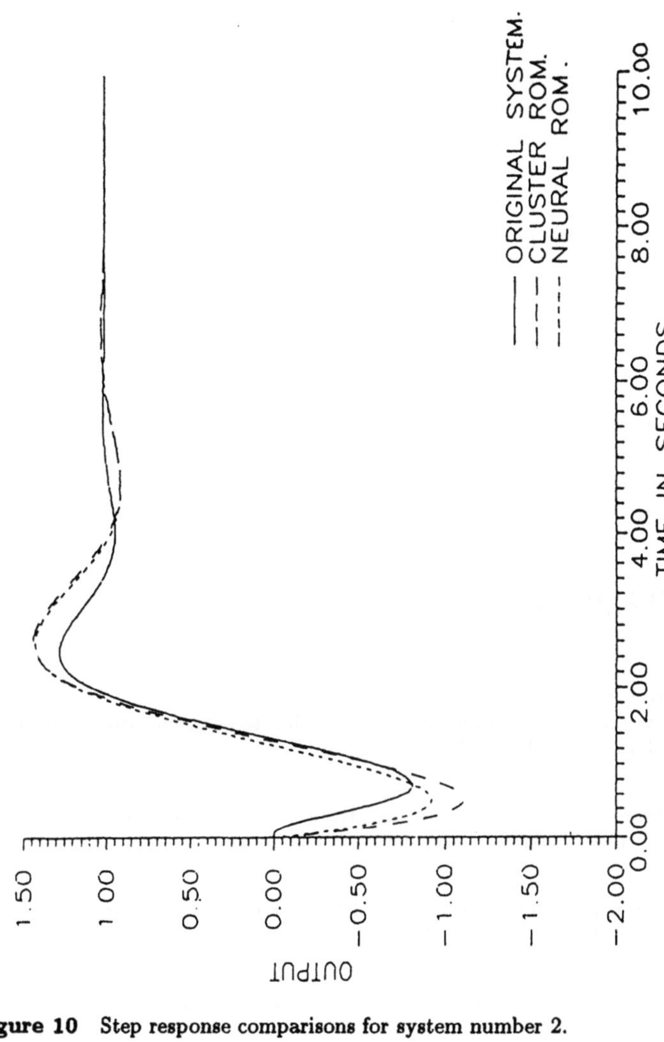

Figure 10 Step response comparisons for system number 2.

8 CONCLUSIONS

In this chapter we have described multistage flash desalinization processes and discussed the potential applicability of ANNs to reduced order modeling and robust control of such processes. A new method of ANN-based reduced order modeling has been proposed where the discrete Fourier components of the plant output are used to yield the parameters of the ROM. Some typical SISO examples are used to illustrate the method.

A connectionist approach to implementing a neural robust controller based on the ANN-based ROM identifier is proposed. Simulation results presented in this chapter

Inputs	Desired outputs	Calculated outputs	Percentage error
9959.902	-1.20679	-1.40803	16.67534
0.00000	-1.25434	-1.238	-1.30284
-798.376	5.06253	5.95337	17.59672
684.144	-0.5609	-0.5526	-1.48008
-365.458	-11.12901	-10.85358	-2.47484
591.598	-0.60392	-0.58722	-2.76599
-201.067	4.54151	4.57026	0.63316

Table 3

confirm the validity of the ANN-based reduced order modeling technique. Application of these concepts to modeling and control of complex multistage flash desalinization processes is currently under investigation.

REFERENCES

[1] Al-Gobaisi D.M.K., A.S. Barakzai, and A.M. El-Nashar, "An overview of modern control strategies for optimizing thermal desalinization plants", in Desalinization and water Re-use, Proceedings of the 12th International Symposium, (Miriam Balaban, ed.), (Malta), 15-18 April 1991.

[2] Davison E.J., "A method for simplifying linear dynamic systems", Transactions of IEEE on Automatic Control, Vol. AC-11, pp. 93, 101, 1966.

[3] Wilson D.A., "Optimum solution of model reduction problem", Proceedings of IEE, Vol. 117, pp. 1161, 1165, 1970.

[4] Sinha N.K. and G.T. Bereznai, "Optimum approximation of high order systems by low order models", International Journal of Control, Vol. 14, pp. 951, 959, 1971.

[5] Luus R. and G.D. Howitt, "Model reduction by optimization", Hungarian Journal of Industrial Chemistry, Vol. 16, pp. 29, 38, 1988.

[6] Kokotovic P.V., R.E. O' Malley, Jr, and P. Sannuti, "Singular perturbations and order reduction in control theory-an overview", Automatica, Vol. 12, pp. 123, 132, 1976.

[7] Moore B.C., "Principal component analysis in linear systems: controllability, observability and model reduction", Transactions of IEEE on Automatic Control, Vol. AC-26, pp. 17, 32, 1981.

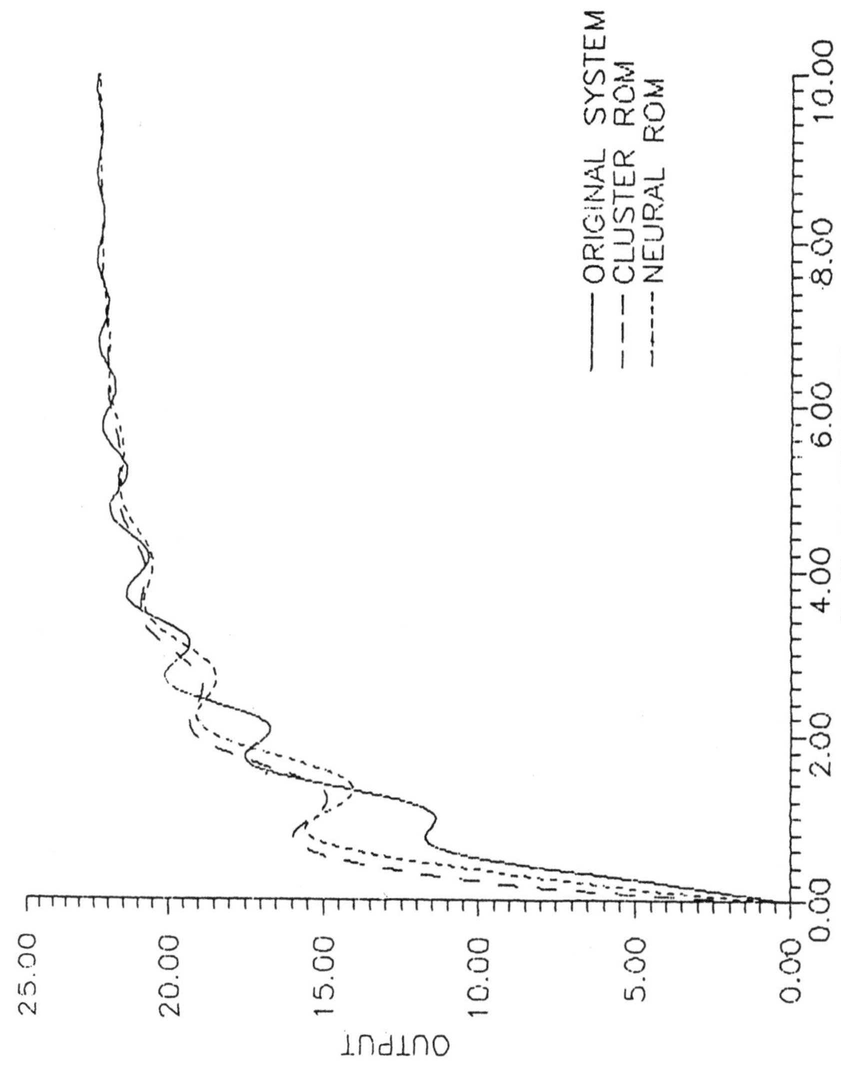

Figure 11 Step response comparisons for system number 3.

[8] Shamash Y., "Stable reduced order models using padé-type approximations", Transactions of IEEE on Automatic Control, Vol. AC-19, pp. 615, 617, 1974.

[9] Pal J., "Improved padé approximants using stability equation method",

Electronics Letters, Vol. 19, pp. 426, 427, 1983.

[10] Glover K., "All optimal hankel-norm approximations of linear multivariable systems and their l^{∞}-error bounds", International Journal of Control, Vol. 39, no. 6, pp. 1115, 1193, 1984.

[11] Chen C.F. and L.S. Shieh, "A novel approach to linear model simplification", International Journal of Control, Vol. 8, pp. 561, 570, 1968.

[12] Pal J., "System reduction by a mixed method", Transactions of IEEE on Automatic Control, Vol. AC-25, no. 5, pp. 973, 976, 1990.

[13] Bosley M.J. and F.P. Lees, "A survey of simple transfer function derivations from high order state-variable models", Automatica, Vol. 8, pp. 765, 775, 1972.

[14] Pal J., "An algorithmic method for the simplification of linear dynamic scalar systems", International Journal of Control, Vol. 43, pp. 257, 269, 1986.

[15] Hutton M.F. and Friedland B., "Routh approximation for reducing order of linear, time-invariant systems", IEEE Transactions of Automatic Control, Vol. AC- 20, pp. 329, 337, 1975.

[16] Pal J., "Stable reduced order padé approximants using the routh-hurwitz array", Electronics Letters, Vol. 15, pp. 225-226, 1979.

[17] Sinha A.K. and J. Pal, "Simulation based reduced order modeling using a clustering technique", Computers and Electrical Engineering, Vol. 16, no. 3, pp. 159, 169, 1990.

[18] D.T. Pham and X. Liu, "State-space identification of dynamic systems using neural networks", Engineering Applications of AI, Vol. 3, pp. 198, 203, 1990.

[19] Chu S.R., R. Shoureshi, and M. Tenorio, "Neural networks for system identification", IEEE Control Systems Magazine, pp. 31, 35, 1990.

[20] Narendra K.S. and K. Parthasarathy, "Identification and control of dynamical systems using neural networks", IEEE Tranasactions of Neural Networks, Vol. 1, no. 1, pp. 4, 27, 1990.

[21] Liang Jin, Peter N. Nikiforuk, and Madan M Gupta, "Model matching control of unknown nonlinear systems using recurrent neural networks", in IFAC 11th Triennial World Congress, (Sydney, Australia), pp. 337, 344, 1993.

[22] Anderson B.D.O. and J.B. Moore, Optimal Control: Linear Quadratic Methods. Prentice-Hall of India Ltd., 1991.

[23] Safanov M., "Future directions in the robust control theory", in IFAC 11th Triennial World Congress, (Tallin, Estoria), pp. 171, 175, 1990.

Hua Li, an associate professor of the Computer Science Department, College of Engineering at Texas Tech University, received his BS degree in Electronics Engineering from Tianjin University, China, and his MS and Ph.D. degrees in Electrical and Computer Engineering from the University of Iowa, USA. His current research interests include vision, neural networks, fuzzy logic and their VLSI implementations. He is the co-author (with Christof Koch) of the book, VISION CHIPS: IMPLEMENTING VISION ALGORITHMS WITH ANALOG VLSI CIRCUITS published by IEEE CS Press. His work of using fuzzy logic for digital image processing was collected in a book, FUZZY MODELS FOR PATTERN RECOGNITION, edited by Bezdek and Pal and published by IEEE Press. He and his graduate students have designed and built a real time fuzzy logic controller for the demonstration system, beam-and-ball system. This system was given an "Industrial Neural Network Award" in the 1994 World Congress Neural Network Conference in San Diego in 1994. Based on the understanding of the biological vision system, he has designed a two-dimensional network suitable for analog VLSI implementation for motion detection. He worked as a guest editor for the special sections of Neural Networks and Fuzzy Logic and their applications in Intelligent Manufacturing in semiconductor industry for IEEE Transactions on Components, Hybrid, and Manufacturing Technology. Dr. Li has worked as session chairman or co-chairman for the 13th, 15th, and 16th IEEE International Symposium on Electronics Manufacturing. Dr. Li is a member of IEEE, IEEE Computer Society, Upsilon Pi Epsilon (Computer Science Honor Society) and chairman of the publication subcommittee of CT-4, IEEE CHMT Society.

Madan M. Gupta (Fellow: IEEE and SPIE) received the B. Eng. (Hons.) and the M.Sc. in Electronics-Communications Engineering, from the Birla Engineering College (now the BITS), Pilani, India, in 1961 and 1962, respectively. He received the Ph.D. degree from the University of Warwick, United Kingdom, in 1967 in adaptive control systems. Dr. Gupta is currently Professor of Engineering and the Director of the Intelligent Systems Research Laboratory and the Centre of Excellence on Neuro-Vision Research at the University of Saskatchewan, Canada. He was elected Fellow of IEEE for his contributions to the theory of fuzzy sets and the adaptive control systems, and the advancement of the diagnosis of cardiovascular disease. He was also elected Fellow of SPIE for his contributions to the field of neuro-vision, neuro-control, and neuro-fuzzy systems.

Dr. Gupta has served the engineering community worldwide in various capacities through societies such as IEEE, IFSA, IFAC, SPIE, NAFIP, UN, CANSFINS, and ISUMA. He has been elected as a visiting professor and a special advisor, in the areas of high technology, to the European Centre for Peace and Development (ECPD), University for Peace, which was established by the United Nations. In addition to publishing over 400 research papers, Dr. Gupta has co-authored two books on fuzzy logic with Japanese translation, and has edited fourteen volumes in the field of adaptive control systems, fuzzy logic/computing, neuro-vision, and neuro-control systems.

Dr. Gupta's present research interests are expanded to the areas of neuro-vision, neuro-control and integration of fuzzy-neural systems, neuronal morphology of biological vision systems, intelligent and cognitive robotic systems, cognitive information, new paradigms in information processing, and chaos in neural systems. He is also developing new architectures of computational neural networks (CNNs), and computational fuzzy neural networks (CFNNs) for applications to advanced robotic systems.

INDEX

A

Active stereo method, 214
Adaptive fuzzy PID, 300
Adaptive sensing strategy, 225
Aggregation, 1
Agility, 86
Angle of attack, 94
Application-specific IC (ASIC), 271
Area method, 10
Arithmetic mean, 6
Automatic design of fuzzy systems, 317
Automatic flight control system (AFCS), 86
Average-of-maxima method, 9

B

Beam-and-ball system, 157
Bode diagrams, 193
Bounded product, 71
Bounded sum, 71

C

Camcorder operations, 409
Camera calibration, 220, 247
Canomical fuzzy set-extension, 73
Canonical isomorphisms, 72
Center-of-gravity method, 8
Center-of-gravity procedure, 89
Clustering, 319
Controllability condition, 373

D

Design procedure 1, 196
Design procedure 2, 197
Direct Adaptive Fuzzy Control, 289
Direct adaptive fuzzy control, 293
Discrete-time, 369
Dual geometric mean, 6
Dual harmonic mean, 6
Dynamic characteristics, 92
Dynamic recurrent neural network (DRNN), 374
Dynamic recurrent neural networks (DRNNs), 368

E

EPROM, 285
Equivalence between Sugeno;s system and the RBF, 302
Evolution algorithms, 320

F

Feedforward neural networks, 23
Flight control systems, 86
Force-torque sensory data, 214
FPGA, 271
Frequency and transient characteristics, 189
Function approximation, 17
Function estimation, 337
Fuzzy basis function networks, 50
Fuzzy order relation, 77
Fuzzy pitch control, 107
Fuzzy real quantities, 69
Fuzzy roll control, 112
Fuzzy yaw control, 110

G

Gaussian network, 46
Gradient method, 291
Graphical analysis, 140

H

Hand-eye robotic manipulator system, 257
Height method, 10
Homogeneous transformation matrix, 252
HON, 37
HONU, 37
Human control, 167
Human-like capabilities, 367

I

Ill-posed inverse mapping problem, 246
Intelligent control, 367

Interface
8255A, 166
Inverse mapping problem, 249
Inverted pendulum, 342

K

Kalman filter, 169

L

Linear Quadratic Gaussian (LQG) problem, 169
Linearization, 159
Linearized system, 373
Linguistic fuzzy control rules, 97
Linguistic rules, 126

M

Maximal area method, 11
Maximal height method, 10
Mean-of-maximum procedure, 89
Median method, 9
Midpoint-of-maxima method, 9
MIMO designs, 424
Moving vector, 411
Multi-input and multi-output (MIMO), 369
Multilayer feedforward neural network, 430
Multiple fuzzy reasoning, 2
Multiple-input multiple-output (MIMO), 274
Multiple-input single- output (MISO), 274
Multistage flash desalinization, 424

N

Neural computing approach, 414
Neural networks, 221
Nonlinear aerodynamic, 106

O

Output regulation, 386
Output tracking control, 392
Output tracking, 371

P

Phase-lead compensation, 191
Photo sensors, 166
Piecewise linear controller, 135
Pitch control surface, 90
Pitch PID control rules, 99
Proportional fuzzy logic controller, 130
Proportional-derivative fuzzy logic controller, 132

R

Real-time nonlinear system control, 166
Reduced order modeling, 428
Regulator problem, 169
Riccatti equation, 169
Ridge polynomial network, 45
Robust controller design, 439
Roll control surface, 90
Roll PI control rules, 103

S

Self-Organizing Controller (SOC), 318
Sensor integration, 214
Sigma-pi networks, 42
Simplified reasoning method, 127
SISO (single-input single-output), 274
Specialized learning, 296
Spot-light, 220
Stability augmentation system (SAS), 86
Stability derivative, 93
Stability, 305
Stabilization, 370
State regulation, 370, 382, 388
Steepest-descent technique, 384
Stochastic fuzzy real numbers, 72
Sugeno's fuzzy systems, 291
Surface analysis, 145
Surface carving system, 213

T

T-conorms, 70
T-norms, 70
Time-varying system, 161
Trigonometric functions, 21

U

Universal approximators, 321

V

VHDL description, 280

W

Wang's Fuzzy System (WFS), 322
World coordinate system, 249

X

X-29 aircraft, 92